Interactive LINEAR ALGEBRA

Interactive LINEAR ALGEBRA

A LABORATORY COURSE USING MATHCAD®

Gerald J. Porter
University of Pennsylvania

David R. Hill
Temple University

Springer

Textbooks in Mathematical Sciences

Library of Congress Cataloging-in-Publication Data

Porter, Gerald J.
Interactive linear algebra : a laboratory course using Mathcad / Gerald J. Porter, David R. Hill.
p. cm. — (Textbooks in mathematical sciences)
Includes bibliographical references (p. –) and index.
ISBN 0-387-94608-X (softcover : alk. paper)
1. Algebras, Linear—Computer-assisted instruction. 2. Mathcad
I. Hill, David R. (David Ross), 1943– . II. Title. III. Series.
QA184.P668 1996 96-1762
512'.5'078—dc20

Printed on acid-free paper.

Production managed by Steven Pisano; manufacturing supervised by Jeffrey Taub.
Photocomposed from the authors' camera copy.
Printed and bound by Hamilton Printing Company, Rensselaer, NY
Printed in the United States of America.

9 8 7 6 5 4 3 2 1

ISBN 0-387-94608-X Springer-Verlag New York Berlin Heidelberg SPIN 10516671

PREFACE

Interactive Linear Algebra: A Laboratory Course Using Mathcad is an electronic book, contained on the two disks at the back of this text. It must be used on a computer with Mathcad. The explorations, calculations, graphics, and reports need to be completed on the computer. This print version is intended to complement the electronic text, not to replace it. We have provided this print version so that you, the student, can prepare for class, keep notes, and rapidly locate items in sections other than the one currently open on your computer screen. It is a great deal quicker to open a section in a book than it is electronically.

This book contains a printout of the electronic book, although here, in this version, we have kept white space to a minimum. Also, in many places we have evaluated expressions that we ask you to evaluate in the electronic book. Obviously, instructions such as "press F9" or "change the value of x" cannot be executed here. Since the electronic book is intended to be viewed on a color monitor we refer to the "red line" or the "green line." In this print version there are no colors and the reader must determine which line is which color.

In the electronic version, the Index and Contents are hypertext documents. Click on a section and that section is opened. Again, however, such instructions cannot work in this version.

We have included here a section entitled "Before You Start." This section has instructions about the environment in which we intend the electronic book to be run. This includes colors, fonts, and evaluation options. Finally, the various palettes in Mathcad 5.0 and 6.0 are illustrated.

ACKNOWLEDGMENTS

Our involvement with interactive mathematics texts began with discussions we had with Jerry Uhl and Horacio Porta, who, with Bill Davis, were the authors of the seminal interactive text *Calculus & Mathematica* (Addison-Wesley). David Smith convinced us of the difference between teaching Linear Algebra (or any course for that matter) with a lab and as a lab. From these discussions, our ideas about the nature of this text emerged. Anyone who is familiar with Alan Tucker's *A Unified Introduction to Linear Algebra* (Macmillan) will see its influence on this text. Many threads that run through this book as well as many of the applications in our text were "derived" from Tucker's book.

The MAA's Interactive Mathematics Text Project provided the impetus for us to author this text. We also appreciate the support that IBM and the NSF gave to that project.

The creation of this text was supported by Award P116B20642 from the Fund for the Improvement of Post Secondary Education (FIPSE) of the U.S. Department of Education. We thank FIPSE and our project officer, Brian Lekander, for their encouragement. In addition, we thank Horacio Porta, Eugene Herman, Alan Tucker, Nancy Baxter Hastings, and David Smith who have served as evaluators on this project, and Jane Day, who commented on an earlier version. Their candid comments and advice helped keep us on the right track.

Each of us taught Linear Algebra five or six times using preliminary editions of this text. We wish to thank our colleagues for allowing us to do so, and our students, whose comments, both positive and negative, helped us hone the final version of this text.

MathSoft, Inc., the publisher of Mathcad, has provided us with assistance in the creation of the dll files that accompany this text as well as giving us access to authoring tools. Frank Purcell has been particularly helpful in this endeavor.

We thank Jerry Hefley for the encouragement that he provided for the Interactive Mathematics Text Project and for writing the rowops.dll that accompanies this text. Without his efforts that dll would not exist.

Jerry Lyons and Ken Dreyhaupt at Springer-Verlag have worked closely with us to actualize our vision of what this text should be. We thank them for their efforts.

Finally, we thank our wives, Judy and Sue, for their encouragement and support of our many activities and projects.

Gerald J. Porter

David R. Hill

CONTENTS

Before you start - Books and Palettes - Mathcad

This interactive text is written in Mathcad. There are some basic processes that you need to know about Mathcad before you begin. The first is how to get started. This is what you must do to get started.

1. Find the Mathcad icon on your screen and double click on it.

This book will run under either Mathcad 5.0 or Mathcad 6.0; however, if you save a file in Mathcad 6.0 you will be unable to open that file in 5.0.

2. Click on **Books** to open the Book menu.

Mathcad - [Untitled:1]

File Edit Text Math Graphics Symbolic Window Books Help

3. If Interactive Linear Algebra appears on the menu, click on it. Otherwise click on **Open Book ...**

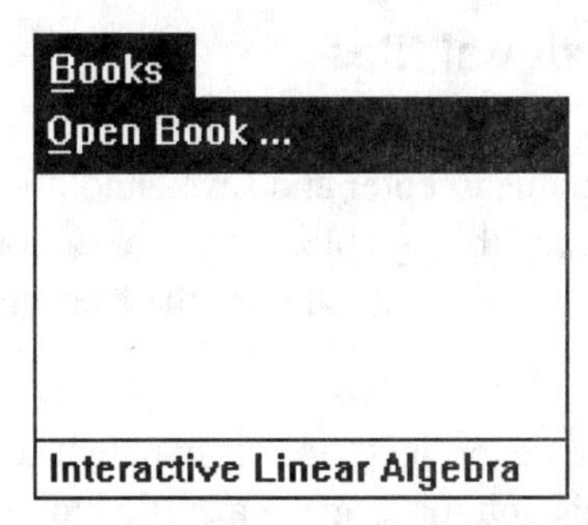

4. If you click on **Open Book ...** look for the file **lin_alg.hbk** and double click on that.

5. Once you are in the **Table of Contents** you can double click to open any section.

Environment

This book is setup up to run with black text, dark blue equations and red annotations on a gray background. You can set these by opening the **Window** menu and selecting **Change Colors**. These colors are saved in the mcad.ini file as follows:

```
[System-Colors]
Color-0=12632256     ← Background color
Color-1=0            ← Text color
Color-2=8388608      ← Equation color
Color-3=128          ← Annotation color
```

The default fonts for Text, Variables and Equations are saved in the file mcad.mcc. To execute this file open the File menu, click on **Execute Configuration File...** click on mcad.mcc from the file list and then the **OK** button.

The default font for Text is 10 pt Arial; for Variables and Constants it is 10 pt Bold Arial. If the defaults are different from these choices you may set them by using the **Change Default Font...** option on the **Text** Menu and the **Modify Font Tag...** option on the **Math** menu. The **Automatic Mode** should be off on the **Math** menu. This configuration may be saved by clicking on **Save Configuration...** on the **File** menu.

Annotations

If you are using this text as an Electronic Book you should enter and save your work as annotated files. Your work will be entered and saved in the color chosen as the annotation color. You should make sure that **Annotate Book** is selected on the **Books** menu.

To save your work you should select **Save Edited Sections** from the **Annotate Options** choice on the **Books** menu. If you do this your edited file will be saved as an *.mcu file. If this is deleted the original (unedited) book is restored. If you are working in an edited file there will be an * on the title bar after the name of the section.

Using the text as individual files

We described above how to use this book as a Mathcad Electronic Book. This has certain advantages such as being able to enter and save annotations in a color different from that in which the book is written and being able to go to a section by double clicking in the Table of Contents. In addition, you will be able to use the Book Index and **Search Book...** on the **Books** menu.

In some environments, however, it is more convneient to use this text as a collection of individual files. To do this you must open each file individually as follows:

1. From the **File** menu choose **Open...**

2. Select the directory winmcad\handbook\lin_alg. (Here winmcad is the directory in which Mathcad is stored. This may may be different on the system you are using.)

3. The Table of Contents is the file contents.mcd. (You will not be able to jump to a Section from the Table of Contents unless you are using the text as an electronic book.)

4. The individual sections have names such as 1_5.mcd. (Chapter 1, Section 5).

5. Select the section that you wish to work on and open the file.

You may then work on that section; however, your work will be in the same color in which the book is written. When you are done you should save the file either to a floppy disk or to your own directory (check with your instructor to find where you save your work). If you wish to access your work to complete the section you should open the file that you saved rather than the unedited file that you began with.

Palettes

Both Mathcad 5.0 and Mathcad 6.0 have icon buttons that make it easy to use various operations and to insert Greek letters. The icons buttons are arranged on palettes that contain related operations.

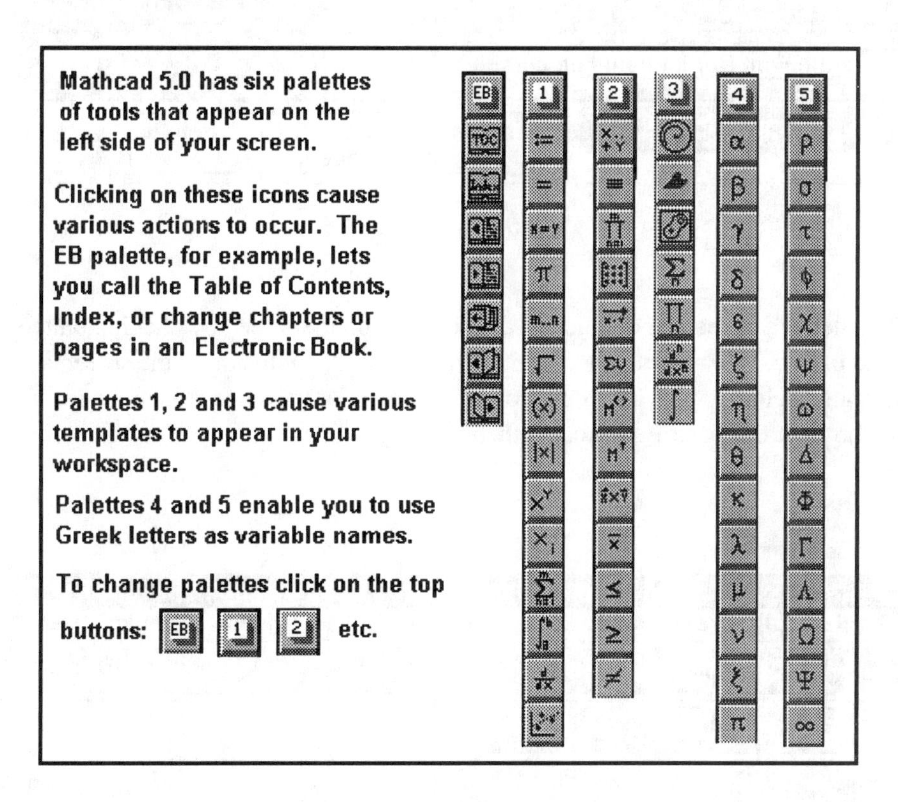

Palettes in Mathcad 5.0

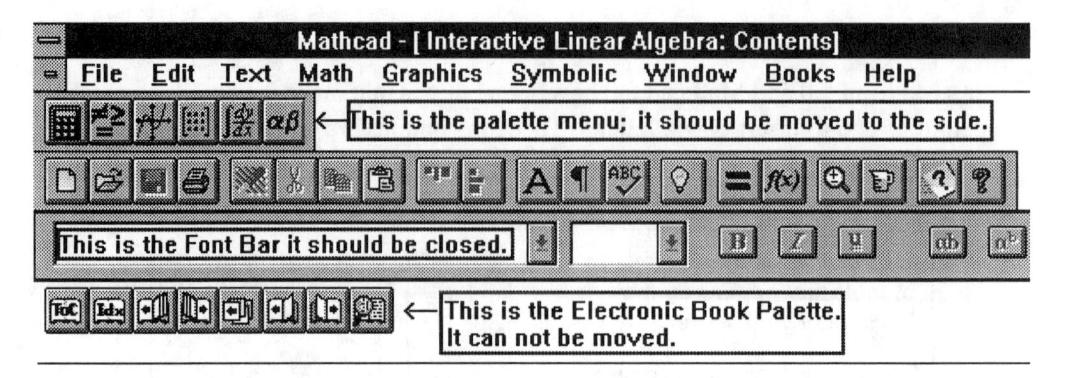

Palettes in Mathcad 6.0

The palette menu can be dragged to any location on the screen by clicking on the far left of the palette. We recommend that it be docked on the left side of the screen as illustrated below to provide more working space. For the same reason we recommend that the Font Bar be closed except when actively used.

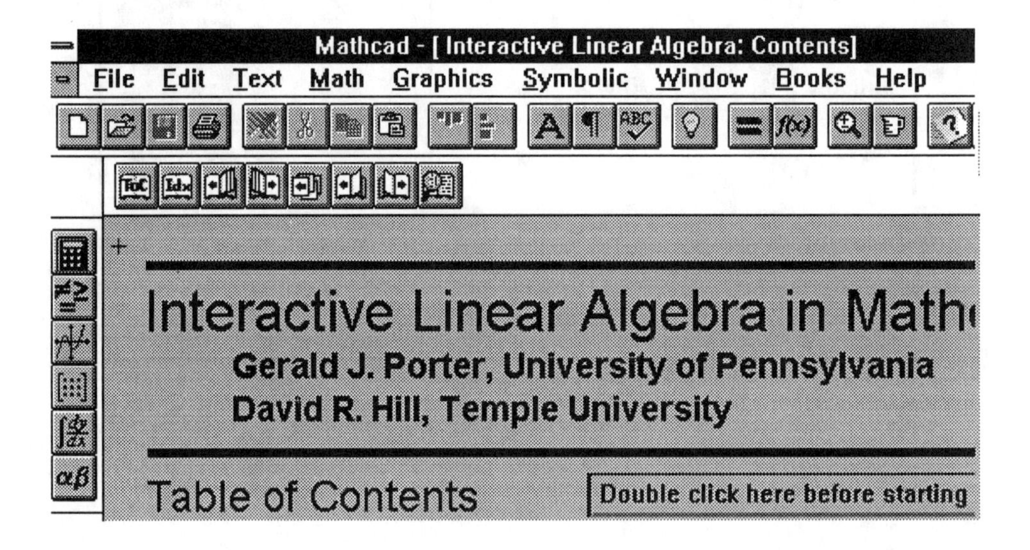

CHAPTER 0

INTRODUCTION

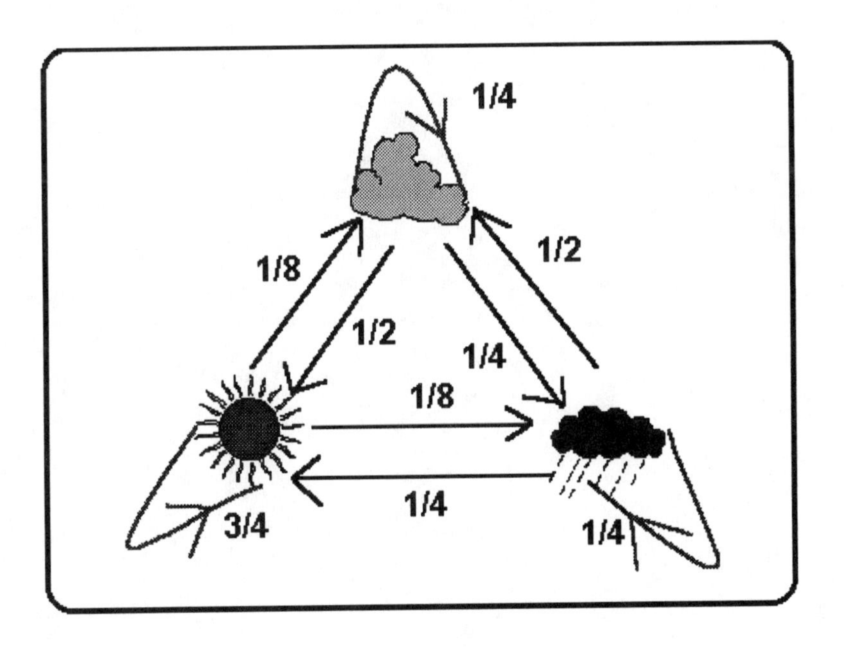

Section 0-1 About This Text

INTERACTIVE LINEAR ALGEBRA: A Laboratory Course with Mathcad is a new type of textbook. We call it a ***computer based interactive mathematics text***. This refers both to the environment in which you read it and the way that you use it to learn mathematics.

The environment

This text has been created in a computer environment called Mathcad. Mathcad is a smart electronic blackboard. By this we mean that mathematical expressions appear in Mathcad the same way that they would in a "standard" mathematics text or as a professor might write them on a blackboard. Here are some calculus examples:

$$f(x) := \frac{x^2 \cdot e^x}{4 + \sin(x)} \qquad g(x) := \int_1^x \frac{t^3}{t^2 + 3 \cdot t + 2}\, dt$$

The difference between a printed text and this text is that the expressions given above are "live." We can evaluate them, graph them, take a derivative, and in general "do mathematics." That is what we mean by smart. Here are some examples:

$f(3) =$ $\qquad$ $g(4) =$

To evaluate these expressions, press the **keyboard** function key

The important thing is that these quantities are calculated when you press the key. You may change the objects as you wish. Use the mouse to move the cursor next to the 3 in f(3) above and click on the left mouse button. A blue vertical line should appear next to the 3, backspace over the 3 and type in another number. Then press F9 and the value of f at the new number will be computed.

We can also graph these functions:

$x := 0, .05 .. 2$ (This expression means: x goes from 0 to 2 in increments of .05.)

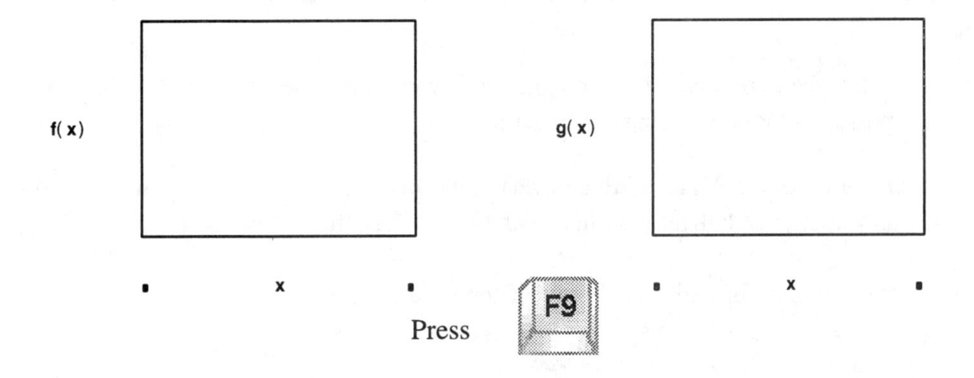

We can vary the range of x in the expression above to view a different part of the graph. As you see, Mathcad has the capacity to evaluate both numerical and graphical expressions. In addition, Mathcad can compute symbolically. Mathcad uses another language - one called Maple - to do its symbolic calculations.

If you are using Mathcad 5.0 click on the maple leaf icon on the tool bar above to load the symbolic processor. This is not required in Mathcad 6.0.

We illustrate the symbolic capability by taking the derivative of the function f given above.

$$\frac{x^2 \cdot e^x}{4 + \sin(x)}$$

Move the cursor next to any one of x's in the expression. Open the **Symbolic** menu by clicking on it, then click on **Differentiate on Variable**.

The expression above is the derivative of f(x). Try it by hand and see if you get the same answer.

Mathcad can also do integration.

$$\frac{x^3}{x^2 - 3 \cdot x + 2}$$

Put your cursor next to x and click.
Now pull down the **Symbolic** menu and click on **Integrate on Variable**.

We can check that this is the correct answer by differentiation. (Remember to put the cursor next to one of the x's, then click on **Differentiate on Variable** in the **Symbolic** menu.)

This doesn't look like what we began with, so let's collect terms: To do this successively click on the + and - signs in the expression until the entire expression is boxed. Then click on **Simplify** in the **Symbolic** menu.

It still isn't quite what we began with so let's expand the denominator. To do this, click on the dot between the two factors. Then click on **Expand Expression** in the **Symbolic** menu.

Finally, we have the derivative of the integral we began with. The symbolic capabilities of Maple are powerful tools for doing mathematics.

There is one more tool in Mathcad that is very important. The tool that is being used to write this introduction, Mathcad's ability to enter text. Try this below.

Initiate a text region by pressing the " key. Then type your name.

In summary, Mathcad provides five important tools that we will use to study Linear Algebra.

1. **Mathematical expressions are displayed in mathematical format.**
2. **Expressions can be evaluated numerically.**
3. **Expressions can be evaluated symbolically.**
4. **Graphs can easily be drawn.**
5. **Text can be entered into the document.**

Since Mathcad operates in the Windows environment, there is another important feature of Mathcad; namely, we can paste material generated by other Windows' programs into our document. That is what we did above with the F9 key image. That image was generated in Paintbrush, copied to the clipboard and pasted into our document. So we can add

6. **The ability to generate materials in other Windows programs and paste them into our document,**

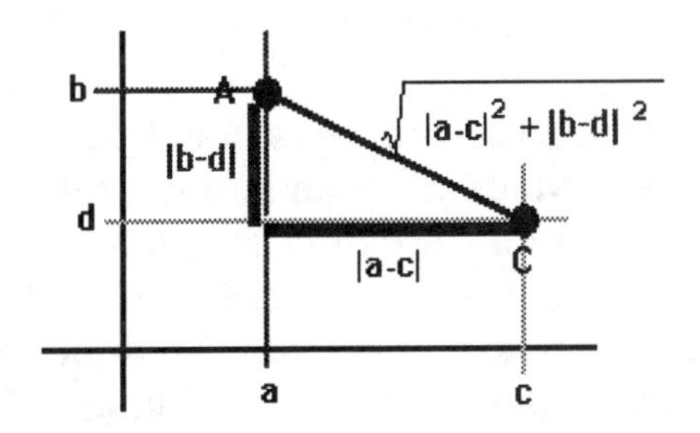

This is what we mean by an interactive environment. There are many ways that such an environment can be used in mathematics learning. The most common is simply as a powerful calculator. In this mode, the material in your course would be presented as it always has and the learner would use Mathcad to solve problems. That is not our goal in this text; our goal is ***discovery based learning***.

Discovery Based Learning

In "traditional" mathematics courses algorithms are presented to the class by the teacher. The students then "learn" these algorithms by doing many examples that are intended to reinforce retention of the algorithm. Long division is a good example of this process but so is the way that you were taught the product rule or the chain rule in calculus. We refer to this as *template learning*. The student is presented a template for solving a particular type of problem. The student then does many problems, the majority of which involve substituting different numbers or different functions into the template. For example there is nothing conceptually different between the following problems:

$$\frac{d}{dx}\left(\frac{x\cdot\sqrt{x^2+1}}{x^2+1}\right) \qquad \frac{d}{dt}\left(\frac{\sin(2\cdot t)\cdot\cos(t^2)}{2+\sin(t)}\right) \qquad \frac{d}{dx}\left(\frac{x^2\cdot\ln(x^3-3)}{e^x}\right)$$

They are all exercises to reinforce the quotient rule and chain rule. You can surely remember the many problems that were used to reinforce learning of long division.

There are two major problems with template learning. The first is that research has shown that memorizing algorithms does not lead to retained knowledge. The second is that problem solving skills are not significantly improved. You have learned how to solve a specific problem and that solution is not easily generalizable to a broader class of problems. Most problems that arise from applications do not fit nicely into one of the small number of classes of problems that appear in books. You must remember that, for the most part, text book problems are limited to those that can be solved by the methods of that text.

The goal of this text is to improve both long range retention of the subject matter and problem solving skills. The basis for discovery learning is given in the following paragraph by Lynn Steen.

> **"Learning takes place when students construct their own representation of knowledge. Facts and formulas will not become part of deep intuition if they are only committed to memory. They must be explored, used, revised, tested, modified, and finally accepted through a process of active investigation, argument and participation. Science (and mathematics) instruction that does not provide these types of activities rarely achieves its objectives."**

Mathcad provides us with a rich laboratory not unlike those of the chemist and biologist in which we can experiment with, explore, and test mathematical concepts. The way that mathematicians do research is in "hand to hand" combat with concepts and ideas. The neat structure of Definition, Lemma, Theorem that eventually emerges is the result of those struggles. Presenting mathematics in that mode hides all the sweat and struggle from the student and deprives the student of the opportunity of understanding the concepts that led to the final structure.

In this text definitions appear only when it becomes clear that they are needed to give some property or object a name. Theorems, perhaps without the word, appear after they have been tested and shown to hold. The presentation is informal and is intended to expose the structural foundations and not simply the beautiful facade. You will be asked questions without being told the answers previously. You are encouraged to guess, conjecture and to just "Try it!." That is the way that science and mathematics work.

Communications are an important skill in mathematics as elsewhere. We encourage you to work on this text with a lab partner or partners. Talk with them about mathematics. When you hit a wall they may have an idea. Team projects are commonplace in business school. Can you imagine getting a new job and your boss instructing you not to talk with anyone about the project to which you have been assigned. Working together is both an important way to learn and a skill that helps you after college. Writing is also important! Not only does it provide a way of communicating your thoughts with others but it also forces you to clarify your thoughts before putting them to paper or computer screen. **Every section in this text concludes with a writing exercise.**

Communications consists of three components:

1. Reading
2. Speaking
3. Writing

All of these are required by this approach to learning.

You might ask what your professor is getting paid for if you have to do all the work yourself and he or she won't even tell you how to do the problems. The role of the instructor is changed in this mode of learning. The professor has changed from "the sage on the stage" to "the guide on the side." We are here to help you, to answer the questions that didn't occur to us when the text was authored, to rescue you when expeditions of discovery take you to shores far removed from your intended destination and, more generally, to help guide your discovery. This text and your instructor are intended as discovery guides.

Finally, we must warn you that this course requires a different level of participation than does the traditional course. Many students study in bursts mostly motivated by the challenge (threat) of examinations. Thus this week is economics week because there is an economics exam on Thursday. Next week will be chemistry and then a week on math before the math exam. Such an approach will be as fatal in this course as sprints are to a long distance runner. What is required is a steady pace. That gives time for the ideas to be absorbed and internalized. There will be times that you fall behind for one reason or another. When that happens try to get back on track as quickly as possible. If the class is talking about a subject you haven't explored then the discussion will not be beneficial to you and it won't be repeated when you need it.

We wish you well on your trip into Linear Algebra. Never be afraid to try something. You will not damage the computer. And if you really foul things up your instructor will be there to help straighten things out. Happy learning!

Gerald J. Porter, University of Pennsylvania
David R. Hill, Temple University

Section 0-2 Mathematical Models

Name:________________

Score:_____________

Overview

A mathematical model is a description in mathematical language of an object that exists in a non-mathematical universe. The mathematical model may or may not accurately reflect the object that it models. We are all familiar with weather forecasts that are based on a mathematical model of the weather and with economic forecasts that are based on a mathematical model of the economy. The success or failure of these models is a reflection of how accurately the mathematical model represents the original object not of the correctness of the mathematics in analyzing the model. Some models are good for some things and bad for others. Newtonian mechanics (a mathematical model) can still be used to predict many events accurately despite the fact that Einstein's relativity theory (another mathematical model) tells us that it is inaccurate.

Generally speaking there are three steps in mathematical modeling.

BUILD THE MODEL
Translate the non-mathematical object into mathematics.

ANALYZE THE MODEL
Study the model mathematically.

INTERPRET THE MATHEMATICAL ANALYSIS
Apply the results of the mathematical study to the original non-mathematical object.

The mathematician is involved primarily at the second stage: studying the mathematical model. The investigator from the original subject discipline and the mathematician can be involved together in formulating the model and in interpreting the mathematical results.

Most of the applications of calculus (e.g., max-min problems) involve mathematical models.

Linear algebra provides many tools for use in mathematical modeling. Applications include economics, demography, engineering, computer graphics, as well as probabilistic models. In this section we present three models. These models will be referred to throughout the remainder of the text so make sure you understand how they are formulated.

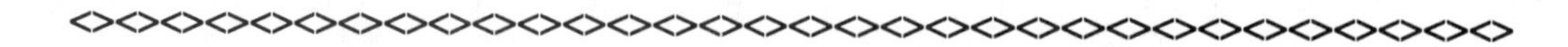

An Oil Refinery Model

A company runs three oil refineries. Each refinery produces three petroleum-based products:

Heating Oil, Diesel Oil, & Gasoline.

Suppose that from one barrel of crude petroleum, the first refinery produces 4 gallons of heating oil, 2 gallons of diesel oil, and 1 gallon of gasoline. The second and third refineries produce different amounts of these three products as described by the array **A** given below.

		Refinery 1	Refinery 2	Refinery 3
	Heating Oil	4	2	2
A =	Diesel Oil	2	5	2
	Gasoline	1	2.5	5

Each column of **A** is a vector of outputs by a refinery. For example, from 1 barrel of crude refinery 3 produces an output vector.

$$\begin{pmatrix} 2 \\ 2 \\ 5 \end{pmatrix}$$

Each row of **A** is a vector of amounts of a particular product produced by different refineries. The row vector for gasoline is

$$\begin{pmatrix} 1 & 2.5 & 5 \end{pmatrix}$$

Let $\mathbf{x}_i$ denote the number of barrels of crude petroleum used by the ith refinery. Suppose there is the following set of demands for the refined products:

600 gals of Heating Oil, 800 gals of Diesel Oil, & 1000 gals of Gasoline.

It follows that the number of barrels of crude required to satisfy this demand for the three refineries must satisfy the linear system of equations

$$4 \cdot x_1 + 2 \cdot x_2 + 2 \cdot x_3 = 600$$

$$2 \cdot x_1 + 5 \cdot x_2 + 2 \cdot x_3 = 800$$

$$1 \cdot x_1 + 2.5 \cdot x_2 + 5 \cdot x_3 = 1000$$

This system can be expressed as a vector equation in terms of the columns of array **A** as follows:

$$x_1 \cdot \begin{pmatrix} 4 \\ 2 \\ 1 \end{pmatrix} + x_2 \cdot \begin{pmatrix} 2 \\ 5 \\ 2.5 \end{pmatrix} + x_3 \cdot \begin{pmatrix} 2 \\ 2 \\ 5 \end{pmatrix} = \begin{pmatrix} 600 \\ 800 \\ 1000 \end{pmatrix}$$

Let **b** be the column vector of demands, **x** the column vector of required inputs to achieve the demands, and **A** the refinery manufacturing data given above. Linear algebra gives us a way to write the system of equations developed above concisely.

This is an example of a mathematical model. It might be the case that when we solved the above system of equations we discovered that one of the variables was negative. That would not imply that the model was a "bad" model, only that there might not be any solution to the model.

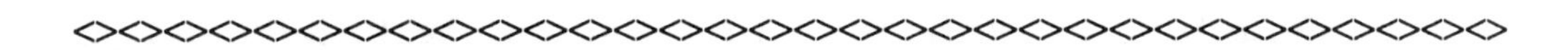

A Weather Model

Suppose we have **three states** of the weather:

SUNNY

CLOUDY

RAINY

Suppose that if it is cloudy today, then the probability is 1/2 that it will be sunny tomorrow; 1/4 that it will be cloudy tomorrow; and 1/4 that it will be rainy tomorrow. The numbers 1/2, 1/4, and 1/4 are called the transition probabilities from today's weather to tomorrow's given that today it is cloudy. Other probabilities for tomorrow's weather apply if it is sunny today or if it is rainy today. It is convenient to display these **transition probabilities** as a matrix **A**. (A matrix is a rectangular table of numbers.) The matrix **A** is called the **transition matrix** (from today to tomorrow).

		Today		
	Tomorrow	Sunny	Cloudy	Rainy
	Sunny	$\frac{3}{4}$	$\frac{1}{2}$	$\frac{1}{4}$
A =	Cloudy	$\frac{1}{8}$	$\frac{1}{4}$	$\frac{1}{2}$
	Rainy	$\frac{1}{8}$	$\frac{1}{4}$	$\frac{1}{4}$

To illustrate the meaning of the matrix entires consider the following:

> The (2,1) entry (row 2, column 1) is 1/8. (We write this as $\mathbf{A}_{2,1} = 1/8$.) This means that given today is sunny the probability that tomorrow will be cloudy is 1/8 = 0.125.

> The (3,2) entry (row 3, column 2) is 1/4. (We write this as $\mathbf{A}_{3,2} = 1/4$.) This means that given today is cloudy the probability that tomorrow will be rainy is 1/4 = 0.25.

Note that **the probabilities in each column of the transition matrix must add up to 1.** (Given that the weather is in a certain state today it is certain that tomorrow it will be in one of the three possible states.)

A convenient way to display the transition information is with a transition diagram or graph. The states represented by our weather icons on the graph are called nodes. There is a directed arrow for each of the transition probabilities.

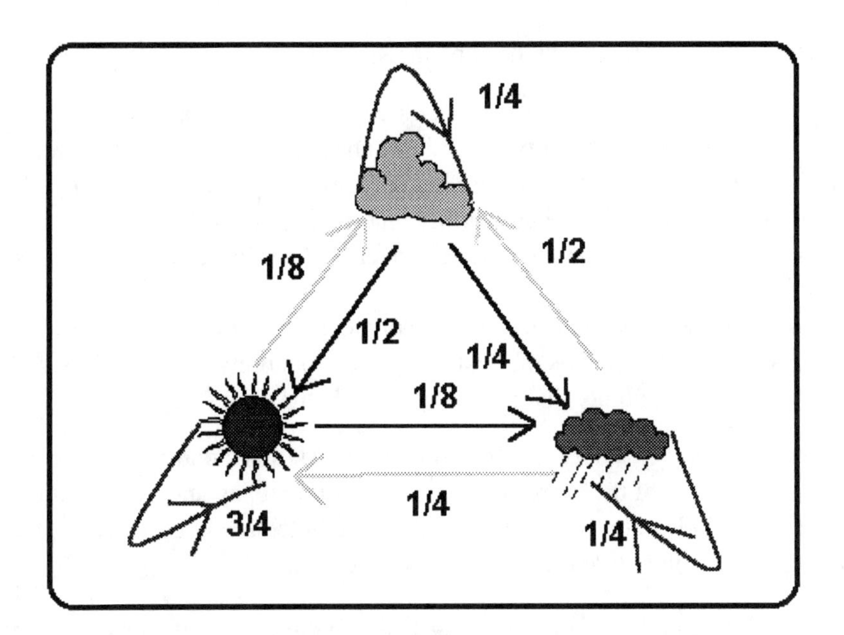

The data in the transition matrix **A** is used to compute the weather probabilities for tomorrow given the weather probabilities today. The input to the computational process is a 3 by 1 vector **p** representing the probabilities for the 3 weather states today. For example suppose that today's probability distribution is

no chance of being sunny ===> p(1) = 0

50% chance of being cloudy ===> p(2) = 1/2

50% chance of being rainy ===> p(3) = 1/2

That is, the vector **p** is given by $\quad \mathbf{p} := \begin{bmatrix} 0 \\ \frac{1}{2} \\ \frac{1}{2} \end{bmatrix}$

The entries of **p** are denoted using subscripts as follows:

$p_1 = 0$ $\quad p_2 = 0.5$ $\quad p_3 = 0.5$ <==== *Press F9.*

Subscripts can be entered by pressing the [key or by clicking on the icon 

Using the transition probabilities from today to tomorrow given in the transition matrix **A** we compute the probability distribution **p_tomorrow** of the weather tomorrow.

The calculation of **p_tomorrow** can be explained as follows:

We must be in one of the three states (sunny, cloudy, rainy) today and we will transition to some state tomorrow. The probability that if it is sunny today it will be sunny tomorrow is the (1,1) entry in the transition matrix. If it is cloudy today, the probability that it will be sunny tomorrow is the (1,2) entry and the probability that if it is rainy today it will be sunny tomorrow is the (1,3) entry. These quantities are called **conditional probabilities.**

The vector **p** gives the probability of each type of weather for today. This means that the probability that it is sunny today is the first entry, the probability that it is cloudy today is the second entry and the probability that it is rainy today is the third entry.

We know from probability theory that the probability that it is cloudy today and cloudy tomorrow is the product of the probability that it is cloudy today and the conditional probability that it will be cloudy tomorrow if it is cloudy today. In terms of our variables the probability that it is cloudy today and tomorrow is $\mathbf{A}_{2,2} \cdot \mathbf{p}_2$.

The fundamental assumption of this model is that each day the weather is in exactly one of the three states. It follows that the probability that it is sunny tomorrow must be the sum of (sunny today and sunny tomorrow) + (cloudy today and sunny tomorrow) + (rainy today and sunny tomorrow). Thus we have the following formula:

$$\text{p\_tomorrow}_1 := A_{1,1} \cdot p_1 + A_{1,2} \cdot p_2 + A_{1,3} \cdot p_3$$

$$\text{p\_tomorrow}_1 := \frac{3}{4} \cdot p_1 + \frac{1}{2} \cdot p_2 + \frac{1}{4} \cdot p_3$$

We can calculate this directly by substituting for $\mathbf{p}_1$, $\mathbf{p}_2$ and $\mathbf{p}_3$. This gives us

$$\frac{3}{4} \cdot 0 + \frac{1}{2} \cdot \frac{1}{2} + \frac{1}{4} \cdot \frac{1}{2} = \frac{3}{8}$$

However, we might as well let Mathcad do the work. $\text{p\_tomorrow}_1 = 0.375$ *<==Press F9.*

Similarly, $\text{p\_tomorrow}_2 := \frac{1}{8} \cdot p_1 + \frac{1}{4} \cdot p_2 + \frac{1}{2} \cdot p_3$ $\text{p\_tomorrow}_2 = 0.375$ *<==Press F9.*

and $\text{p\_tomorrow}_3 := \frac{1}{8} \cdot p_1 + \frac{1}{4} \cdot p_2 + \frac{1}{4} \cdot p_3$ $\text{p\_tomorrow}_3 = 0.25$ *<==Press F9.*

The probability distribution of tomorrow's weather given that today's weather had the distribution given in vector **p** is the vector **p_tomorrow** which is

$$\text{p\_tomorrow} = \begin{pmatrix} 0.375 \\ 0.375 \\ 0.25 \end{pmatrix}$$ *<==Press F9.*

Another fundamental assumption of this model is that the transition doesn't depend upon the particular day. It will be the same from the kth day to the (k+1)st day (for any value of k) as it was from today to tomorrow. We use this assumption in our example to compute the weather of day 3 using the transition matrix and the probability distribution for the weather on day 2 = **p_tomorrow**. Call the answer **p_day_3**. Fill in the formulas needed for the next 3 calculations. Then press F9 to compute **p_day_3**.

$\text{p\_day\_3}_1 := \blacksquare$

$\text{p\_day\_3}_2 := \blacksquare$

$\text{p\_day\_3}_3 := \blacksquare$

$\text{p\_day\_3} =$

You should find that **p_day_3** has components 34/64, 17/64, and 13/64. You can continue to use the transition matrix to compute future probability distributions. This process is called a **MARKOV CHAIN**.

Markov chains have wide applicability and are an important area of mathematics. An interesting question about successive predicting of probability distributions is: **Does the distribution become the same after many repetitions?** That is, does the input vector match the output vector? If it does, then we say we have a ***stable distribution***. We will learn how to determine if this occurs by analyzing the properties of the transition matrix **A**. For the transition matrix **A** in our model there is a stable distribution vector

$$\begin{bmatrix} \frac{14}{23} \\ \frac{5}{23} \\ \frac{4}{23} \end{bmatrix}$$

<><><><><><><><><><><><><><><><><><><><><><><><><><><><><><><><>

Leontief Models (Input-Output Analysis)

This model has as its goal the balance of supply and demand throughout an economy. For each industry there is an equation which relates supply to demand. A Leontief model for an entire society can have thousands of such equations.

We consider a model with three industries:
Energy, Construction and Transportation.

	Supplies		Energy	Constr	Transp	consumer demand
Energy	x	=	0.4 x	+ 0.2 y	+ 0.1 z	+ 100
Construction	y	=	0.2 x	+ 0.4 y	+ 0.1 z	+ 50
Transportation	z	=	0.15 x	+ 0.2 y	+ 0.2 z	+ 100

We call this portion D for industrial demand.

In the table above, the ***left side** of each equation represents the **supply*** of the commodity named. We assume that the supply is measured in dollars (it could as easily be measured in millions of dollars).

x = the supply of energy,
y = the supply of construction services,
z = the supply of transportation services.

The ***right side** of each equation consists of the **demands***: there are two types of demands, the demands by the industries themselves (i.e., the amount of resources required to produce the supply) and the consumer demands. Each of these is again measured in dollar units.

To understand the relationships expressed in the table consider the following.

> To make $1 of energy requires 40 cents of energy, 20 cents of construction services and 15 cents of transportation services.

> The consumer demand for energy is $100 worth of energy.

> The industrial demand for energy is the sum of the energy required to produce x units of energy, y units of construction, and z units of transportation.

Complete the following sentences:

To provide $1 of transportation services requires ___ cents of energy, ___ cents of construction services and ___ cents of transportation services.

The total amount of construction supply required by the industrial demands in this model is $_____.

$_____ is allocated to energy.

$_____ is allocated to construction.

$_____ is allocated to transportation.

The amount of the transportation services needed to meet consumer demand is $_____.

<><><><><><><><><><><><><><><><><><><><><><><><><><><><><><><><>

The fundamental question of the Leontief model is: **How many units of each commodity must be produced to ensure that consumer demand is satisfied?**

Explain what a solution **(x,y,z)** would mean if consumer demand were 0 for each commodity.

ANSWER:

<><><><><><><><><><><><><><><><><><><><><><><><><><><><><><><><>

If there exists a solution to the above model, we say that the model is in

EQUILIBRIUM.

Leontief Input Constraint:

The sum of the coefficients in each column of **D** is < 1.

The entries in the jth column of **D** represent the number of units of the various input commodities needed to produce one output unit of commodity j. For uniformity these are all measured in dollars. Thus the Input Constraint says that it costs less than one dollar to produce one dollar of commodity j.

We can write the equations of the Leontief system in the form

$$x = .4 \cdot x + .2 \cdot y + .1 \cdot z + 100$$

$$y = .2 \cdot x + .4 \cdot y + .1 \cdot z + 50$$

$$z = .15 \cdot x + .2 \cdot y + .2 \cdot z + 100$$

We can solve these using Mathcad's Solve Block as follows:

First make a guess at the answer.

$x := 100 \qquad y := 50 \qquad z := 100$

Set up the solve block

Given

$$x = .4 \cdot x + .2 \cdot y + .1 \cdot z + 100$$

$$y = .2 \cdot x + .4 \cdot y + .1 \cdot z + 50$$

$$z = .15 \cdot x + .2 \cdot y + .2 \cdot z + 100$$

$$u := \mathrm{Find}(x, y, z)$$

$$u = \begin{pmatrix} 276.316 \\ 213.816 \\ 230.263 \end{pmatrix}$$

Press F9 to see the answer.

Explain what the answer means in terms of the model.

ANSWER:

Fill in new coefficients for the equations given below consistent with the requirements that they be ≥ 0 and that the sum of the coefficients in each column be less than 1. Does the system have a solution for your choice of coefficients?

$x := 50 \qquad y := 50 \qquad z := 50$

Given

$$x = \blacksquare \cdot x + \blacksquare \cdot y + \blacksquare \cdot z + 100$$

$$y = \blacksquare \cdot x + \blacksquare \cdot y + \blacksquare \cdot z + 50$$

$$z = \blacksquare \cdot x + \blacksquare \cdot y + \blacksquare \cdot z + 100$$

$$u := \mathrm{Find}(x, y, z)$$

ANSWER:

◇◇◇◇◇◇◇◇◇◇◇◇◇◇◇◇◇◇◇◇◇◇◇◇◇◇◇◇◇◇◇◇◇◇◇◇◇◇◇

Exercise: (Individual) Describe the three models introduced in this section in your own words.

CHAPTER 1

VECTORS
AND
MATRICES

$$\begin{bmatrix} \mathbf{a}_{1,1} & \mathbf{a}_{1,2} & \mathbf{a}_{1,3} & \mathbf{a}_{1,4} \\ \mathbf{a}_{2,1} & \mathbf{a}_{2,2} & \mathbf{a}_{2,3} & \mathbf{a}_{2,4} \\ \mathbf{a}_{3,1} & \mathbf{a}_{3,2} & \mathbf{a}_{3,3} & \mathbf{a}_{3,4} \\ \mathbf{a}_{4,1} & \mathbf{a}_{4,2} & \mathbf{a}_{4,3} & \mathbf{a}_{4,4} \end{bmatrix}$$

Section 1-1 Matrices

Name:_______________

Score:____________

Overview

A **matrix** is a *data structure* consisting of rows and columns. Its entries, the positions in the matrix, represent information. Since the information we usually deal with is numeric in nature, we view a matrix as a **rectangular array of numbers** (or symbols representing numbers).

$$\begin{pmatrix} -3 & 0 & 0 & 7 & 0 \\ 4 & 11 & -9 & 8.1 & 0 \\ 2 & 1 & 2.31 & 0.5 & 1 \end{pmatrix}$$

This is a 3 by 5 matrix.

A **matrix** is a rectangular array of numbers. The contents of a matrix are called **entries**, **elements**, or **components**. We refer to its **dimension** or **size** as the number of rows by the number of columns. A 2 by 3 matrix has two rows and three columns. (The word matrix is Latin for womb. Originally the word referred to a matrix of type used by a printer.)

Create a 2 by 3 matrix called **M**.

To create a matrix we must first open a dialog box similar to the one on the right. This can be done in three different ways.
(1) Press Ctrl M.
(2) Click on Math menu above and then on Matrix.
(3) Click on the button on the matrix palette.

Matrices
Rows: 2 Columns: 3
Create Insert Delete

Assignment is done by using := which is entered either by typing a colon or by clicking on the := button on the evaluation palette.

Enter **M** followed by the assignment symbol:

Click on the small square to the right of := and then on **Create** in the dialog box.

You now should have a two by three matrix of placeholders. To put a value in a particular placeholder, click on it with the mouse and then enter the variable. You may step your way through the matrix by using the Tab key.

Choose numbers for the entries. Choose at least one of the entries to be negative.

Create **M** here ==>

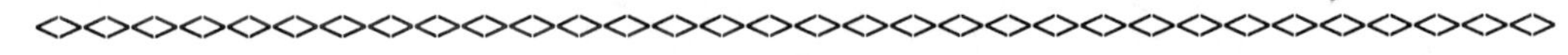

We can always make more space in a document by pressing Ctrl F9. To close up space, press Ctrl F10.

<><><><><><><><><><><><><><><><><><><><><><><><><><><><><><><><><><><><><><>

Create a second matrix called **N** and make it 3 by 3. Choose numbers for the entries.

Create **N** here ==>

<><><><><><><><><><><><><><><><><><><><><><><><><><><><><><><><><><><><><><>

Add **M** and **N**.
To evaluate an expression, you must enter the equal key (=), then press F9.

Show your work here ==>

Explain what happened.

ANSWER:

Create a Text Region to enter your response. Press " to create the text region.

Go back and edit the matrix **N** so that it is 2 by 3.

To change the size of a matrix:

(1) Open the Matrix dialog box.

(2) Enter the number of rows and columns you wish to add or delete.

(3) Place your cursor next to an element in the matrix. Mathcad inserts rows above the element chosen and columns to the left; it deletes rows below and columns to the right.

(4) Click on **Insert** to insert elements or **Delete** to delete elements.

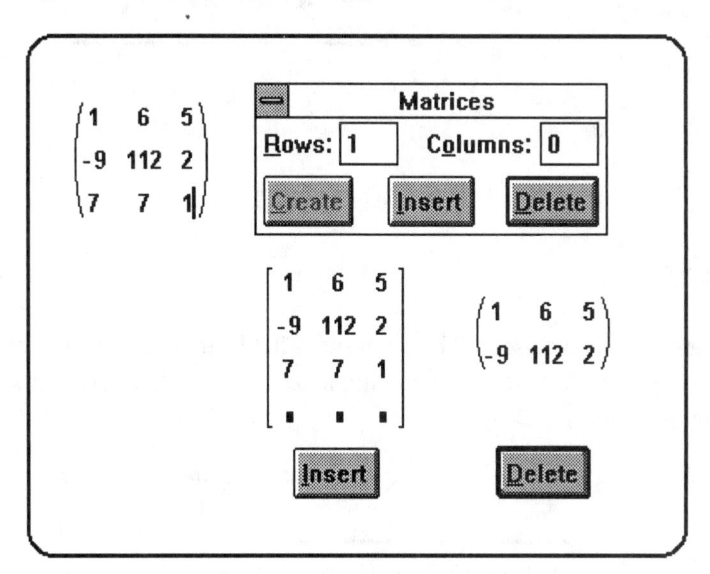

Now try to add **M** and **N** below.

Explain how addition of matrices is performed. Are there any restrictions?

ANSWER:

Create a Text Region to enter your response by pressing ".

<><><><><><><><><><><><><><><><><><><><><><><><><><><><><><><><><>

Determine which of the following sums are possible. If it is possible, display the result.

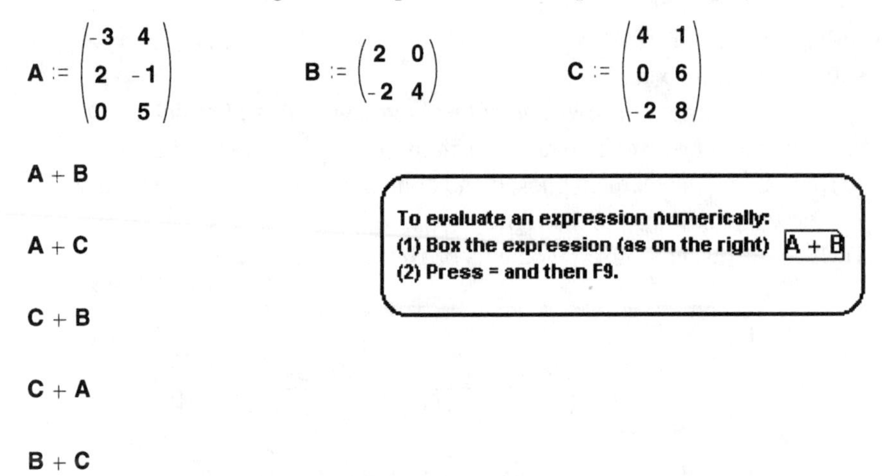

$$\mathbf{A} := \begin{pmatrix} -3 & 4 \\ 2 & -1 \\ 0 & 5 \end{pmatrix} \qquad \mathbf{B} := \begin{pmatrix} 2 & 0 \\ -2 & 4 \end{pmatrix} \qquad \mathbf{C} := \begin{pmatrix} 4 & 1 \\ 0 & 6 \\ -2 & 8 \end{pmatrix}$$

$\mathbf{A} + \mathbf{B}$

$\mathbf{A} + \mathbf{C}$

$\mathbf{C} + \mathbf{B}$

$\mathbf{C} + \mathbf{A}$

$\mathbf{B} + \mathbf{C}$

$\mathbf{B} + \mathbf{A}$

Many of the matrices that we use will be numeric. That is, their entries will be numbers. However, we can also use matrices that contain letters that represent numbers. We will call such matrices **symbolic**.

Symbolic matrices are created the same way you created numeric matrices.

To work with symbolic matrices we do not name them; instead, we work directly with the matrix.

Create a pair of 2 by 2 symbolic matrices below; make the entries different letters.

Your matrices ==>

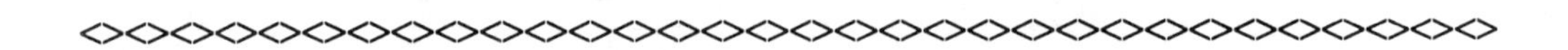

If you are using Mathcad 5.0, click on the Maple Leaf icon **on the toolbar** to load the Symbolic Processor. This is not necessary in Mathcad 6.0.

To add the symbolic matrices you created above:

1. Press +.
2. Copy and paste the first into the first placeholder.
3. Copy and paste the second matrix into the second placeholder.
4. Click on the + to box the expression.
5. Press Shift F9.

ANSWER:

<><><><><><><><><><><><><><><><><><><><><><><><><><><><><><><><><><><><><><><>

Numbers like 2 , 3.2157, and e are called **scalars**. Multiply a matrix by a scalar. (Multiply is the Shift 8 key, *.)

$$3 \cdot \begin{pmatrix} 4 & -2 & 3 \\ 1 & 0 & -1 \end{pmatrix}$$

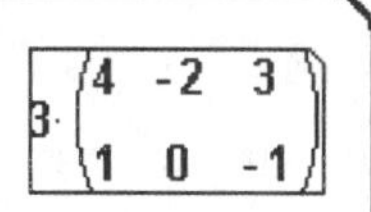

Explain how *scalar multiplication* is performed.

ANSWER:

Perform scalar multiplication symbolically.

$$k \cdot \begin{pmatrix} a & -5 \\ b & 8 \end{pmatrix}$$

To evaluate an expression symbolically:
(1) Box the expression (as on the right)
(2) Either press Shift F9 or click on Evaluate Symbolically on the Symbolic Menu.

$$k \cdot \begin{pmatrix} a & c \\ b & d \end{pmatrix}$$

<><><><><><><><><><><><><><><><><><><><><><><><><><><><><><><><><><><><><><><>

Let **A** and **B** be m by n matrices. Explain carefully what we mean when we say **A** equals **B** (written as **A** = **B**).

ANSWER:

<><><><><><><><><><><><><><><><><><><><><><><><><><><><><><><><><><><><><><><>

Which of the following *laws* hold for matrices and scalars? Matrices will be denoted in this text by **bold** capital letters; scalars, by small letters (either italic or plain, never bold). Assume below that all matrices have the same dimension. Write a short description explaining why the law holds or fails in each case.

1. Commutative Law: $\mathbf{M} + \mathbf{N} = \mathbf{N} + \mathbf{M}$.
2. Associative Law: $(\mathbf{M} + \mathbf{N}) + \mathbf{P} = \mathbf{M} + (\mathbf{N} + \mathbf{P})$.
3. First Distributive Law: $a \cdot (\mathbf{M} + \mathbf{N}) = a \cdot \mathbf{M} + a \cdot \mathbf{N}$.
4. Second Distributive Law: $(a + b) \cdot \mathbf{M} = a \cdot \mathbf{M} + b \cdot \mathbf{M}$.
5. Additive Identity: There is a matrix $\mathbf{0}$ such that $\mathbf{0} + \mathbf{M} = \mathbf{M} = \mathbf{M} + \mathbf{0}$.
6. Additive Inverse: $(-1) \cdot \mathbf{M} + \mathbf{M} = \mathbf{0}$.

ANSWERS:

<><><><><><><><><><><><><><><><><><><><><><><><><><><><><><><><><><><><><>

Subtraction of matrices adds the first matrix to the negative of the second. We write $\mathbf{A} - \mathbf{B}$, but this is to be interpreted as $\mathbf{A} + (-1) \cdot \mathbf{B}$. Let

$$A := \begin{pmatrix} 3 & 1 \\ 4 & 6 \\ -2 & 0 \end{pmatrix} \qquad B := \begin{pmatrix} -2 & 3 \\ 5 & -4 \\ 6 & 8 \end{pmatrix}$$

$A - B$

To evaluate an expression numerically:
(1) Box the expression (as on the right) $\boxed{A - B}$
(2) Press = and then F9.

Evaluate the following expressions symbolically.

$$\begin{pmatrix} s & 8 \\ t & a \end{pmatrix} - \begin{pmatrix} b & 8 \\ 3 & c \end{pmatrix}$$

To evaluate an expression symbolically:
(1) Box the expression (as on the right)
(2) Either press Shift F9 or click on
Evaluate Symbolically on the Symbolic Menu.

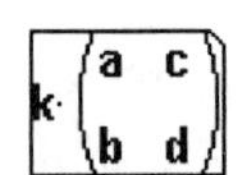

<><><><><><><><><><><><><><><><><><><><><><><><><><><><><><><><><><><><><>

$\mathbf{P}_{i,j}$ (called the *(i,j)th element of the matrix* **P**) is the element in the ith row and jth column of **P**. The row and column subscripts must be separated by a comma.

To enter a subscript, either press [on the keyboard or click on the X_i key on the matrix palette. (This is on menu 1 in Mathcad 5.0.)

To exit subscript mode press the space bar.

Enter a 2 by 4 matrix, **P**. Write and evaluate the expression for the element in the 2nd row and 3rd column in terms of the matrix name **P**.
ANSWER:

Set the element in the first row and fourth column equal to 55. (Use the $\mathbf{P}_{i,j}$ notation.)

ANSWER:

Are the following symbols recognized by Mathcad? If so, what do they mean? $\mathbf{P}_2$? $\mathbf{P}^{<2>}$?
ANSWER:

To enter $P^{<2>}$ click on the Matrix Column button $M^{<>}$ on the matrix palette.

<><><><><><><><><><><><><><><><><><><><><><><><><><><><><><><><><><><>

The transpose of a matrix is denoted by $\mathbf{M}^T$.

To enter M^T click on the M^T icon on the matrix palette or Enter Ctrl 1 from the keyboard.

Examine the properties of the transpose. How is $\mathbf{M}^T$ related to **M**? Use Mathcad to compute each of the following. Then explain the relationship between **M** and the result of the computation.

$\mathbf{M}^T$ $\qquad\qquad\qquad\qquad$ ${\mathbf{M}^T}^T$

ANSWER:

Compare $\mathbf{M}^T + \mathbf{N}^T$ with $(\mathbf{M} + \mathbf{N})^T$.

ANSWER:

Write a Mathcad expression whose value is the 2nd row of the matrix **M**. (Hint: Use transposes and column extraction <> together on the matrix **M**.)

ANSWER:

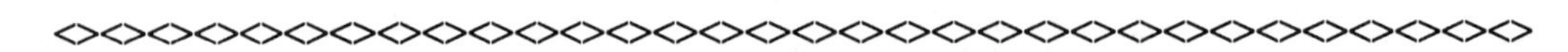

Exercises

1. For each of four months the FLIP Box Company made the following purchases:

pulp:	120,	142,	93,	136	(tons)
glue:	14,	17,	10,	16	(tons)
laminate:	1.5,	2.1,	.85,	1.72	(tons)

a) Construct a matrix **M** representing this information where each row corresponds to a product and each column to a month.
ANSWER:

b) Construct a matrix **R** representing this information where each row corresponds to a month and each column to a product.
ANSWER:

c) An audit showed that in month 2 only 15.5 tons of glue were purchased. Correct matrices **M** and **R**, but rename these new data structures **MM** and **RR** respectively.
ANSWER:

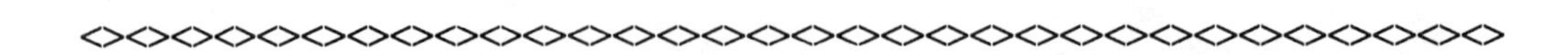

2. Both Beth and LuAnn started work at $750 per month. After the first month, performance incentives gave them a raise each month. Over a period of 6 months Beth's raises were 2%, 3%, 0%, 4%, 2%, 3% while LuAnn's were 0%, 4%, 2%, 3%, 3%, 2%.

a) Create a matrix **S** giving Beth's salary for each of the seven months in column 1 and LuAnn's in column 2. (Do this directly, do not try to be clever.)

ANSWER:

b) Describe in a short paragraph how you computed the salaries of each woman. Who earned more for the seven month period? (To type this paragraph below, position the mouse where you want to start and click it. Then go to the **Text** menu and select **Create Text Region**. Begin to type.)

ANSWER:

<><><><><><><><><><><><><><><><><><><><><><><><><><><><><><><><><><><><><><><><>

3. There are four different 'equals signs' in Mathcad. If you click on the icon (on the palette) with the right mouse button you will see the name of the operation. In Mathcad 6.0 all of these operations are located on the evaluation palette. For help about the meanings, click on the **Help** menu, then click on **Index**, and **Search** for **Equal signs**. When you see text in green, you may click on that text for more information. Explain below, in your own words, what each of the operations means. (Follow the directions in 2 b) to create a text region.)

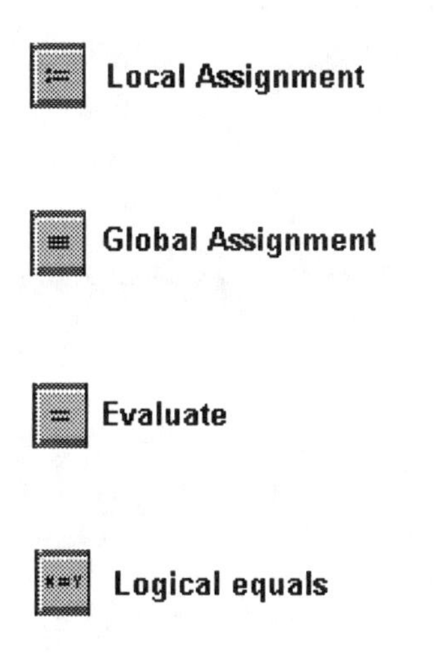

<><><><><><><><><><><><><><><><><><><><><><><><><><><><><><><><><><><><><><><><>

4. (Individual) Write a summary of the mathematics you learned in this section.

Save your work before closing this file.

The procedure for saving a file depends on the installation of the handbook. Ask your instructor how the files should be saved.

Section 1-2 Vectors

Name:________________

Score:_____________

Overview

There is a special subset of matrices called **vectors**. We investigate three **equivalent** interpretations of vectors with two components: ordered pairs of real numbers, points in the plane, and directed line segments. We explore a geometric model for addition and scalar multiplication of such vectors. Intuitively these models extend to vectors with more than two components. Finally we examine a way to build new vectors from other vectors using addition and scalar multiplication.

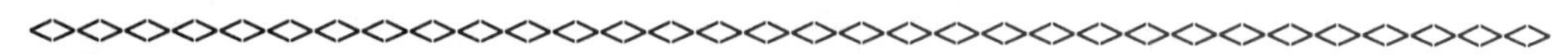

Matrices with exactly two components are called **2-vectors**. Each of the following is a 2-vector:

$$\mathbf{x} := \begin{pmatrix} 2 \\ 5 \end{pmatrix} \quad \mathbf{y} := \begin{pmatrix} 4 \\ 3 \end{pmatrix} \quad \begin{pmatrix} a \\ b \end{pmatrix} \qquad \mathbf{s} := \begin{pmatrix} 1 & -4 \end{pmatrix} \quad \mathbf{t} := \begin{pmatrix} 0 & 3 \end{pmatrix} \quad \begin{pmatrix} c & d \end{pmatrix}$$

The first three vectors are called **column vectors**; the last three, **row vectors**. Be careful to distinguish between row and column vectors. For certain mathematical studies the difference is unimportant, but Mathcad is column oriented, hence for much of our work with vectors we will use columns.

The set of all 2-vectors is denoted by $\mathbf{R}^2$. As you can guess, the **R** stands for real and the 2 indicates the number of components. In general, $\mathbf{R}^k$ is the set of all k-vectors (i.e., there are k components) with real components and $\mathbf{C}^k$ is the set of all k-vectors with complex components.

There are many different ways to look at $\mathbf{R}^2$.

1) $\mathbf{R}^2$ is the collection of all **ordered pairs** of real numbers. We say *ordered* because there is a top number and a bottom number and if we interchange the first and second component we get a different pair.

2) $\mathbf{R}^2$ is the **plane**. Here we identify the first component with the x-coordinate of the point and the second component with the y-coordinate.

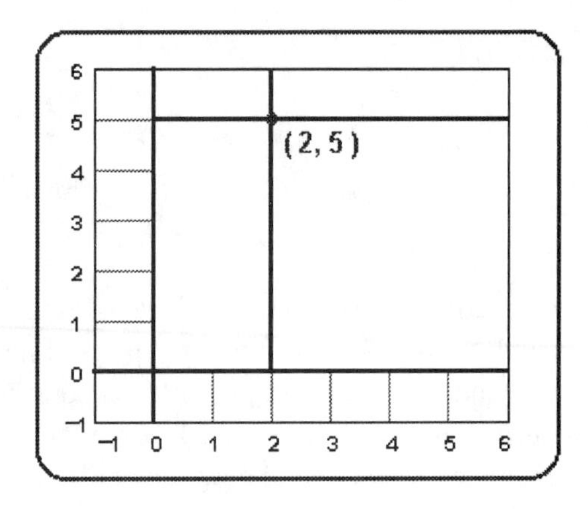

3) $\mathbf{R}^2$ is the set of **vectors** in the plane. Here we identify the pair (x,y) with the line segment from (0,0) = the origin to the point with coordinates x and y.

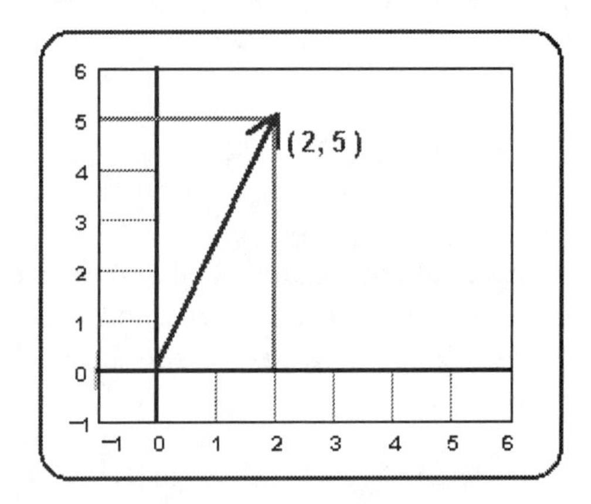

This may sound confusing, but the beauty of it all is that with different representations of the same object we can always choose the representation that makes the most sense in a particular application without worrying that we are jumping back and forth between interpretations.

Let's begin by trying to understand what addition means for each of the three interpretations of $\mathbf{R}^2$ given above. Recall that a 2-vector is a special type of matrix hence addition of 2-vectors is the same as addition of matrices, which is performed by adding corresponding components. Thus both addition of ordered pairs and addition of points in the plane is done the same way. We illustrate using Mathcad's symbolic capabilities.

If you are using Mathcad 5.0, click on the Maple Leaf icon **on the toolbar** to load the Symbolic Processor. This is not necessary in Mathcad 6.0.

Box the expression on the right ======>
and type Shift F9 or click on Evaluate
Symbolically on the Symbolic Menu.

$$\begin{pmatrix} x1 \\ x2 \end{pmatrix} + \begin{pmatrix} y1 \\ y2 \end{pmatrix}$$

The result that appears should confirm the statement about adding corresponding components.

We list below the interpretation of addition for each of the three equivalent representations of $\mathbf{R}^2$.

1) In the first interpretation (ordered pairs) there is no meaning other than the obvious meaning that we add component-wise. This is purely an algebraic function.

2) In the second meaning, points in the plane, addition means that we add the x-components and the y-components to get a new point.

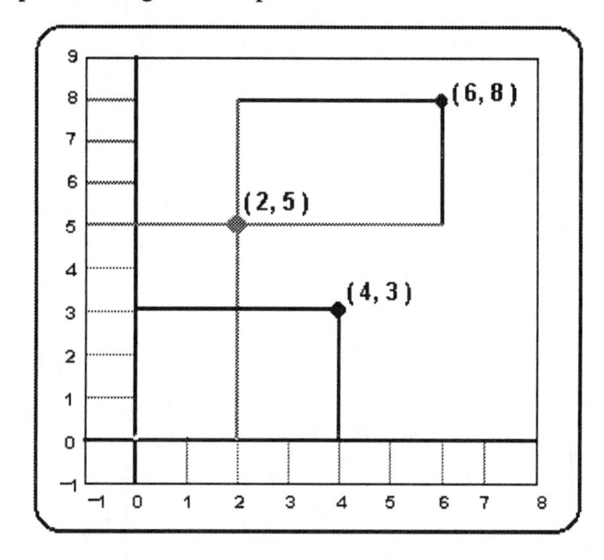

3) In the vector interpretation, the sum means that we translate the second vector so that it begins at the end of the first vector. The vector from the origin to the 'end' of the second vector is then the sum.

$$\begin{pmatrix} 4 \\ 3 \end{pmatrix} + \begin{pmatrix} 2 \\ 5 \end{pmatrix} = \begin{pmatrix} 6 \\ 8 \end{pmatrix}$$

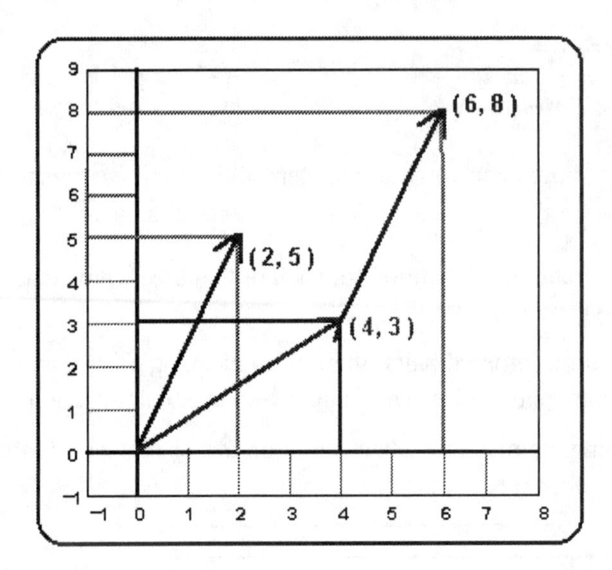

We could have 'added' the vectors in the opposite order.

$$\begin{pmatrix} 2 \\ 5 \end{pmatrix} + \begin{pmatrix} 4 \\ 3 \end{pmatrix} = \begin{pmatrix} 6 \\ 8 \end{pmatrix}$$

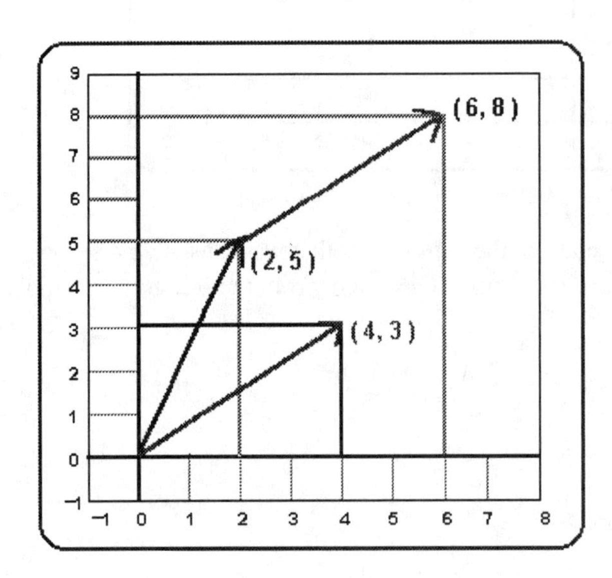

Combining the preceding figures we have the following.

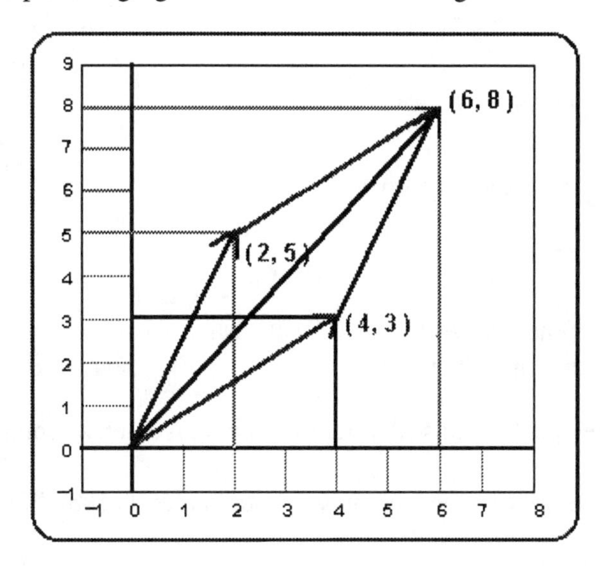

Since the purple vectors (4, 3) are parallel and the red vectors (2, 5) are parallel, they form a parallelogram in which ***the sum is the main diagonal***. This is sometimes referred to as the **parallelogram rule for addition**.

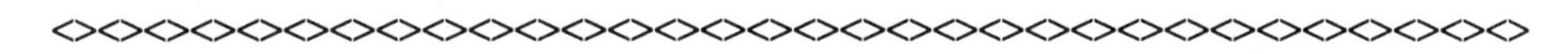

We can write any 2-vector $\begin{pmatrix} x \\ y \end{pmatrix}$ as the sum of two vectors as follows:

$$\begin{pmatrix} x \\ y \end{pmatrix} = \begin{pmatrix} x \\ 0 \end{pmatrix} + \begin{pmatrix} 0 \\ y \end{pmatrix}$$

This is an important idea that we will use later.

a) Describe this in terms of addition of points in $\mathbf{R}^2$.

ANSWER:

b) Describe this in terms of addition of vectors in $\mathbf{R}^2$. (Hint: draw a picture.)

ANSWER:

<><><><><><><><><><><><><><><><><><><><><><><><><><><><><><><><><><><><><><><><><><><><><><><>

The product of a scalar times a vector is called **scalar multiplication**. To see how this is defined evaluate the following expression.

Box and evaluate symbolically. ===> $k \cdot \begin{pmatrix} a \\ b \end{pmatrix}$

Describe how scalar multiplication is defined.

ANSWER:

Using the equivalence of the three forms for $\mathbf{R}^2$ established above we have the following interpretations of scalar multiplication.

1) For ordered pairs the operation of scalar multiplication gives a new point (ka, kb).

2) For points in the plane we get a new point with coordinates (ka, kb). For example if

$a := 2 \qquad b := 4 \qquad k := 2$

then

Box and evaluate numerically. ===> $k \cdot \begin{pmatrix} a \\ b \end{pmatrix}$

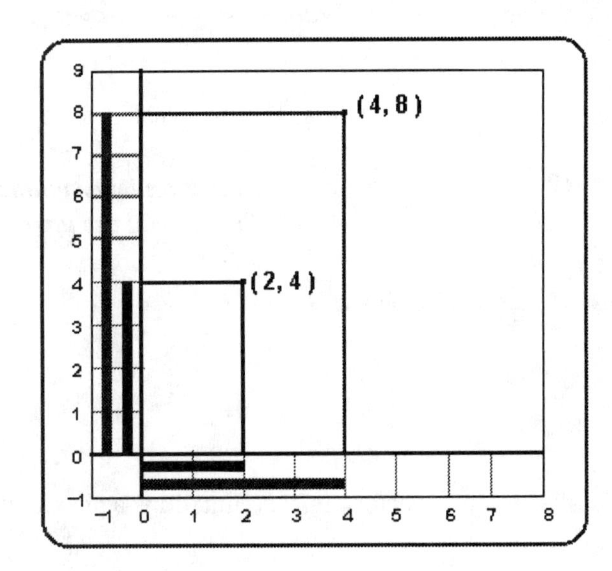

3) For 2-vectors, scalar multiplication changes the length of the vector represented by the line segment connecting the origin to the endpoint.

We illustrate for the vector $\mathbf{v} := \begin{pmatrix} 1 \\ 3 \end{pmatrix}$ and various values of scalar k.

This is the scalar value k that we use below ==> $k := 2$ $i := 1..2$ **<==Indices**

The line from the origin to the vector **v**, can be represented as the matrix

$$\text{original} := \begin{pmatrix} 0 & v_1 \\ 0 & v_2 \end{pmatrix} \qquad k \text{ times this is} \qquad \text{new} := \begin{pmatrix} 0 & k \cdot v_1 \\ 0 & k \cdot v_2 \end{pmatrix}$$

To graph these vectors we compute the maximum and minimum values of the x and y coordinates.

$$\text{xmin} := -1 + \min\left(\begin{pmatrix} 0 & v_1 \\ 0 & k \cdot v_1 \end{pmatrix}\right) \qquad \text{xmax} := 1 + \max\left(\begin{pmatrix} 0 & v_1 \\ 0 & k \cdot v_1 \end{pmatrix}\right)$$

$$\text{ymin} := -1 + \min\left(\begin{pmatrix} 0 & v_2 \\ 0 & k \cdot v_2 \end{pmatrix}\right) \qquad \text{ymax} := 1 + \max\left(\begin{pmatrix} 0 & v_2 \\ 0 & k \cdot v_2 \end{pmatrix}\right)$$

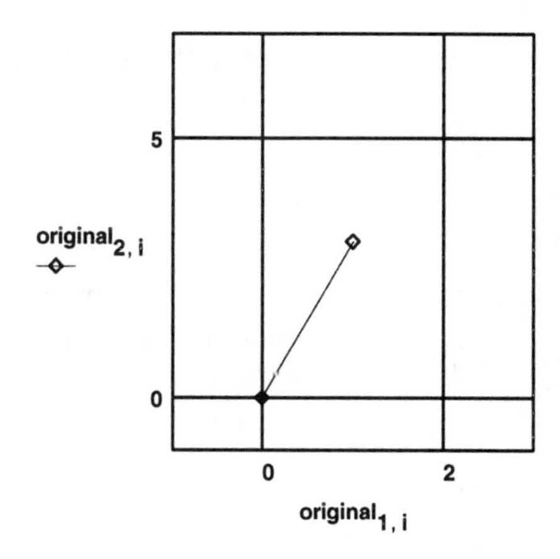

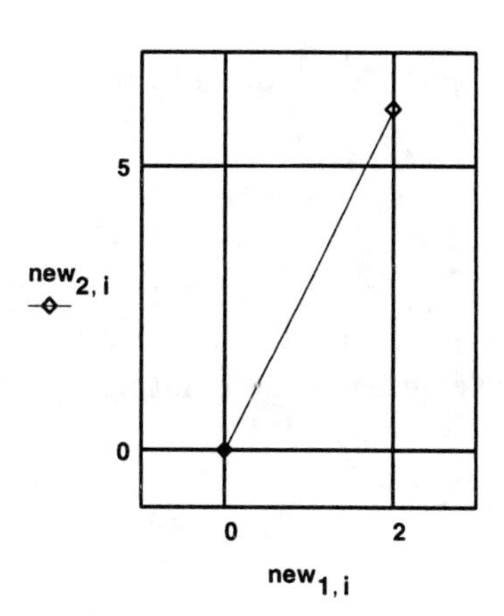

Execute these graphs by pressing F9.

Compare the lengths of the vectors pictured when the scalar $k = 2$.

ANSWER:

Change k to be -1 in the preceding statements and redraw the graphs by pressing F9. Compare the lengths of the vectors pictured.
ANSWER:

Change k to be -3 in the preceding statements and redraw the graphs by pressing F9. Compare the lengths of the vectors pictured.

ANSWER:

Change k to be 1/2 in the preceding statements and redraw the graphs by pressing F9. Compare the lengths of the vectors pictured.
ANSWER:

Change k to be -1/2 in the preceding statements and redraw the graphs by pressing F9. Compare the lengths of the vectors pictured.
ANSWER:

If $\mathbf{v}$ is a vector in $\mathbf{R}^2$, describe the vector $k \cdot \mathbf{v}$ for any scalar k not zero. (Hint: Consider cases.)
ANSWER:

<><><><><><><><><><><><><><><><><><><><><><><><><><><><><><><>

Let n be an integer (1, 2, 3, ...) then we can think about *taking the sum of n copies of a vector* (i.e., $\mathbf{v} + \mathbf{v} + \ldots + \mathbf{v}$ n-times). Is this the same as multiplying (multiplying means *scalar multiplication*) by n?
ANSWER:

Geometrically what does it mean to *take the sum of n copies of a vector*?
ANSWER:

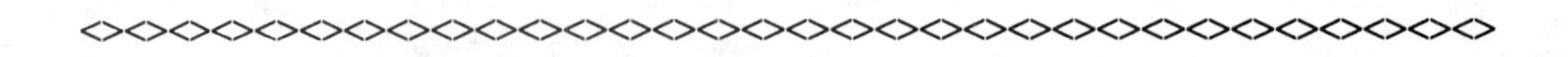

The **subtraction of vectors, v - w**, is the sum of vector **v** and vector (-1)·**w**. That is

$$\mathbf{v} - \mathbf{w} = \mathbf{v} + (-1)\cdot\mathbf{w}$$

Pictorially we have

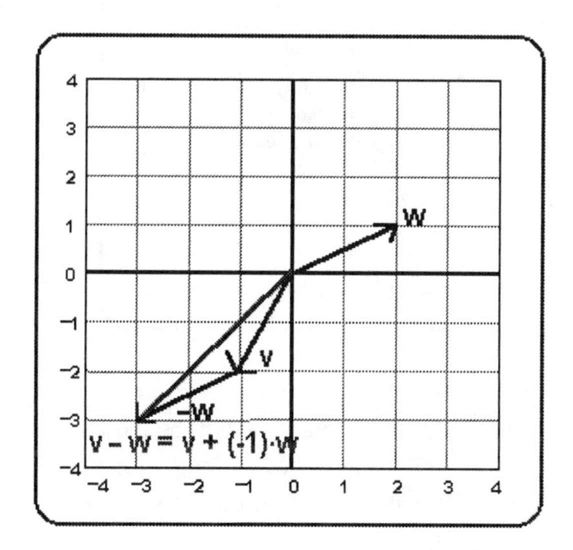

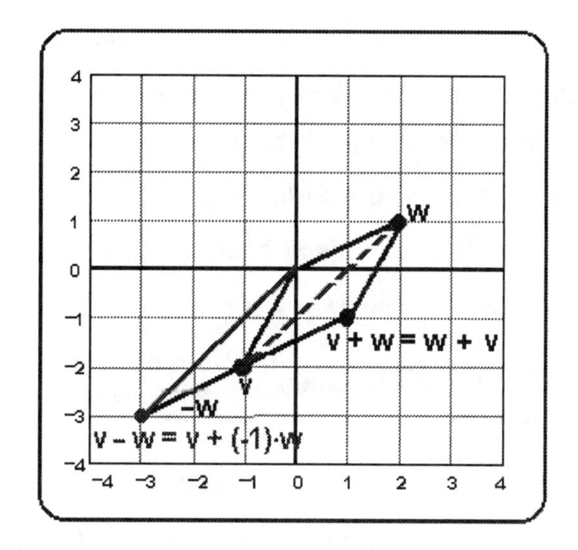

In this picture **w** = (2, 1), and **v** = (-1, 2). The blue line from the origin to (2, 1) is the vector **w** while the red line from the origin to (-1, 2) is the vector **v**. The purple line from the origin to (-3, -3) is then the vector (**v** - **w**). This has the same length and direction as the dotted purple line from **w** to **v**. This is because **w** + (**purple vector**) = **v**; so, the **purple vector** equals (**v** - **w**).

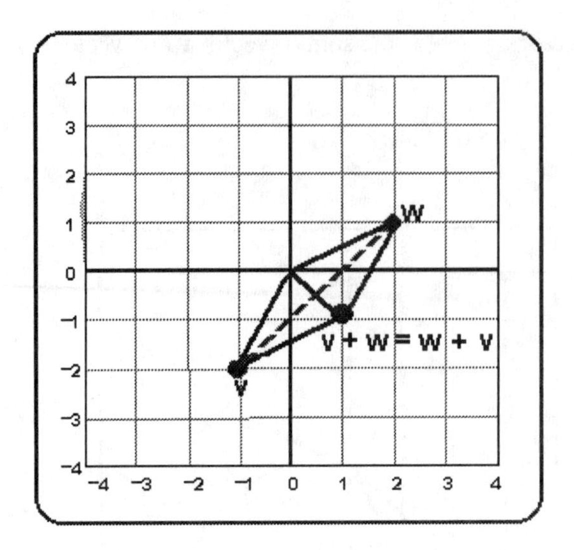

If we add the vector ($\mathbf{v} + \mathbf{w}$) to the previous picture, we can then see that if we draw the parallelogram with sides $\mathbf{v}$ and $\mathbf{w}$, the main diagonal will be the vector sum ($\mathbf{v} + \mathbf{w}$) while the minor diagonal will be either ($\mathbf{v} - \mathbf{w}$) or ($\mathbf{w} - \mathbf{v}$) depending upon the direction of the corresponding vector. (From $\mathbf{v}$ to $\mathbf{w}$ is ($\mathbf{w} - \mathbf{v}$); from $\mathbf{w}$ to $\mathbf{v}$ is ($\mathbf{v} - \mathbf{w}$).)

<><><><><><><><><><><><><><><><><><><><><><><><><><><><><><><><><><><><><><>

The preceding geometric model for vectors in R^2 provides an intuitive foundation for vectors in R^k even though for $k > 3$ we can not visualize a similar model. Since any two vectors in R^k lie in a plane, addition in R^k can also be interpreted in terms of the parallelogram model; similarly scalar multiplication changes the length and possibly reverses the direction of k-vectors.

<><><><><><><><><><><><><><><><><><><><><><><><><><><><><><><><><><><><><><>

Exercise: (Individual) Write a summary of the mathematics that you learned in this section.

Save your work before closing this file.

Section 1-3 Linear Combinations

Name:________________

Score:_____________

Overview

A **linear combination of vectors** is a new vector formed by sums and differences of scalar multiples of the original vectors. Let **v** and **w** be vectors in $\mathbf{R}^n$. Their sum **v** + **w** and their difference **v** - **w** are linear combinations of **v** and **w**. More generally, any expression of the form $k \cdot \mathbf{v} + t \cdot \mathbf{w}$, where k and t are scalars is a linear combination of the vectors **v** and **w**.

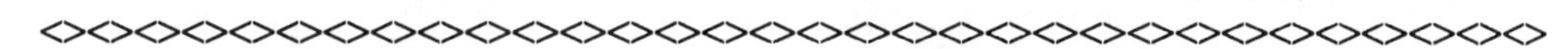

The notion of linear combinations extends to more than two vectors. Let $\mathbf{v}_1, \mathbf{v}_2, \ldots, \mathbf{v}_n$ be a set of n vectors and let a_i, i = 1, 2, ..., n be scalars, then the expression

$$a_1 \cdot v_1 + a_2 \cdot v_2 + \ldots + a_n \cdot v_n = \sum_{i=1}^{n} a_i \cdot v_i$$

is a linear combination of the vectors $\mathbf{v}_i$, i = 1, 2, ..., n.

We are not restricted to linear combinations of vectors. We can form linear combinations of matrices (all the matrices must be the same size) or functions. The well known example in terms of functions is the polynomial:

$$a_0 + a_1 x + a_2 x^2 + \ldots + a_n x^n$$

which is a linear combination of 1, x, x^2, ..., x^n.

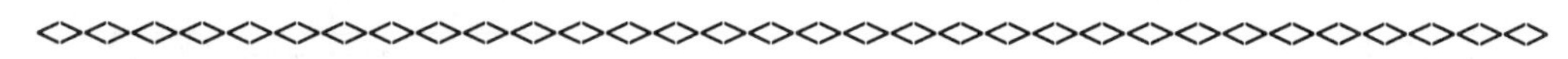

Examples

Linear combination of column vectors. =====> $5 \cdot \begin{pmatrix} -1 \\ 2 \end{pmatrix} - 3 \cdot \begin{pmatrix} 2 \\ 4 \end{pmatrix}$
Determine the result of this combination.

Linear combination of matrices. ====> $3 \cdot \begin{pmatrix} 1 & -4 \\ 2 & 5 \end{pmatrix} - 2 \cdot \begin{pmatrix} 17 & .5 \\ -11 & 1 \end{pmatrix} - \frac{1}{2} \cdot \begin{pmatrix} 1 & 0 \\ 0 & 1 \end{pmatrix}$
Evaluate this expression.

A polynomial. ====> $2 - 4 \cdot x + 5 \cdot x^2$

A linear combination of trig functions. ====> $-1 \cdot \cos(x) + 4 \cdot \sin(x)$

Linear combinations play an important role in the study of linear algebra.

- Given a set of vectors we can construct other vectors by forming linear combinations.
- Given a particular vector, can we write it as a linear combination of a specified set of vectors? (This is a *construction question*; that is, is it possible to build a particular vector from a specified set of other vectors?)

Convex Linear Combinations

For a set of vectors $\mathbf{v}_1, \mathbf{v}_2, \ldots, \mathbf{v}_n$, the **family** of all linear combinations

$$t_1 \cdot v_1 + t_2 \cdot v_2 + \ldots + t_n \cdot v_n \qquad \text{where} \qquad t_1 + t_2 + \ldots + t_n = 1$$

is called **the set of all convex linear combinations** of the vectors $\mathbf{v}_1, \mathbf{v}_2, \ldots, \mathbf{v}_n$.

Here we think of both the vectors and the linear combinations as **points**.

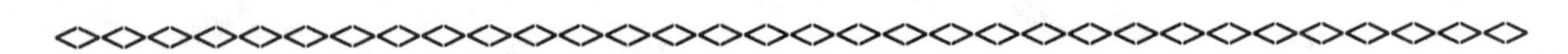

Two points

We begin by considering the case of two points $\mathbf{v}$ and $\mathbf{w}$ and all points

$$z(t) = (1 - t) \cdot v + t \cdot w \qquad \text{where} \qquad 0 \le t \le 1$$

It is easy to see that $\mathbf{z(0)} = \mathbf{v}$ and $\mathbf{z(1)} = \mathbf{w}$.
As t changes from 0 to 1, $\mathbf{z(t)}$ *moves* from $\mathbf{v}$ to $\mathbf{w}$.

We examine the points $(1\text{-}t)\cdot\mathbf{v} + t\cdot\mathbf{w}$ for t between 0 and 1. We do this with an animation below. First look at the fixed picture on the right. This is the picture for t = 0.3. The point, $\mathbf{v}$, is indicated by an open red diamond. The point $\mathbf{w}$ is shown as an open blue diamond and the point $(1 - t)\cdot\mathbf{v} + t\cdot\mathbf{w}$ as a magenta (purple) diamond. What is the point A? What is the point B?

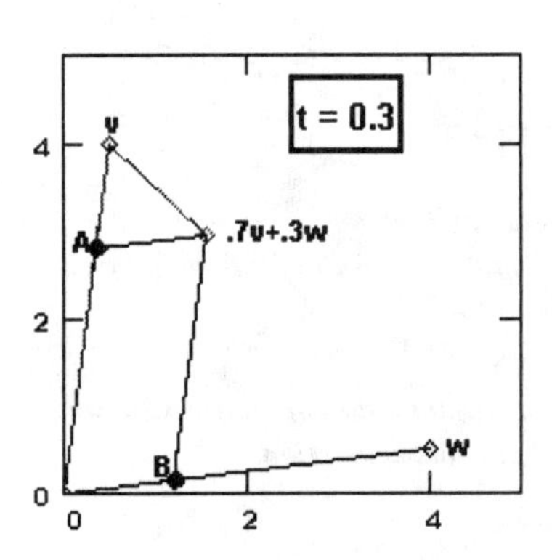

ANSWER:

What is the magenta line between $\mathbf{v}$ and $0.7\cdot\mathbf{v} + 0.3\cdot\mathbf{w}$?

ANSWER:

We now animate the picture given above as t goes from 0 to 1. The Mathcad statements below are required for the animation and the graphics. You may ignore them.

The points ==> $\quad \mathbf{v} := \begin{pmatrix} 1 \\ 3 \end{pmatrix} \quad \mathbf{w} := \begin{pmatrix} 2 \\ 1 \end{pmatrix} \quad \mathbf{o} := \begin{pmatrix} 0 \\ 0 \end{pmatrix} \quad \mathbf{k} := 1..2 \quad \mathbf{kk} := 1..3$

The points z(t) ==> $\quad \mathbf{z(t)} := (1 - t)\cdot \mathbf{v} + t\cdot \mathbf{w} \quad \mathbf{ZZ_1} := 1 \quad \mathbf{ZZ_2} := -0.02 \quad$ **WRITEPRN(QQ) := ZZ**

◎ $\mathbf{ZZ_2} := \mathbf{READPRN(QQ)_2} + .02$ **WRITEPRN(QQ) := ZZ** **t := min(ZZ)**

x := augment(v·(1 – t), augment(z(t), w·t)) **ww := augment(w, o)**

vv := augment(augment(o, v), z(t)) **zz := augment(v, z(t))**

Current value of t.

t = 0.52

◇ v
◇ w

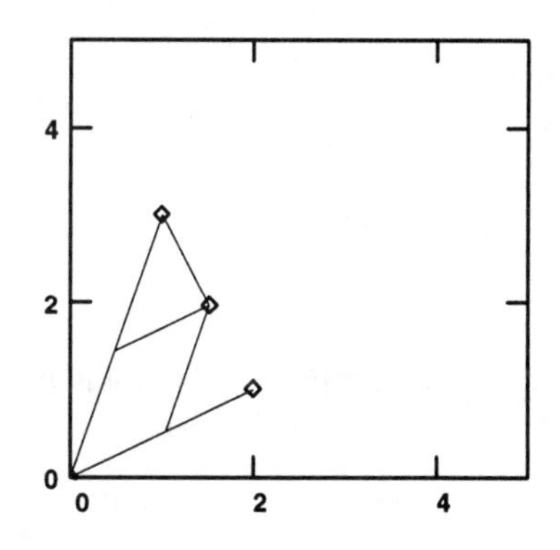

To view this animation. Place your cursor between the Z's at the start of the line marked by ◎ above. Then press F9 repeatedly until t = 1.

Now try other values for **v** and **w**. To keep the points in the graph window choose the coordinates between 0 and 5. To initialize the animation place your cursor between the **Z**'s in the expression $\mathbf{ZZ_1}$:=1 above and press F9. Then repeat the animation procedure given above.

Based on your observations complete the following sentence:

The set of points (1 - t)·**v** + t·**w** for $0 \le t \le 1$ describes the __________________ between the points **v** and **w**.

Describe the set of points (1 - t)·**v** + t·**w** for $0 \le t \le 1$ if **v** and **w** are points in $\mathbf{R}^3$.

$$\mathbf{v} := \begin{pmatrix} 2 \\ 0 \\ 9 \end{pmatrix} \quad \text{and} \quad \mathbf{w} := \begin{pmatrix} 7 \\ -3 \\ 2 \end{pmatrix}$$

ANSWER:

What is **z(t)** if **v** = **w**?

ANSWER:

<><><><><><><><><><><><><><><><><><><><><><><><><><><><><><><><><><><><><><><><>

Three points

Let $\mathbf{v} := \begin{pmatrix} 0.5 \\ 4 \end{pmatrix}$ $\mathbf{w} := \begin{pmatrix} 4 \\ .5 \end{pmatrix}$ and $\mathbf{u} := \begin{pmatrix} 2 \\ 4 \end{pmatrix}$

We wish to determine all convex linear combinations of **u**, **v** and **w**. We write this as $\mathbf{r \cdot u + s \cdot v + t \cdot w}$ where $\mathbf{0 \le r, s, t \le 1}$ and $\mathbf{r + s + t = 1}$

We can rewrite this as $\mathbf{(1 - t) \cdot (q \cdot u + (1 - q) \cdot v) + t \cdot w}$ where $\mathbf{q = \frac{r}{r + s}}$

when $r + s \neq 0$. If $r + s = 0$, $t = 1$ and $1 - t = 0$ and it doesn't matter what q is.

It follows that this set is equal to all convex linear combinations of the point **w** and the points $q \cdot \mathbf{u} + (1 - q) \cdot \mathbf{v}$ where $0 \le q \le 1$.

The set $q \cdot \mathbf{u} + (1 - q) \cdot \mathbf{v}$ is, as we saw above, the line segment between **u** and **v**. Complete the following sentence.

The set of convex linear combinations of three points in the plane is ________________.

We check this graphically. $\mathbf{v} := \begin{pmatrix} 0.5 \\ 4 \end{pmatrix}$ $\mathbf{w} := \begin{pmatrix} 4 \\ .5 \end{pmatrix}$ $\mathbf{u} := \begin{pmatrix} 2 \\ 4 \end{pmatrix}$ $\mathbf{t := 0, .02 .. 1}$ $\mathbf{q := 0, .01 .. 1}$

$$\mathbf{z(t, q) := (1 - t) \cdot (q \cdot u + (1 - q) \cdot v) + t \cdot w}$$

$$\mathbf{zz(q) := augment(z(0, q), z(1, q))} \qquad \mathbf{i := 1 .. 2}$$

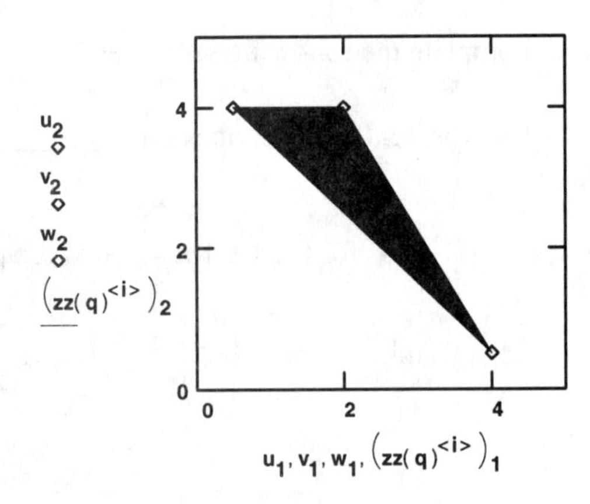

What would the result be if **u**, **v** and **w** were collinear?

ANSWER:

◇◇◇◇◇◇◇◇◇◇◇◇◇◇◇◇◇◇◇◇◇◇◇◇◇◇◇◇◇◇◇◇◇◇◇◇◇

Exercise

Rather than look at convex linear combinations let's examine **the set of all possible linear combinations** of a pair of vectors **u** and **v** in $\mathbf{R}^2$. That is, the set of all

$$s \cdot \mathbf{u} + t \cdot \mathbf{v}, \quad \text{where} \quad -\infty \leq s, t \leq \infty$$

Of course there is no way to do this since we can only look at a portion of the plane at one time. To investigate this problem change the values that t and s run through. The graph below will plot the various points.

$u := \begin{pmatrix} 3 \\ 5 \end{pmatrix}$ $\quad v := \begin{pmatrix} -1 \\ 4 \end{pmatrix}$

$t := 0, .1 .. 5$

$s := 0, .1 .. 5$

These expressions define a range of values. For details use Help above. Then select headings Index, Iteration, Range variables. When you are finished click on File and then on Exit.

$w(s,t) := u \cdot s + v \cdot t$

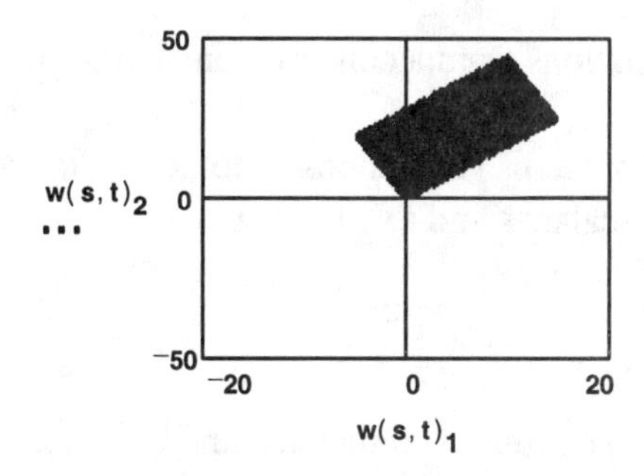

Change the range of t by letting the maximum value decrease from 5 to 4, then 3, then 1, and then set t = 0. Describe how the picture changes.

ANSWER:

Now set the maximum of t back to 5 and decrease the values of s as above. Describe how the picture changes.

ANSWER:

Now set s = 0,-.1..-5 and explain how the picture changes.

ANSWER:

Change the values for **u** and **v** and repeat the investigation.

Describe the values of **w**(s,t) = s·**u** + t·**v** in each of the following cases:

a) **u** = k·**v**.

ANSWER:

b) **u** = **0** the zero vector.

ANSWER:

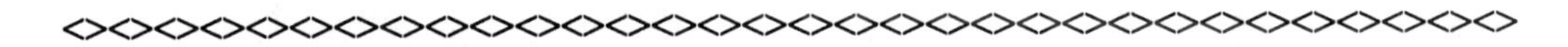

We next look at linear combinations from a constructionist point-of-view.

Choose vectors **u** and **v** and fix them. Then choose another vector **p** in $\mathbf{R}^2$. Experiment to see if it is possible to find the scalars s and t so that s·**u** + t·**v** = **p**. For example, let

$$u := \begin{pmatrix} 4 \\ 3 \end{pmatrix} \qquad v := \begin{pmatrix} 2 \\ -1 \end{pmatrix} \qquad p := \begin{pmatrix} 2 \\ 9 \end{pmatrix}$$

In the following graph we have chosen the range of s and t so that the blue diamond contains the corner opposite to the origin.

$s := 0, .05 .. 2$

$t := 0, -.05 .. -3$

$w(s,t) := u \cdot s + v \cdot t$

These expressions define a range of values. For details use Help above. Then select headings Index, Iteration, Range variables. When you are finished click on File and then on Exit.

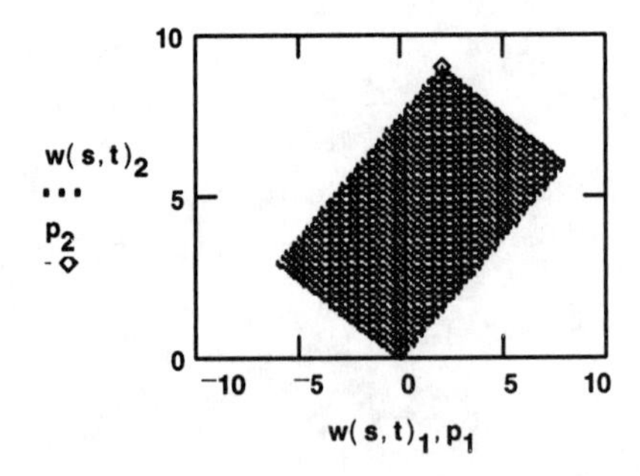

What can you conclude about the relationship between **u**, **v** and **p**?

ANSWER:

<><><><><><><><><><><><><><><><><><><><><><><><><><><><><><><><><><><>

$\mathbf{u} := \begin{pmatrix} 4 \\ 3 \end{pmatrix} \quad \mathbf{v} := \begin{pmatrix} 2 \\ -1 \end{pmatrix}$ $\qquad$ $\mathbf{s} := 0, .05 .. 2$

$\mathbf{t} := 0, -.05 .. -3$

Now let $\qquad \mathbf{p} := \begin{pmatrix} -4 \\ 8 \end{pmatrix}$

Adjust the range variables so that the blue diamond is at the vertex opposite the origin.
Hint: Change one range at a time.

$\mathbf{w(s,t)} := \mathbf{u \cdot s + v \cdot t}$

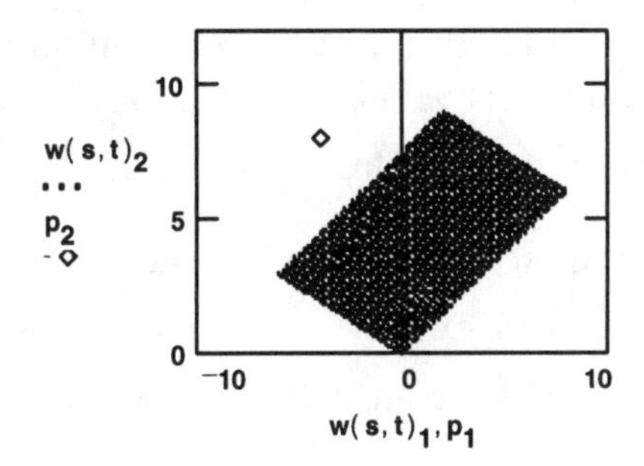

Check your answer by substituting the range limit in the following

$\mathbf{w(\blacksquare,\blacksquare)} =$

Explain what the previous line is computing.

ANSWER:

Explain how to compute the exact values for s and t algebraically. Then compute s and t.

ANSWER:

Under what conditions on **u** and **v** is the following statement true?

Let **p, u** and **v** be vectors in the plane, then there are scalars, s and t such that $\mathbf{p} = s \cdot \mathbf{u} + t \cdot \mathbf{v}$.

ANSWER:

<><><><><><><><><><><><><><><><><><><><><><><><><><><><><><><><><><><>

Exercise: Convex sets

A set of points in n-space ($\mathbf{R}^n$) is said to be convex if whenever $\mathbf{v}$ and $\mathbf{w}$ are in the set, all the points on the line segment between $\mathbf{v}$ and $\mathbf{w}$ are also in the set.

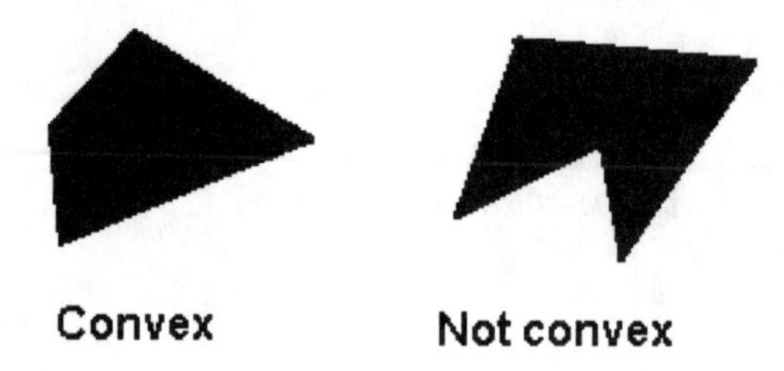

The goal of this exercise is to show that the **the set of all convex linear combinations** of the set $\mathbf{v}_1, \mathbf{v}_2, \ldots, \mathbf{v}_n$ is the smallest convex set that contains the points, $\mathbf{v}_1, \mathbf{v}_2, \ldots, \mathbf{v}_n$.

To see this we must show (1) **the set of all convex linear combinations** of the set $\mathbf{v}_1, \mathbf{v}_2, \ldots, \mathbf{v}_n$ is a convex set (in the above sense); and (2) any point in this set must also be in any other convex set that contains $\mathbf{v}_1, \mathbf{v}_2, \ldots, \mathbf{v}_n$.

Let CLC($\mathbf{v}_1, \mathbf{v}_2, \ldots, \mathbf{v}_n$) stand for **the set of all convex linear combinations** of the set $\mathbf{v}_1, \mathbf{v}_2, \ldots, \mathbf{v}_n$. To show that this is a convex set we must show that whenever $\mathbf{p}$ and $\mathbf{q}$ are in the set, all points of the form $(1-t)\cdot\mathbf{p} + t\cdot\mathbf{q}$ are also in the set for $0 \le t \le 1$.

$$p = r_1\cdot v_1 + r_2\cdot v_2 + \ldots + r_n\cdot v_n \qquad r_1 + r_2 + \ldots + r_n = 1 \qquad 0 \le r_i$$

$$q = s_1\cdot v_1 + s_2\cdot v_2 + \ldots + s_n\cdot v_n \qquad s_1 + s_2 + \ldots + s_n = 1 \qquad 0 \le s_i$$

To show that $(1-t)\cdot\mathbf{p} + t\cdot\mathbf{q}$ **is in** CLC($\mathbf{v}_1, \mathbf{v}_2, \ldots, \mathbf{v}_n$), calculate the sum of the coefficients of the $\mathbf{v}$'s in the expression below.

$$(1 - t)\cdot\left(r_1\cdot v_1 + r_2\cdot v_2 + \ldots + r_n\cdot v_n\right) + t\cdot\left(s_1\cdot v_1 + s_2\cdot v_2 + \ldots + s_n\cdot v_n\right)$$

ANSWER:

To prove (2) above we must show that each point in CLC($\mathbf{v}_1, \mathbf{v}_2, \ldots, \mathbf{v}_n$) must be in every convex set containing the $\mathbf{v}$'s. If n = 2 this is trivially true.

Section 1-4 Dot Product (Scalar Product)

Name:________________

Score:______________

Overview

The **dot product** of two vectors is a function that associates with a pair of vectors a scalar quantity. A dot product is also called a **scalar product** or an **inner product**. The dot product computation makes it easy to calculate the distance between vectors, the length of a vector, and the angle between two vectors. These concepts are related to the geometry of vectors, especially in $\mathbf{R}^2$ and $\mathbf{R}^3$ where we can construct geometric models.

<><><><><><><><><><><><><><><><><><><><><><><><><><><><><><>

Define a pair of numeric 4-vectors (column vectors). Call them **u** and **v**.

u := ▪ v := ▪

Evaluate **u**·**v**. (Multiplication is the Shift 8 key.) ==>

The result is a scalar, a number. Our objective is to discover how the output from this operation is computed.

Make a conjecture here! ==>
(*Conjecture* is a fancy word for *guess*.)

Change the entries of **u** and **v** and see if your conjecture is valid. Experiment to find a pattern that involves the use of the components of the vectors **u** and **v**. Revise your conjecture as needed.

<><><><><><><><><><><><><><><><><><><><><><><><><><><><><><>

If you are using Mathcad 5.0, click on the Maple Leaf icon **on the toolbar** to load the Symbolic Processor. This is not necessary in Mathcad 6.0.

Enter symbolic values in the vectors below (e.g., x, y, t). Box the product by clicking on the multiplication symbol between the vectors. Then type Shift F9 or click on **Evaluate Symbolically** on the **Symbolic** Menu to evaluate the product. (This product will not work with row vectors.)

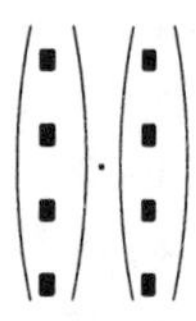

Carefully inspect the output from the symbolic dot product calculation. Write a sentence that tells how the dot product of a pair of vectors is computed.
ANSWER:

Can you take the dot product of two vectors that have different sizes (e.g., a 5-vector and a 3-vector)? Try it and see. Explain carefully.

ANSWER:

Suppose that in a given course there are five grades (three tests, a term paper, and a final exam). Each test counts for 15% of the course grade, the paper counts 25 % and the final exam 30%. Use the language of vector algebra, including dot product, to compute the course grade. Compute the course grade for a student who earns 75, 85, and 88 on the three tests; 92 on the term paper and 83 on the final.

ANSWER:

Let **g** be the vector of grades.

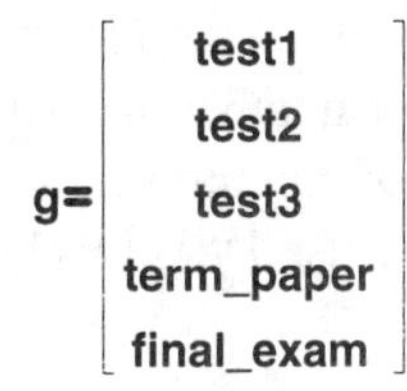

Use the dot product to express the formula for the course grade in terms of the vector **g**.

ANSWER:

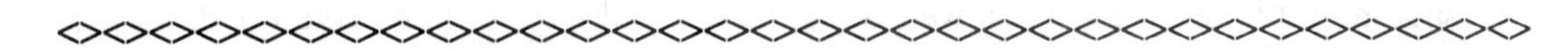

Let **v** be a vector with entries $\mathbf{v}_1, \mathbf{v}_2, \ldots, \mathbf{v}_k$. Write a brief description of the result of taking the dot product of **v** with itself.

ANSWER:

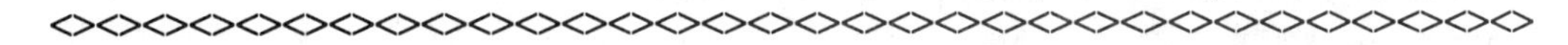

Up to this point, we have considered the dot product of two column vectors. From a mathematical point of view, it makes just as much sense to consider the dot product of two row vectors. Try this in Mathcad and report on the results of your experimentation below.

ANSWER:

<><><><><><><><><><><><><><><><><><><><><><><><><><><><><><><><><><><><><>

Now try to take the dot product of a row vector with a column vector. Does this yield the same result as the dot product of the transpose of the row vector with the column vector? Can you take the dot product of a row vector with a column vector when they have a different number of entries?

ANSWER:

<><><><><><><><><><><><><><><><><><><><><><><><><><><><><><><><><><><><><>

At first glance it might appear that the dot product of a row vector and a column vector is the same as the dot product of two column vectors. Check this further by taking the reciprocal of each of these numbers (i.e., 1 over each of these objects; **do not take the number to the (-1) power!**).

ANSWER:

If **M** is a matrix, the Mathcad function rows(**M**) returns the number of rows in **M** and the function cols(**M**) returns the number of columns. For each of the objects in the previous exercise (i.e., dot product of column vectors and dot product of row vector with column vector) compute rows and cols. Does this help to explain the previous result?

ANSWER:

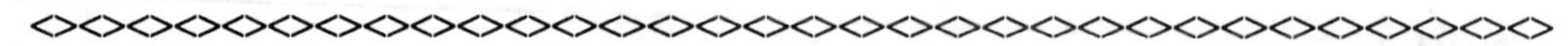

The moral of the previous exercise is that sometimes objects are other than what they appear to be. In particular, a one by one matrix and a scalar 'look' the same on the screen. We shall return to this topic when we discuss matrix multiplication in Section 1-8.

<><><><><><><><><><><><><><><><><><><><><><><><><><><><><><><><><><><><><>

For completeness, you should take the dot product of a column vector with a row vector. In this case the two vectors do not need to have the same dimension. The result of this multiplication is neither a scalar nor a one by one matrix instead it is a matrix called the *outer product*. Use symbolic entries and multiplication to derive the formula for the outer product. Explain how each row of the outer product is computed.

ANSWER:

<><><><><><><><><><><><><><><><><><><><><><><><><><><><><><><><><><><><><>

Before we proceed further we should check the formal properties of the dot product to be sure that the product behaves as we expect it *should*. Check each of the following properties symbolically for column vectors in $\mathbf{R}^4$. Determine which hold and, for those that fail to hold, explain why they fail. Show your work.

Let **u**, **v** and **w** be vectors; *a* and *b* be scalars.

$$\begin{bmatrix} u_1 \\ u_2 \\ u_3 \\ u_4 \end{bmatrix} \quad \begin{bmatrix} v_1 \\ v_2 \\ v_3 \\ v_4 \end{bmatrix} \quad \begin{bmatrix} w_1 \\ w_2 \\ w_3 \\ w_4 \end{bmatrix}$$

1) $\mathbf{u}\cdot\mathbf{v} = \mathbf{v}\cdot\mathbf{u}$.

2) $(\mathbf{u}\cdot\mathbf{v})\cdot\mathbf{w} = \mathbf{u}\cdot(\mathbf{v}\cdot\mathbf{w})$.

3) $a\cdot(\mathbf{u}\cdot\mathbf{v}) = (a\cdot\mathbf{u})\cdot\mathbf{v} = \mathbf{u}\cdot(a\cdot\mathbf{v})$.

4) $(\mathbf{u} + \mathbf{v})\cdot\mathbf{w} = \mathbf{u}\cdot\mathbf{w} + \mathbf{v}\cdot\mathbf{w}$.

5) $\mathbf{u}\cdot\mathbf{u} = 0$ can only be true if $\mathbf{u} = \mathbf{0}$ (the zero vector).

6) $\mathbf{u}\cdot(a\cdot\mathbf{v} + b\cdot\mathbf{w}) = a\cdot(\mathbf{u}\cdot\mathbf{v}) + b\cdot(\mathbf{u}\cdot\mathbf{w})$.

ANSWERS:

Applications to geometry

Here we relate the distance between two vectors and the length of a vector to the dot product calculation.

<><><><><><><><><><><><><><><><><><><><><><><><><><><><><><><><><><><><><>

The distance between two points in $\mathbf{R}^n$ is defined in terms of the absolute value function in Mathcad. (There is an 'absolute value' icon on either the calculator palette or the matrix palette or you can use shift \ .) We first check this on the line, $\mathbf{R}^1$.

Use Mathcad to compute the distance on the line, $\mathbf{R}^1$, between

a) 5 and 11

b) -5 and 11

c) -5 and -11

d) in general between a and b.

ANSWERS:

<><><><><><><><><><><><><><><><><><><><><><><><><><><><><><><><><><><><><>

In the plane, $\mathbf{R}^2$, distance is computed using the Pythagorean Theorem as indicated in the following diagram:

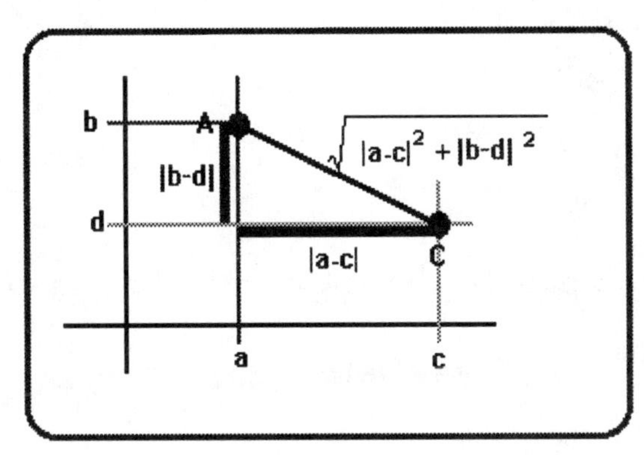

The distance from the point A = (a,b) to the point C = (c,d) above is

$$\sqrt{(a-c)^2 + (b-d)^2}$$

(The icon for square root is on the calculator palette or we can use the backward slash, \ .)

Verify, by symbolically evaluating the expression below, that if A = (a,b) and C = (c,d) then the distance between A and C can be calculated in Mathcad as |A - C|.
ANSWER:

$$\left|\begin{pmatrix} a \\ b \end{pmatrix} - \begin{pmatrix} c \\ d \end{pmatrix}\right|$$

Before proceeding further, we need to talk about the length of a vector.

The length of a vector

The length of a vector is the distance from the origin to the terminal point of the vector. (Here we think about a vector as the directed line segment from the origin to the point. Thus the length of the vector is the length of the corresponding directed line segment.)

Properties of length

1. Obviously, the length should be non-negative. (What sense would a vector of length of -1 make?)

2. The only vector with length = 0 should be the zero vector.

3. There must be a relation between the length of a vector **v** and the length of $c \cdot \mathbf{v}$ where c is a scalar. Discuss this relationship below.
ANSWER:

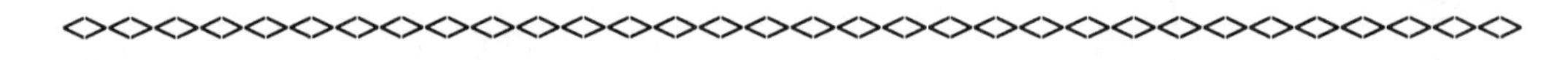

Compute $\left|3 \cdot \begin{pmatrix} a \\ b \\ c \\ d \end{pmatrix}\right|$ and $\left|-3 \cdot \begin{pmatrix} a \\ b \\ c \\ d \end{pmatrix}\right|$

Does the computation agree with your answer in part 3 above? If it does not, go back and revise your answer.

<><><><><><><><><><><><><><><><><><><><><><><><><><><><><><><><><><><><>

The way to calculate the length of a vector in Mathcad is to use the absolute value (also called magnitude) of the vector. (**Beware!** There is a Mathcad function called length that returns the number of components of the vector. This is not what we are speaking about here.) Check this symbolically for vectors in $\mathbf{R}^n$, for n = 1 and n = 2.

ANSWER:

<><><><><><><><><><><><><><><><><><><><><><><><><><><><><><>

Make a conjecture about the algebraic expression for the length of a vector, $\mathbf{v} = (\mathbf{v}_1, \mathbf{v}_2, \mathbf{v}_3)$, in $\mathbf{R}^3$. Explain.

ANSWER:

<><><><><><><><><><><><><><><><><><><><><><><><><><><><><><>

Verify that your answer is the same as: $\left|\begin{pmatrix} v_1 \\ v_2 \\ v_3 \end{pmatrix}\right|$

ANSWER:

<><><><><><><><><><><><><><><><><><><><><><><><><><><><><><>

The function for the length of a vector is written using the absolute value symbol, **|v|**. (The icon for | | is located on both the calculator and matrix palette or you can enter the | key.) Verify that this is the case for each of the following vectors. Verify that this works for column vectors and not for row vectors.

$$x := \begin{pmatrix} 2 \\ -1 \\ 1 \end{pmatrix} \qquad y := (1\ \ -1\ \ 0\ \ 5)^T \qquad \begin{pmatrix} a \\ 2 \\ s \end{pmatrix}$$

ANSWER:

Let $\mathbf{v}=\begin{bmatrix} v_1 \\ v_2 \\ \dots \\ v_n \end{bmatrix}$ be a vector in $\mathbf{R}^n$. (The ... indicates that there are other components.)

How would you define the length of **v**? Explain why each of the three required properties of length given above are satisfied by your definition of length in $\mathbf{R}^n$.

ANSWER:

Definition:

1. $|\mathbf{v}|$ is non-negative because ____________________________________.

2. $|\mathbf{v}| = 0$ only if $\mathbf{v} = \mathbf{0}$ because __________________________________ .

3. $|k \cdot \mathbf{v}| = |k||\mathbf{v}|$ because ____.

<><><><><><><><><><><><><><><><><><><><><><><><><><><><><><><><>

What is the relationship between the length of a vector, **v**, and the dot product $\mathbf{v} \cdot \mathbf{v}$?
Hint: Examine the algebraic expression for **v** and the expression for $\mathbf{v} \cdot \mathbf{v}$.

ANSWER:

<><><><><><><><><><><><><><><><><><><><><><><><><><><><><><><><><>

Distance in $\mathbf{R}^n$

We saw above that the distance between two points in the plane was $|A - C|$. We may interpret this in two ways. The first is simply an algebraic formalism. The second is to think of $|A - C|$ as the length of the vector A - C. From a vector viewpoint, we can see that the vector, A - C, has the same length as the line segment from A to C (or C to A).

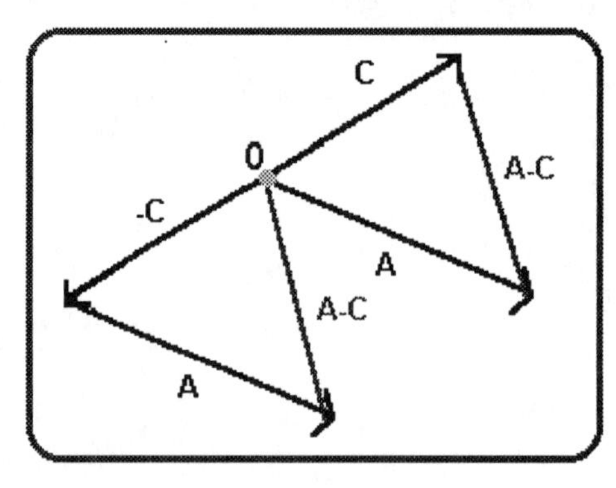

Generalize these definitions to define the distance between two points in three (or more) dimensional space. In particular what is the distance between (x_1,x_2,x_3,x_4) and (y_1,y_2,y_3,y_4)? Express the result as

a) an algebraic formula: ________________________________

b) in terms of the absolute value function and the vectors: __________

c) in terms of the dot product: ___________________________

<><><><><><><><><><><><><><><><><><><><><><><><><><><><><><>

Consider the distance between a pair of vectors (points) **x** and **y** in $\mathbf{R}^n$ as a real valued function, d(**x**, **y**).

a) How should d(**x**, **y**) and d(**y**, **x**) be related?

ANSWER:

b) What should d(**x**, **x**) = ?

ANSWER:

c) What should d(**x**, **y**) = 0 mean?

ANSWER:

<><><><><><><><><><><><><><><><><><><><><><><><><><><><><><>

There is another important property called the **triangle inequality**. (The triangle inequality states that the sum of the lengths of any two sides of a triangle is greater than the length of the third side.) Explain what this has to do with distance and how it is expressed mathematically. (Hint: Inspect the vector diagram below.)

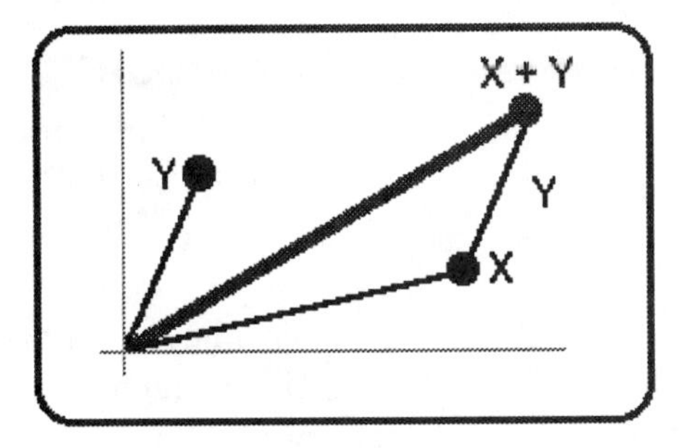

ANSWER:

d) The triangle inequality must be valid. State in words the relation between the lengths of **x** and **y** and the length of **x** + **y**. Then express the relationship in terms of the dot product.
ANSWER:

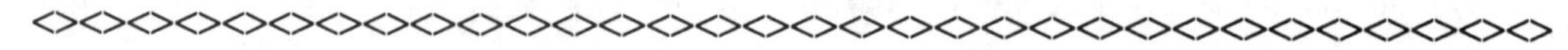

Define vectors **x** and **y** and verify for different components of **x** and **y** that the triangle inequality holds. Use the absolute value function.

ANSWER:

Unit Vectors

A unit vector is one that has length = 1. Examples: $\begin{pmatrix} -1 \\ 0 \end{pmatrix}$ $\begin{bmatrix} \frac{\sqrt{2}}{2} \\ \frac{\sqrt{2}}{2} \end{bmatrix}$ $\begin{pmatrix} \cos(t) \\ -\sin(t) \end{pmatrix}$

<><><><><><><><><><><><><><><><><><><><><><><><><><><><><><><><><><><><><><><>

Plot the vectors (cos(t), sin(t)) for t = 0 to 2 pi in steps of pi/10. Enter the Greek letter pi, π, by typing Ctrl p or using the Greek letter palette.

$i := 1..20$ $j := 1..40$ $k := 1..2$ $t_i := i \cdot \frac{\pi}{10}$

$G_{k,2 \cdot i} := 0$ $G_{1,2 \cdot i-1} := \cos(t_i)$ $G_{2,2 \cdot i-1} := \sin(t_i)$

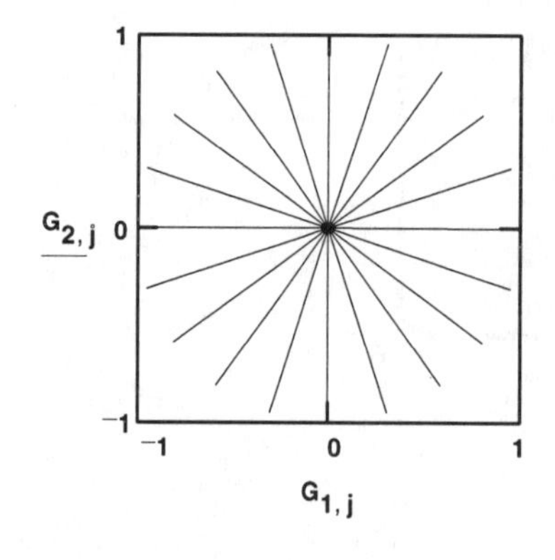

You can adjust the aspect ratio (width to height) of a graph by selecting the graph (i.e., putting a dotted box around it) and then "pulling" on the bottom and/or right side. Try it!

What is the length of each these vectors?

ANSWER:

What geometric curve do the end points of the vectors lie on?

ANSWER:

Which of the following are unit vectors?

ANSWER:

$$\begin{bmatrix} 1 \\ -1 \\ 1 \\ -1 \\ 1 \end{bmatrix} \quad \begin{bmatrix} .2 \\ .2 \\ .2 \\ .2 \\ .2 \end{bmatrix} \quad \begin{pmatrix} .5 \\ .5 \\ .5 \\ .5 \end{pmatrix} \quad \begin{bmatrix} -.3 \\ .3 \\ .5 \\ -.4 \\ -.5 \\ .4 \end{bmatrix}$$

Give three examples of 3-vectors and 4-vectors that are unit vectors.

ANSWER:

<><><><><><><><><><><><><><><><><><><><><><><><><><><><><><>

Exercises

1. Find a scalar c, such that $|c \cdot \mathbf{v}| = 1$ for each of the following:

$$\begin{pmatrix} -4 \\ 0 \end{pmatrix} \quad \begin{pmatrix} 0 \\ 0 \\ -1 \end{pmatrix} \quad \begin{bmatrix} 0 \\ 0 \\ 7 \\ 0 \\ 0 \end{bmatrix}$$

Express the scalar c in terms of the length of the vector.

ANSWER:

2. Find a scalar c such that $|c \cdot \mathbf{v}|$ is 1 for each of the following. How many such c's are there for each vector?

$$\begin{pmatrix} 1 \\ 2 \\ 3 \\ -4 \end{pmatrix} \qquad \begin{pmatrix} 1 \\ 1 \\ 1 \\ 1 \end{pmatrix} \qquad \begin{bmatrix} -2 \\ 7 \\ 3 \\ 10 \\ -5 \\ 1 \end{bmatrix} \qquad \begin{pmatrix} a \\ b \\ c \end{pmatrix}$$

ANSWER:

3. (Individual) Write a summary of the MATHEMATICS that you learned in this section. Include a list of the properties of the dot product, distance, and length.

Section 1-5 Angles and Orthogonality

Name:_______________

Score:____________

Overview

In this section we explore the relationship between the dot product of two vectors and the angle between the vectors. Our goal is to compute the angle as a function of the two vectors. Using this function we can then determine when two vectors are orthogonal (perpendicular). Orthogonality is very important for many of the applications of linear algebra.

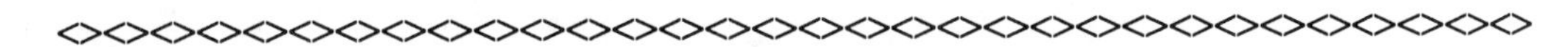

Angles

We start with the idea of an angle between vectors in the plane and extend the concept to the angle between vectors in k-space, $\mathbf{R}^k$. Our primary tool is the dot product.

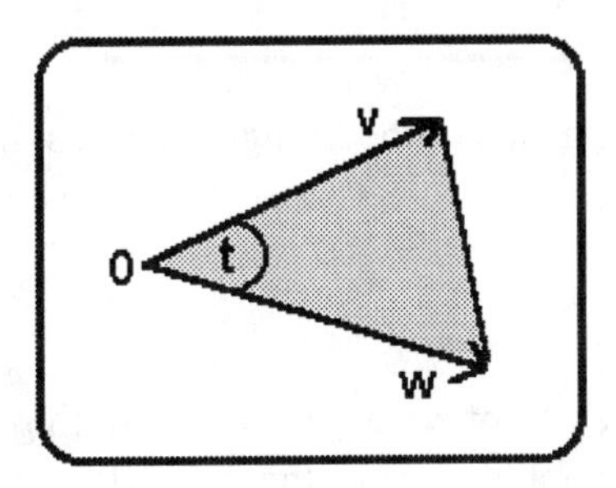

Our goal is to describe the angle t in terms of vectors **v** and **w**.

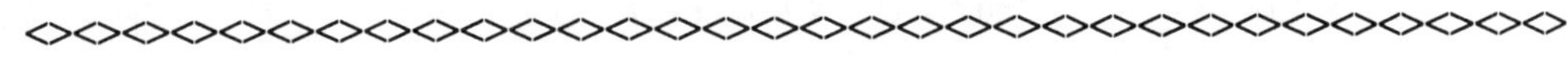

Law of cosines

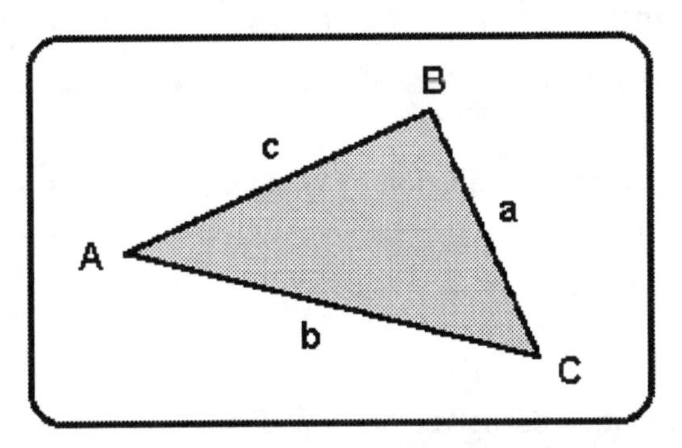

a is the length of the side opposite angle A
b is the length of the side opposite angle B
c is the length of the side opposite angle C

If angle(A) were a right angle we would have $a^2 = b^2 + c^2$. Why?

ANSWER:

The **law of cosines** is the *correction* when angle(A) is not a right angle.

$$a^2 = b^2 + c^2 - 2\cdot b\cdot c\cdot \cos(A)$$

Our goal is to express the law of cosines as a vector equation. We begin by assuming that vertex A is the origin and that the sides of the triangle that form angle t are the vectors **v** and **w**.

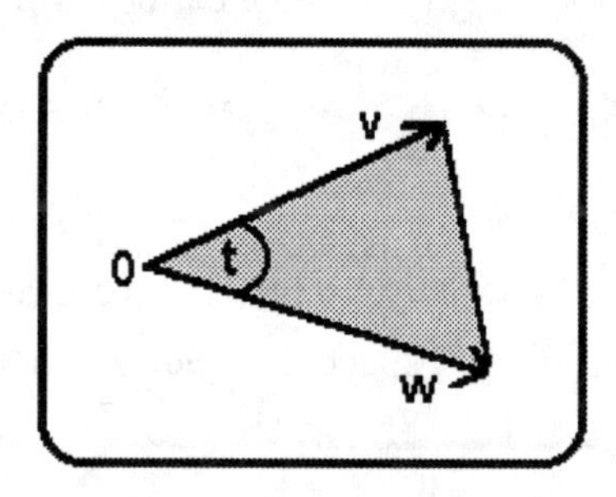

Express the length of the third side of the triangle in terms of the vectors **v** and **w**.

ANSWER:

Write the law of cosines as a vector equation by substituting the previous expression on the left hand side of the equation that gives the vector law of cosines.

Law of Cosines: $a^2 = b^2 + c^2 - 2\cdot b\cdot c\cdot \cos(A)$

Vector Law of Cosines: $\blacksquare = v\cdot v + w\cdot w - 2\cdot |v|\cdot |w|\cdot \cos(t)$

^^^^

substitute here

Expand the left hand side of this equation by using the properties of the dot product. Then simplify the equation.

ANSWER:

Solve for cos(t) as a function of the vectors, **v** and **w**.

ANSWER:

By taking the arccosine of the above expression determine the angle t as a function of the vectors **v** and **w**. (Note: The arccosine function is called acos in Mathcad.)
ANSWER:

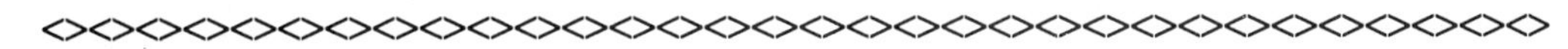

Exercises

1. Use the expression developed above to show that the angle between the following pairs of vectors is as shown. If you do get a different answer you should check your work above.

		YOUR ANSWER:	CORRECT ANSWER:
a)	$\begin{pmatrix} 1 \\ -1 \end{pmatrix}$ $\begin{pmatrix} 1 \\ 1 \end{pmatrix}$		angle$=\frac{\pi}{2}$
b)	$\begin{pmatrix} 1 \\ 1 \end{pmatrix}$ $\begin{pmatrix} -1 \\ -1 \end{pmatrix}$		angle$=\pi$
c)	$\begin{pmatrix} 1 \\ 1 \end{pmatrix}$ $\begin{pmatrix} 1 \\ 2 \end{pmatrix}$		angle=0.322...

2. Is the angle between **v** and **w** the same as the angle between **w** and **v**? Explain.
ANSWER:

3. What is the relationship between the angle between **v** and **w** and the angle between **v** and **-w**?
ANSWER:

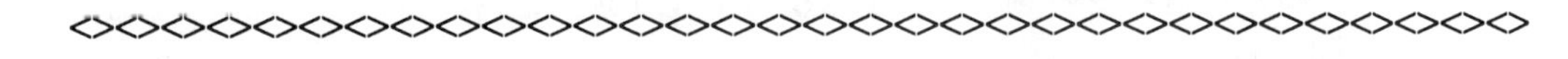

There are two angles between each pair of vectors. Which angle are we speaking of when we refer to the angle between two vectors? Does it make sense to speak of THE ANGLE between two vectors?
Hint: Examine the graph of the arccosine function.
ANSWER:

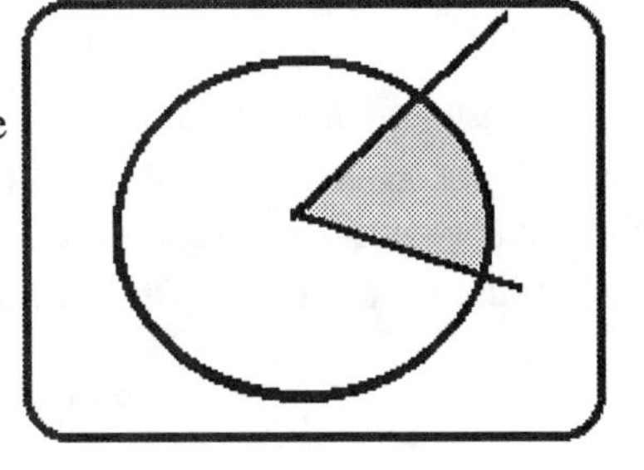

The angle between two vectors in $\mathbf{R}^n$ for n = 2, 3, 4, 5, ... can be defined since two vectors always lie in a plane. The dot product is used to compute this angle in the same way as above. In particular, if t is the angle between **v** and **w** then

$$\cos(t) = \frac{v \cdot w}{|v| \cdot |w|}$$

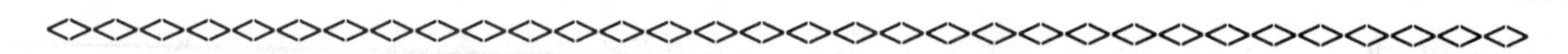

Compute the angle between the following pairs of vectors in $\mathbf{R}^n$ for n = 3 ,4 ,5 , ...

a) $\begin{bmatrix} 1 \\ 2 \\ 0 \\ -3 \\ 1 \end{bmatrix}$ $\begin{bmatrix} 2 \\ 4 \\ 5 \\ 2 \\ -4 \end{bmatrix}$ b) $\begin{pmatrix} 1 \\ 2 \\ 0 \\ 1 \end{pmatrix}$ $\begin{pmatrix} 3 \\ 1 \\ 1 \\ 2 \end{pmatrix}$ c) $\begin{pmatrix} 0 \\ 1 \\ 3 \end{pmatrix}$ $\begin{pmatrix} 1 \\ 2 \\ -2 \end{pmatrix}$

ANSWERS:

a) b) c)

<><><><><><><><><><><><><><><><><><><><><><><><><><><><><><><><>

Correlation Coefficient

Suppose that two sets of data are collected in an experiment. For example, for each student in this class we look at his or her **GPA** and the score on the first exam. We want to know if there is a link or relationship relation between these quantities.

$$GPA := (3.5\ \ 2.8\ \ 3.1\ \ 3.9\ \ 2\ \ 3.3\ \ 2.3\ \ 3.5\ \ 2.1\ \ 2.9)^T$$

$$EXAM := (88\ \ 72\ \ 78\ \ 84\ \ 60\ \ 77\ \ 60\ \ 91\ \ 70\ \ 90)^T$$

The first thing we need to do is ask what *relationship* means mathematically. An easy answer might be that there is some function, f, such that f(**GPA**) ~ **EXAM**. Here the ~ means 'is approximately equal to.' The trouble with this meaning is that the function might be a very complicated one; for example:

$$f(x) := 2 \cdot \log(x)^8 + 5 \cdot \log(x)^3 + \frac{1}{x^2 + 2}$$

It would be very difficult (1) to determine such a function and (2) to give it any meaning. In the example above we are really asking if there is a simple relationship between **GPA** and the score on the first exam. (*Simple* means linear relationship.) Our first attempt might be to see if there is some constant k such that **EXAM** = k·**GPA**.

Experiment below with different values of k to see the relationship between **EXAM** and k·**GPA**.

$j := 1 .. \mathrm{rows}(GPA)$ $\qquad k := 20$ $\qquad i := 1 .. 2$

$u := k \cdot (0 \;\; 4)^T$ $\qquad v := (0 \;\; 4)^T$

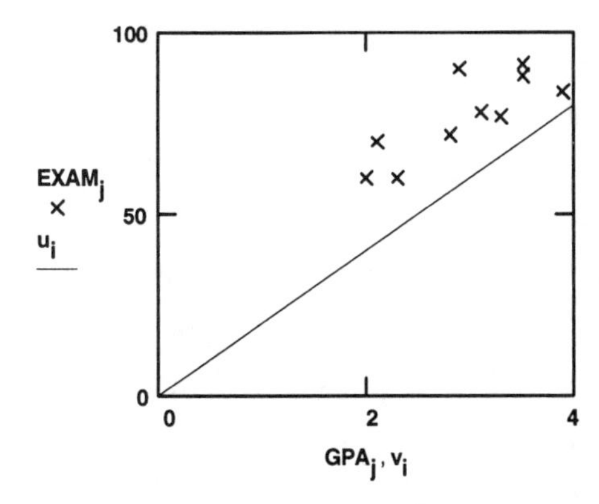

v = $(0,4)^T$ was chosen so that the line from (0,0) to (4, 4k) will go through the graphics box. Other vectors **v** could have been used.

What value of k generates the line that comes closest to the data?

ANSWER:

There are many ways to measure the *strength* of the above relationship. One is to compute the sum of the distances from the data points to the line $k \cdot (0,4)^T$. Another way is to measure the angle between the two vectors of data, **EXAM** and k·**GPA**. The latter way has the advantage, as we see below, that we do not need to know k.

Write the expression for the cosine of the angle between **EXAM** and k·**GPA** in terms of the dot product. Simplify this expression directly using the properties of the dot product and length. Do not use the computer to simplify. How is this related to the cosine of the angle between the vectors **EXAM** and **GPA**? (Hint: For non-zero values of k there are two cases: $k > 0$ and $k < 0$.)

ANSWER:

Compute the cosine of the angle between the vectors **GPA** and **EXAM**.

ANSWER:

What is the geometric relationship between **GPA** and **EXAM** if the angle between them is 0? What if the angle is π? What if the angle is $\pi/2$?

ANSWERS:

We have insisted above that the line go through the origin. Removing this restriction allows us to experiment with lines of the form: EXAM = k·GPA + b.

$\mathbf{k} := 10 \qquad \mathbf{b} := 20$

$\mathbf{u} := (\mathbf{b} \quad 4\cdot\mathbf{k} + \mathbf{b})^T \qquad \mathbf{v} := (0 \quad 4)^T$

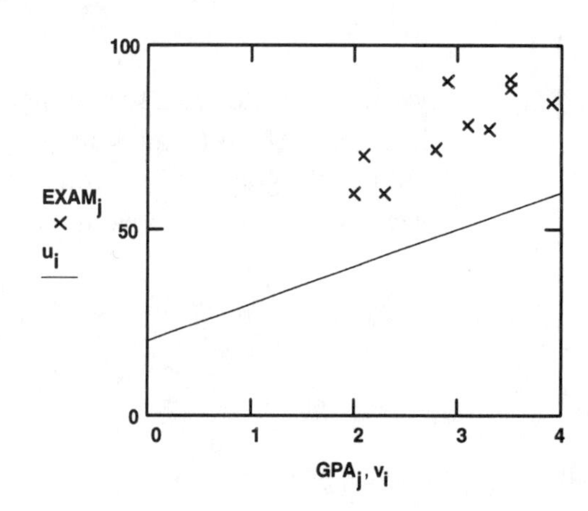

$\mathbf{k}\cdot\begin{pmatrix}0\\4\end{pmatrix}$ is a line L through the origin.

$\mathbf{k}\cdot\begin{pmatrix}0\\4\end{pmatrix} + \begin{pmatrix}\mathbf{b}\\\mathbf{b}\end{pmatrix} = \begin{pmatrix}\mathbf{b}\\4\cdot\mathbf{k}+\mathbf{b}\end{pmatrix}$

is a line parallel to L.

Experiment with different values of k and b to generate a line that comes closest to the data. What values generate such a line?

ANSWER:

The determination of the values of k and b that provide the best approximation to the data is known as the **LEAST SQUARE APPROXIMATION** or **LINEAR REGRESSION** and is discussed in Chapter 5. Our goal here is to get an estimate of the strength of the fit without actually computing the line that is the best fit. To do this we make the assumption that the line will pass through the *center of mass* of the data. This is the point (mean(**GPA**), mean(**EXAM**)). We then compute the cosine of the angle between the new vectors:

$\mathbf{G} := \mathbf{GPA} - \mathbf{mean}(\mathbf{GPA}) \qquad \mathbf{E} := \mathbf{EXAM} - \mathbf{mean}(\mathbf{EXAM})$

What is the **mean** of a vector **v**? Write a Mathcad expression (other than mean(**v**)) to compute this number. Verify your answer by calculating the mean of **GPA**. The correct answer is 2.94. (The number of entries in a vector, **v**, is given by the function, length(**v**).)

ANSWER:

Suppose that **v** and **w** are vectors of the same size and that mean(**v**) = 8.7 and mean(**w**) = -2.1.

a) What is mean(**v** + **w**)?

ANSWER:

b) What is mean(3·**v** - 2)?

ANSWER:

c) In general, what is mean(**u** - mean(**u**))?

ANSWER:

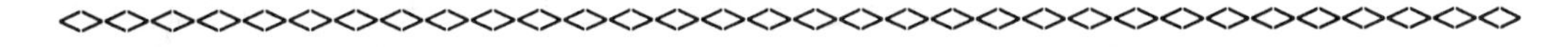

Let **u** and **v** be two vectors having the same size. The cosine of the angle between (**u** - mean(**u**)) and (**v** - mean(**v**)) is called the **Correlation Coefficient** of **u** and **v**.

The Correlation Coefficient of the vectors X and Y is defined as

$$\mathrm{COR}(X, Y) = \frac{(X - \mathrm{mean}(X)) \cdot (Y - \mathrm{mean}(Y))}{|X - \mathrm{mean}(X)| \cdot |(Y - \mathrm{mean}(Y)|}$$

The correlation coefficient measures the strength of the linear relationship between the two vectors. If the value of the coefficient is close to either plus or minus one it means that the relationship is strong. (In this case the angle between the two vectors is either close to zero or close to π.) If the value of the coefficient is close to zero it means the relationship is weak.

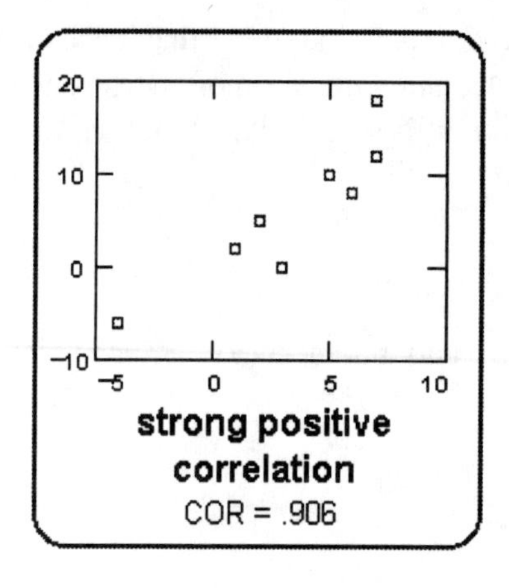

strong positive correlation
COR = .906

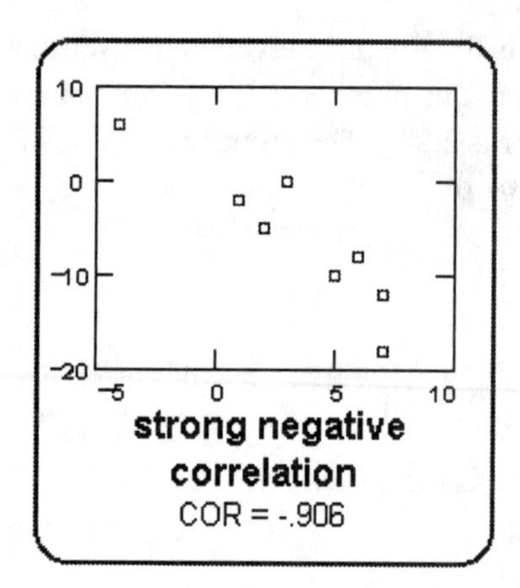

strong negative correlation
COR = -.906

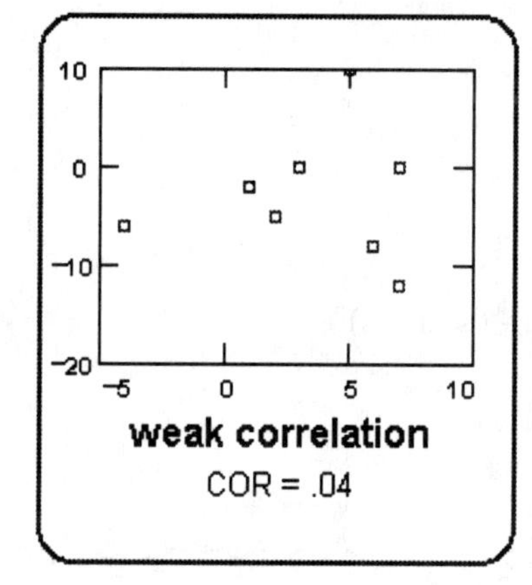

weak correlation
COR = .04

If the correlation coefficient is positive it means that the best fit line has a positive slope. In this case we say the variables are positively correlated. If the correlation coefficient is negative it means that the best fit line has a negative slope. In this case we say the variables are negatively correlated.

Since the correlation coefficient is the cosine of the angle between the vectors the values range between -1 and +1.

Compute the correlation coefficient of vectors **w1** and **w2**.

$$\mathbf{w1} := (53\ \ 49\ \ 50\ \ 55\ \ 57\ \ 48\ \ 42\ \ 46)^T$$

$$\mathbf{w2} := (72\ \ 70\ \ 71\ \ 73\ \ 75\ \ 70\ \ 75\ \ 74)^T$$

ANSWER:

Let $\mathbf{ww1} := \mathbf{w1} - \mathrm{mean}(\mathbf{w1})$ $\mathbf{ww2} := \mathbf{w2} - \mathrm{mean}(\mathbf{w2})$

Plot the points $(\mathbf{ww1}_i, \mathbf{ww2}_i)$ $i := 1..8$

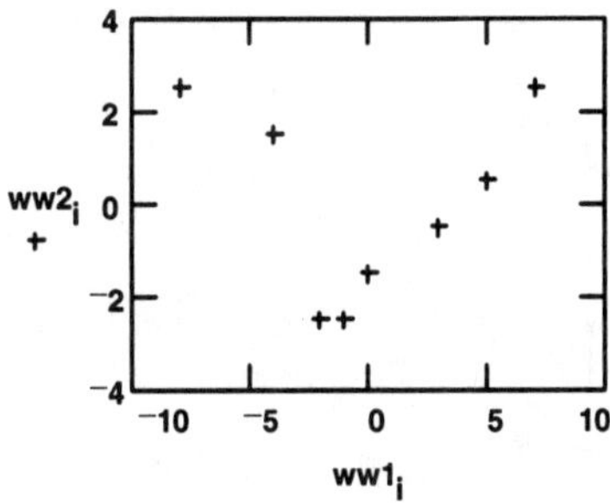

Omit the last two points from the graph by changing the range of i. How do you think that omitting these points will change the correlation coefficient?

ANSWER:

Check your answer to the previous question by computing the correlation coefficient of **w3** and **w4** which are defined below.

$i := 1..6$

ANSWER: $\mathbf{w3}_i := \mathbf{w1}_i$ $\mathbf{w4}_i := \mathbf{w2}_i$

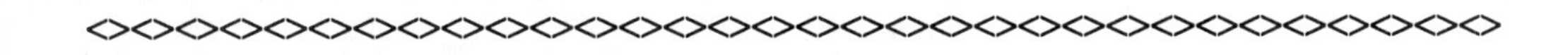

Experiment with the previous data.

$\mathbf{GPA} := (3.5\ \ 2.8\ \ 3.1\ \ 3.9\ \ 2\ \ 3.3\ \ 2.3\ \ 3.5\ \ 2.1\ \ 2.9)^T$

$\mathbf{EXAM} := (88\ \ 72\ \ 78\ \ 84\ \ 60\ \ 77\ \ 60\ \ 91\ \ 70\ \ 90)^T$

$\mathbf{G} := \mathbf{GPA} - \mathrm{mean}(\mathbf{GPA})$ $\mathbf{E} := \mathbf{EXAM} - \mathrm{mean}(\mathbf{EXAM})$

Before doing any calculations, *eyeball* the data and estimate how highly correlated the data is by guessing the value of the correlation coefficient.

ANSWER:

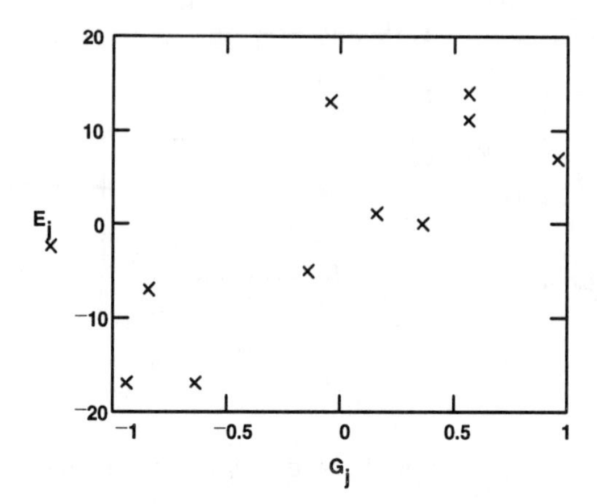

Now compute the correlation coefficient. Compare your eyeball estimate and the correlation coefficient.

ANSWER:

Suppose the fourth student was going through emotional turmoil on the day of the test and received a 10 (instead of an 84) on the exam. How would that affect the correlation coefficient? How could you take anomalies like this into consideration when evaluating relationships between data?

ANSWER:

Next we use the dot product to define a function involving sines and cosines. We first ask you to estimate its maximum and minimum values and their location by observation. Then we solve the problem using techniques from calculus.

Fix some vector (a,b). Plot the dot product of the vectors: $(\cos(t), \sin(t))$

with (a, b) as t goes from 0 to 2π. $t := 0, .1 .. 2\cdot\pi$

Current setting for vector (a,b). ===> $a := -3$ $b := 8$

The function to be studied. ===> $D(t) := \begin{pmatrix} \cos(t) \\ \sin(t) \end{pmatrix} \cdot (a \;\; b)^T$

As you change the vector (a,b) the graph of D(t) will change. We are interested in properties of the graph that depend on the selection of vector (a,b). As you change (a,b) look for patterns of behavior for the properties listed below. Fill in the table given on the next page where each column corresponds to a choice of (a,b). Use the following values for (a,b): (1,0), (0,1), (1,1), (-2,1), (2,-2), and (3,4).

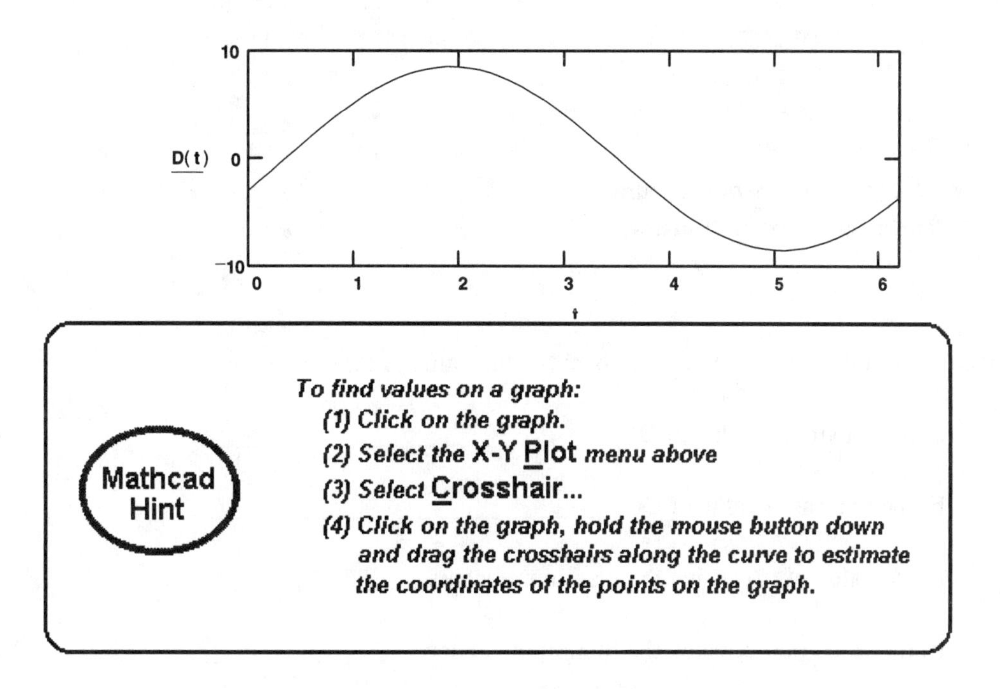

Estimate the maximum value of D(t). (Hint: It is related to vector (a,b).)

ANSWER:

Estimate the value(s) of t at which D(t) is a maximum.

ANSWER:

Estimate the minimum value of D(t).

ANSWER:

Estimate the value(s) of t at which D(t) is a minimum.

ANSWER:

Estimate the value of t at which D(t) is zero.

ANSWER:

Try several different choices for vector (a,b) to determine if there is a pattern for the information requested above. Fill in the chart below.

Trial values for (a,b)

PROPERTY	(1,0)	(0,1)	(1,1)	(-2,1)	(2,-2)	(3,4)
MAX_OF_D(t)	■	■	■	■	■	■
VALUE_of_t_where_max_occurs	■	■	■	■	■	■
MIN_OF_D(t)	■	■	■	■	■	■
VALUE_of_t_where_min_occurs	■	■	■	■	■	■
ESTIMATE_of_horizontal_intercepts	■	■	■	■	■	■

Make conjectures, in terms of a and b, about the value of the following:

a) the maximum value of D(t) ______________________________ .

b) the minimum value of D(t) ______________________________ .

c) the value of t at which the maximum occurs ________________ .

d) the value of t at which the minimum occurs ________________ .

e) the value(s) of t where D(t) = 0 ___________________________ .

<><><><><><><><><><><><><><><><><><><><><><><><><><><><><><><><><><><><><>

The function $\mathbf{D(t)} := \begin{pmatrix} \cos(t) \\ \sin(t) \end{pmatrix} \cdot (a \;\; b)^T$ is the linear combination

$a \cdot \cos(t) + b \cdot \sin(t)$ Use max-min techniques of calculus to determine the location and the values of the maximums and minimums in $[0, 2\pi]$. (Hint: Use Maple.)

ANSWER:

<><><><><><><><><><><><><><><><><><><><><><><><><><><><><><><><><><><><><>

Recall that for vectors **x** and **y** the triangle inequality states that the sum of the lengths of **x** and **y** is greater than or equal to the length of **x** + **y**.

$$|\mathbf{x}+\mathbf{y}| \le |\mathbf{x}| + |\mathbf{y}|$$

Since both sides of the inequality are non-negative we may square them and preserve the inequality.

$$(|\mathbf{x}+\mathbf{y}|)^2 \le (|\mathbf{x}| + |\mathbf{y}|)^2$$

in terms of the dot product:

$$(\mathbf{x}+\mathbf{y})\cdot(\mathbf{x}+\mathbf{y}) \le (|\mathbf{x}| + |\mathbf{y}|)^2$$

Expand this equation and simplify using the fact that $\mathbf{x}\cdot\mathbf{x}=(|\mathbf{x}|)^2$ and $\mathbf{y}\cdot\mathbf{y}=(|\mathbf{y}|)^2$.

ANSWER:

If you did the algebra correctly you should be able to conclude that

$$\mathbf{x}\cdot\mathbf{y} \le |\mathbf{x}|\cdot|\mathbf{y}| \qquad (1)$$

Since this inequality holds for all vectors **x** and **y** it must also hold for the vectors -**x** and **y**. Making this substitution doesn't change the right hand side of the equation but the left hand side become -**x·y.**

$$-\mathbf{x}\cdot\mathbf{y} \le |\mathbf{x}|\cdot|\mathbf{y}| \quad \text{or} \quad -|\mathbf{x}|\cdot|\mathbf{y}| \le \mathbf{x}\cdot\mathbf{y} \qquad (2)$$

putting (1) and (2) together you now have

$-|\mathbf{x}|\cdot|\mathbf{y}| \le \mathbf{x}\cdot\mathbf{y} \le |\mathbf{x}|\cdot|\mathbf{y}|$ which is the same as $|\mathbf{x}\cdot\mathbf{y}| \le |\mathbf{x}|\cdot|\mathbf{y}|$

This called the Cauchy-Schwarz inequality.

Express this relationship in words and tell how it is related to the graphing exercise for D(t) above. (Hint: Set **x** = (cos t, sin t) and **y** = (a,b).) When is there equality in this relationship?

ANSWER:

<><><><><><><><><><><><><><><><><><><><><><><><><><><><><><><>

Orthogonality

Two vectors, **u** and **v**, are called **orthogonal** or **perpendicular** if the angle between them equals $\pi/2$. Express the condition for orthogonality in terms of the dot product.

ANSWER:

What vector is orthogonal to every other vector?

ANSWER:

Find a nonzero vector orthogonal to (a,b). Can there be more than one? Explain.

ANSWER:

Find three different nonzero vectors orthogonal to (2,-5).

ANSWER:

Find a nonzero vector orthogonal to (1,-3,5).

ANSWER:

<><><><><><><><><><><><><><><><><><><><><><><><><><><><><><><><><>

Assume that (x, y, z) is a vector in $\mathbf{R}^3$. Give a condition that ensures that (x, y, z) is perpendicular to (1,-3,5). What sort of geometric object is *the set of all vectors (x,y,z) such that* (x,y,z) *is perpendicular to* (1,-3,5)?

ANSWER:

<><><><><><><><><><><><><><><><><><><><><><><><><><><><><><><><><>

Estimate the value of t such that the distance from the point, t·(2,-4,6) to the point (0,5,8) is a minimum by using the crosshairs on the graph of f(t).

$$f(t) := \left| t\cdot\begin{pmatrix}2\\4\\-6\end{pmatrix} - \begin{pmatrix}0\\5\\8\end{pmatrix}\right|$$

Change these values of t as needed ==> $t := -3, -2.9 .. 3$

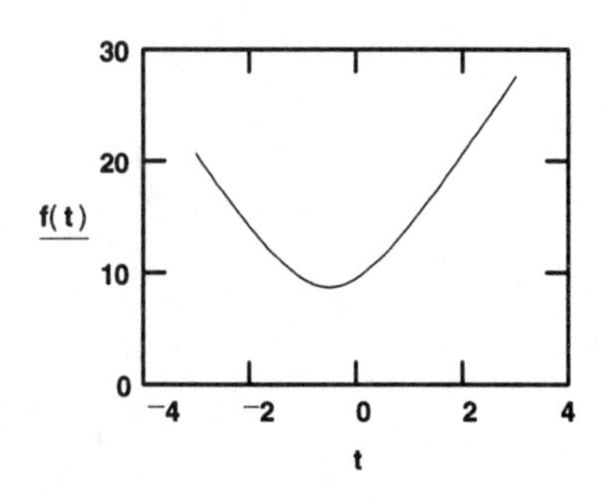

To find values on a graph:
(1) Click on the graph.
(2) Select Crosshairs...
from the X-Y Plot Menu.

ANSWER:

Calculate the exact value of t by using calculus as follows.

1. Evaluate this expression symbolically. $\mathbf{f(t)} := \left| t \cdot \begin{pmatrix} 2 \\ 4 \\ -6 \end{pmatrix} - \begin{pmatrix} 0 \\ 5 \\ 8 \end{pmatrix} \right|$

2. **D**ifferentiate on variable t from the **S**ymbolic menu.

3. Use the root function to find the zero of the derivative. For information about the root function use **H**elp on the toolbar above. Use your estimate from above as the first approximation.

ANSWER:

Compute:

$$\left[a \cdot \begin{pmatrix} 2 \\ 4 \\ -6 \end{pmatrix} - \begin{pmatrix} 0 \\ 5 \\ 8 \end{pmatrix} \right] \cdot \begin{pmatrix} 2 \\ 4 \\ -6 \end{pmatrix}$$

where a is the value of t that minimizes f(t). Explain your result geometrically by drawing a picture.

ANSWER:

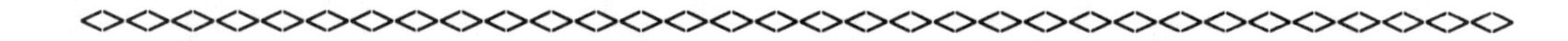

Exercise: (Individual) Write a summary of the MATHEMATICS that you learned in this section.

Section 1-6 Matrix-Vector Product

Name:________________

Score:_____________

Overview

The definition of the dot product of two vectors is extended to a definition for the product of a matrix with a vector.

The Refinery Model gave us the system of equations

$$\begin{aligned} 4\cdot x_1 + 2\cdot x_2 + 2\cdot x_3 &= 600 \\ 2\cdot x_1 + 5\cdot x_2 + 2\cdot x_3 &= 800 \\ 1\cdot x_1 + 2.5\cdot x_2 + 5\cdot x_3 &= 1000 \end{aligned} \tag{1}$$

Inspecting the left side of each equation we see that it looks like a dot product.

$$4\cdot x_1 + 2\cdot x_2 + 2\cdot x_3 = \begin{pmatrix} 4 \\ 2 \\ 2 \end{pmatrix} \cdot \begin{pmatrix} x_1 \\ x_2 \\ x_3 \end{pmatrix}$$

<== Click on the dot; then evaluate symbolically (Shift F9). (If you are using Mathcad 5.0, Click on the Maple Leaf icon on the toolbar first.)

$$2\cdot x_1 + 5\cdot x_2 + 2\cdot x_3 = \begin{pmatrix} 2 \\ 5 \\ 2 \end{pmatrix} \cdot \begin{pmatrix} x_1 \\ x_2 \\ x_3 \end{pmatrix} \tag{2}$$

<== Click on the dot; then evaluate symbolically (Shift F9).

$$1\cdot x_1 + 2.5\cdot x_2 + 5\cdot x_3 = \begin{pmatrix} 1 \\ 2.5 \\ 5 \end{pmatrix} \cdot \begin{pmatrix} x_1 \\ x_2 \\ x_3 \end{pmatrix}$$

<== Click on the dot; then evaluate symbolically (Shift F9).

The left side of each equation in the system in (1) is a dot product of a vector of coefficients with the vector of unknowns. From (1) it is convenient to view the coefficients as coming from a row, one for each equation. We try, below, to write the dot products given above as a row times a column.

$$\begin{pmatrix}4 & 2 & 2\end{pmatrix}\cdot\begin{pmatrix}x_1\\x_2\\x_3\end{pmatrix}$$

<== Click on the dot; then evaluate symbolically (Shift F9).

$$\begin{pmatrix}2 & 5 & 2\end{pmatrix}\cdot\begin{pmatrix}x_1\\x_2\\x_3\end{pmatrix} \qquad (3)$$

$$\begin{pmatrix}1 & 2.5 & 5\end{pmatrix}\cdot\begin{pmatrix}x_1\\x_2\\x_3\end{pmatrix}$$

<><><><><><><><><><><><><><><><><><><><><><><><><><><><><><><><><><><><>

Compare the output from the row times column expressions in (3) to the left sides in (2) and report on the comparison below.

ANSWER:

For the product of a row times a column to be defined what must the row and column have in common? How does the product of a row times a column relate to dot products of vectors?

ANSWER:

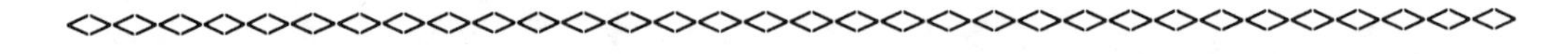

We discovered earlier that the product of a row vector and column vector in Mathcad (assuming they each have the same number of entries) turns out to be a 1-vector rather than a scalar. If we evaluate the same expression using Maple, we will discover that the product is a scalar. This is simply the reflection of different design decisions by the creators of the software. There is no 'right' answer in this case.

<><><><><><><><><><><><><><><><><><><><><><><><><><><><><><><><><><><><>

The row by column products in (3) all have the column in common. So it seems that we should be able to organize the rows to present a more compact expression for the information displayed in (3). Naturally, we organize them into a matrix a row at a time.

$$\begin{pmatrix} 4 & 2 & 2 \\ 2 & 5 & 2 \\ 1 & 2.5 & 5 \end{pmatrix} \cdot \begin{pmatrix} x_1 \\ x_2 \\ x_3 \end{pmatrix}$$ **<== Click on the dot; then evaluate symbolically (Shift F9).** (4)

<><><><><><><><><><><><><><><><><><><><><><><><><><><><><><><><>

Note the result in (4) is a column vector. Describe the entries of this vector in terms of the information in (2); in terms of the information in (3).

ANSWER:

The expression in (4) is called a **matrix-vector product**. In general we have a matrix **A** times a vector **z**

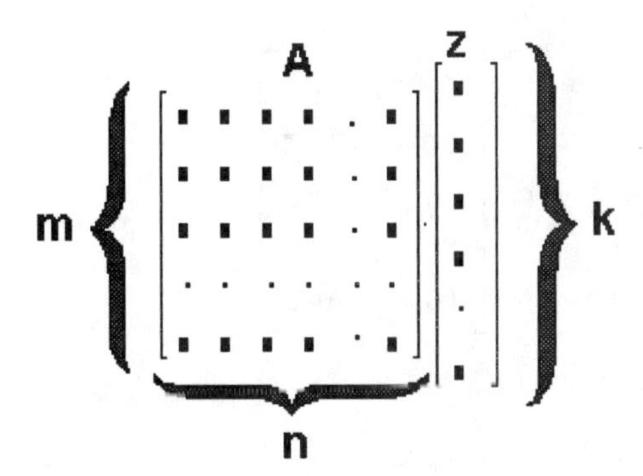

What must the relationship between k and n be for such a product to make sense? Explain why.

ANSWER:

<><><><><><><><><><><><><><><><><><><><><><><><><><><><><><><><>

Example 1

Determine which of the following matrix-vector products make sense using Mathcad. (Before having Mathcad try the matrix-vector product, you should predict success or failure in each case.)

a) $\begin{pmatrix} 2 & -3 \\ 4 & 1 \\ 3 & 5 \end{pmatrix} \cdot \begin{pmatrix} x \\ y \end{pmatrix}$ b) $\begin{pmatrix} 2 & -3 \\ 4 & 1 \\ 3 & 5 \end{pmatrix} \cdot \begin{pmatrix} x \\ y \\ z \end{pmatrix}$

c) $\begin{pmatrix} 1 & 2 & 3 & 4 \\ 5 & 6 & 7 & 8 \end{pmatrix} \cdot \begin{bmatrix} x_1 \\ x_2 \\ x_3 \\ x_4 \end{bmatrix}$ d) $\begin{pmatrix} 1 & 2 & 3 & 4 \end{pmatrix} \cdot \begin{pmatrix} 0 \\ 1 \\ 0 \\ 0 \end{pmatrix}$

ANSWER:

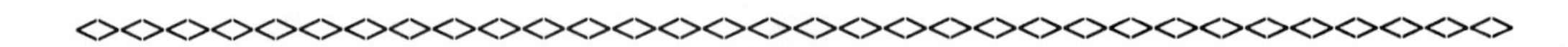

The matrix-vector product is obtained by taking the dot products of the rows of the matrix with the vector.

$$A \cdot z = \begin{bmatrix} \text{row}_1(A) \cdot z \\ \text{row}_2(A) \cdot z \\ \\ \text{row}_m(A) \cdot z \end{bmatrix}$$

This product can be viewed another way.

From the oil refinery model we have the system in (1) which we show again below.

$$\begin{aligned} 4 \cdot x_1 + 2 \cdot x_2 + 2 \cdot x_3 &= 600 \\ 2 \cdot x_1 + 5 \cdot x_2 + 2 \cdot x_3 &= 800 \\ 1 \cdot x_1 + 2.5 \cdot x_2 + 5 \cdot x_3 &= 1000 \end{aligned} \qquad (1)$$

Arrange the coefficients into a matrix **A.** We call this matrix the **coefficient matrix**.

$$A := \begin{pmatrix} 4 & 2 & 2 \\ 2 & 5 & 2 \\ 1 & 2.5 & 5 \end{pmatrix}$$

Arrange the unknowns into a vector **x**.

$$x := \begin{pmatrix} x_1 \\ x_2 \\ x_3 \end{pmatrix}$$

Display the product **A·x**:

$$\begin{pmatrix} 4 & 2 & 2 \\ 2 & 5 & 2 \\ 1 & 2.5 & 5 \end{pmatrix} \cdot \begin{pmatrix} x_1 \\ x_2 \\ x_3 \end{pmatrix}$$

<== Click on the dot, then evaluate symbolically (Shift F9).

The result is a matrix with entries the same as the left side of (1). But looking at (1) we observe that the left side can be viewed as a 'factoring into columns':

$$\begin{pmatrix} 4 \\ 2 \\ 1 \end{pmatrix} \cdot x_1 + \begin{pmatrix} 2 \\ 5 \\ 2.5 \end{pmatrix} \cdot x_2 + \begin{pmatrix} 2 \\ 2 \\ 5 \end{pmatrix} \cdot x_3 = A \cdot x$$

We summarize this as follows.

The product of the matrix A and the vector x is the linear combination of the columns of A with coefficients the corresponding entries of x.

<><><><><><><><><><><><><><><><><><><><><><><><><><><><><><><><><><><><><><><><>

Since the matrix-vector product, **A·z**, is defined in terms of dot products of the rows of **A** with the column **z**, we expect that the matrix-vector product will inherit certain properties from dot products. Let *k* and *r* be scalars; let **u**, **v**, and **w** be n-vectors; let **A** be an m by n matrix and **x** and **y** be n-vectors.

Dot Product Properties

$\mathbf{v}\cdot(\mathbf{u}+\mathbf{w}) = \mathbf{v}\cdot\mathbf{u} + \mathbf{v}\cdot\mathbf{w}$

$\mathbf{v}\cdot(k\cdot\mathbf{u}) = k\cdot(\mathbf{v}\cdot\mathbf{w})$

$\mathbf{v}\cdot(\, k\cdot\mathbf{u} + r\cdot\mathbf{w}\,) = k\cdot(\mathbf{v}\cdot\mathbf{u}) + r\cdot(\mathbf{v}\cdot\mathbf{w})$

Matrix-Vector Product Properties

$\mathbf{A}\cdot(\mathbf{x}+\mathbf{y}) = \mathbf{A}\cdot\mathbf{x} + \mathbf{A}\cdot\mathbf{y}$

$\mathbf{A}\cdot(k\cdot\mathbf{x}) = k\cdot(\mathbf{A}\cdot\mathbf{x})$

$\mathbf{A}\cdot(k\cdot\mathbf{x} + r\cdot\mathbf{y}) = k\cdot(\mathbf{A}\cdot\mathbf{x}) + r\cdot(\mathbf{A}\cdot\mathbf{y})$

<><><><><><><><><><><><><><><><><><><><><><><><><><><><><><><><><><><><>

Verify that $\mathbf{A}\cdot(k\cdot\mathbf{x}) = k\cdot(\mathbf{A}\cdot\mathbf{x})$ as follows.

$$\begin{pmatrix} a & c \\ b & d \end{pmatrix} \cdot \left[k \cdot \begin{pmatrix} x_1 \\ x_2 \end{pmatrix} \right]$$

<== Click on first multiplication symbol, press Shift F9; click on k in the first entry and choose Collect on Subexpression from the Symbolic menu. Repeat on the second entry.

$$k \cdot \left[\begin{pmatrix} a & c \\ b & d \end{pmatrix} \cdot \begin{pmatrix} x_1 \\ x_2 \end{pmatrix} \right]$$

<== Box, then evaluate symbolically by pressing Shift F9.

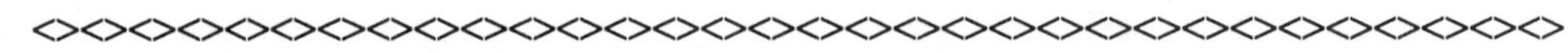

Recall that $k\cdot\mathbf{x} + r\cdot\mathbf{y}$ is a *linear combination* of the vectors **x** and **y**. The property

$$\mathbf{A}\cdot(k\cdot\mathbf{x} + r\cdot\mathbf{y}) = k\cdot(\mathbf{A}\cdot\mathbf{x}) + r\cdot(\mathbf{A}\cdot\mathbf{y})$$

is called the **linearity property**. It is a combination of the first two properties of the matrix-vector product listed above. In effect the linearity property says that the matrix-vector product preserves sums and scalar products. By this we mean it doesn't matter whether you first take the matrix-vector product and then take the sum of the vectors *or* add first and then take the matrix-vector product. Similarly, it doesn't matter whether one takes a scalar product first and then performs the matrix-vector product *or* takes the matrix-vector product first and then takes the scalar product.

The linearity property extends to more than two terms in the linear combination. Let $\mathbf{v}_1, \mathbf{v}_2, ..., \mathbf{v}_p$ be n-vectors, $k_1, k_2, ..., k_p$ scalars, and **A** an m by n matrix, then

$$A\cdot\left(k_1\cdot v_1 + k_2\cdot v_2 + \cdots + k_p\cdot v_p\right) = k_1\cdot\left(A\cdot v_1\right) + k_2\cdot\left(A\cdot v_2\right) + \cdots + k_p\cdot\left(A\cdot v_p\right)$$

The linearity property for a matrix times vectors is an important feature of linear algebra which we use in upcoming work.

<><><><><><><><><><><><><><><><><><><><><><><><><><><><><><><><><><><><>

Exercises

1. Refer to the Weather Model. (Open the file weather.mcd to recall this model.) Enter the transition matrix **A** below.

Display the transition matrix. ===>

Suppose that the probability distribution for today's weather is given by the vector **p** which has components 0, 1/2, 1/2.

Enter p. ===>

Next compute the probability distribution for tomorrow's weather and name it **q**.

Compute q. ===>

Now compute the probability distribution for the weather for the day after tomorrow. Name the vector **r**.

Compute r.===>

Write below a matrix equation for **r** in terms of **p**.

ANSWER:

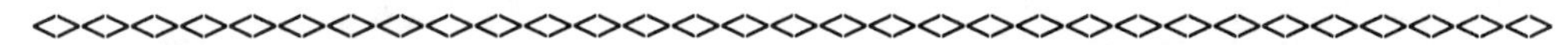

2. Continuing with Exercise 1, we want to predict the probability distribution for the weather a number of days into the future. Given that **A** contains the transition probabilities and that **p** contains the probability distribution for today's weather, then 1 day hence (tomorrow) the probability distribution vector is **A·p**; for 2 days hence the probability vector is **A·(A·p)**; for 3 days hence the probability vector is **A·(A·(A·p))**; and so for k days hence the probability vector is given by

$$\underbrace{\mathbf{A}\cdot(\mathbf{A}\cdot(\mathbf{A}\cdot\ldots(\mathbf{A}}_{\text{k factors of A}}\cdot\mathbf{p}))\ldots)$$

This pattern of repetitive calculation is commonly called an iteration procedure and the result at each stage is called an **iteration**. We can use Mathcad to compute these iterations as follows:

Let **W** be a matrix such that $\mathbf{W}^{\langle i+1 \rangle} = \mathbf{A} \cdot \mathbf{W}^{\langle i \rangle}$. That is, the (i + 1)st column of **W**, $\mathbf{W}^{\langle i+1 \rangle}$, is the probability distribution vector for the weather i days hence.

Defining the transition matrix. ===>

$$A := \begin{pmatrix} .75 & .5 & .25 \\ .125 & .25 & .5 \\ .125 & .25 & .25 \end{pmatrix}$$

Setting the number of iterations. ===> $i := 2 .. 11$

Defining the probability vector for today. ===> $W^{\langle 1 \rangle} := \begin{pmatrix} 0 \\ .5 \\ .5 \end{pmatrix}$

Performing the iteration procedure. ===> $W^{\langle i \rangle} := A \cdot W^{\langle i-1 \rangle}$

Press F9.===> W =

	5	6	7	8	9	10	11
1	0.6	0.606	0.608	0.608	0.609	0.609	0.609
2	0.223	0.219	0.218	0.218	0.217	0.217	0.217
3	0.177	0.175	0.174	0.174	0.174	0.174	0.174

Click on the display to get a scroll bar.

We change the initial probability vector $\mathbf{W}^{\langle 1 \rangle}$ above to have components 0.3, 0.5, 0.2 and recalculate the probability vectors 10 days into the future.

$$W^{\langle 1 \rangle} := \begin{pmatrix} .3 \\ .5 \\ .2 \end{pmatrix}$$

$$W^{\langle i \rangle} := A \cdot W^{\langle i-1 \rangle}$$

W =

	5	6	7	8	9	10	11
1	0.605	0.607	0.608	0.609	0.609	0.609	0.609
2	0.22	0.218	0.218	0.217	0.217	0.217	0.217
3	0.175	0.174	0.174	0.174	0.174	0.174	0.174

Discuss how the prediction is affected.

ANSWER:

Change the initial probability vector **W** above to have components 0.6087, 0.2174, 0.1739 and then recalculate the probability vectors 10 days into the future.

ANSWER:

$$W^{<1>} := \begin{pmatrix} .6087 \\ .2174 \\ .1739 \end{pmatrix} \qquad W^{<i>} := A \cdot W^{<i-1>}$$

W =

	4	5	6	7	8	9	10	11
1	0.609	0.609	0.609	0.609	0.609	0.609	0.609	0.609
2	0.217	0.217	0.217	0.217	0.217	0.217	0.217	0.217
3	0.174	0.174	0.174	0.174	0.174	0.174	0.174	0.174

Describe the day-to-day change in the probability vector.

ANSWER:

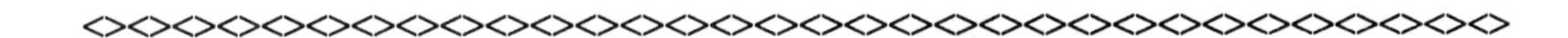

Discuss how the probability vector for the weather in 10 days predicted by this model depends upon the weather today.

ANSWER:

<><><><><><><><><><><><><><><><><><><><><><><><><><><><><><><>

We can use this model to predict the average probability distribution vector for the weather over the next n days as follows. (**p** is the probability distribution today.)

$$\text{average} = \frac{1}{n} \cdot \sum_{k=1}^{n} A^k \cdot p$$

(The average is the sum of the probability tomorrow ($\mathbf{A} \cdot \mathbf{p}$), the probability the day after tomorrow ($\mathbf{A}^2 \cdot \mathbf{p}$), and so forth up to the probability in n days divided by n.)

Discuss how this depends upon the probability distribution for today, **p**.

ANSWER:

<><><><><><><><><><><><><><><><><><><><><><><><><><><><><><><>

$$\mathbf{A}\cdot\mathbf{average}=\mathbf{A}\cdot\left(\frac{1}{n}\cdot\sum_{k=1}^{n}\mathbf{A}^{k}\cdot\mathbf{p}\right)=\frac{1}{n}\cdot\sum_{k=1}^{n}\mathbf{A}^{k+1}\cdot\mathbf{p}=\frac{1}{n}\cdot\sum_{k=1}^{n}\mathbf{A}^{k}\cdot(\mathbf{A}\cdot\mathbf{p})$$

The product of **A** and the average vector is the the average over n days beginning with a different starting probability. If you believe that the average doesn't depend on the starting value as n gets large, you will see that **A·average = average**. This average value is called the ***stable probability distribution*** associated with the given Markov chain. Verify that the vector given below is the stable distribution associated with the Markov chain with transition matrix **A** by evaluating the product symbolically.

$$\begin{bmatrix} \frac{3}{4} & \frac{1}{2} & \frac{1}{4} \\ \frac{1}{8} & \frac{1}{4} & \frac{1}{2} \\ \frac{1}{8} & \frac{1}{4} & \frac{1}{4} \end{bmatrix}\cdot\begin{bmatrix} \frac{14}{23} \\ \frac{5}{23} \\ \frac{4}{23} \end{bmatrix}$$

<><><><><><><><><><><><><><><><><><><><><><><><><><><><><><><><><><><><><><><><><><>

3. Let **u** be a row m-vector and let **A** be an m by n matrix. We can define a vector-matrix product **u·A** in a manner similar to the way we defined the matrix-vector product. The easiest way to do this is to note that $\mathbf{u}^{\mathrm{T}}$ is a column m-vector and $\mathbf{A}^{\mathrm{T}}$ is an n by m matrix. This means that $\mathbf{A}^{\mathrm{T}}\cdot\mathbf{u}^{\mathrm{T}}$ is defined and is a column n-vector. Set **u·A** equal to $(\mathbf{A}^{\mathrm{T}}\cdot\mathbf{u}^{\mathrm{T}})^{\mathrm{T}}$. Then **u·A** is a row n-vector.

a) Verify that the definition given above is consistent with Mathcad as follows:

1. Box the matrix on the right and select **Transpose** from the **Symbolic** menu.

2. Box the vector on the right and select **Transpose** from the **Symbolic** menu.

$$\begin{bmatrix} k & p & u \\ l & q & v \\ m & r & w \\ n & s & x \\ o & t & y \end{bmatrix}\cdot(a\ b\ c\ d\ e)$$

3. Box the product and select **Evaluate Symbolically** from the **Symbolic** menu.

4. Box the result and select **Transpose** from the **Symbolic** menu.

5. Evaluate the vector-matrix product symbolically and show that the result is the same as calculated above.

$$(a\ b\ c\ d\ e)\cdot\begin{bmatrix} k & p & u \\ l & q & v \\ m & r & w \\ n & s & x \\ o & t & y \end{bmatrix}$$

b) Explain how to define the vector-matrix product directly without using the transpose.

ANSWER:

4. (Individual) Write a summary of the MATHEMATICS that you learned in this section.

Section 1-7 Linear Functions

Name:________________

Score:______________

Overview

A function is a mapping from one set to another. We think of it in terms of the following diagram.

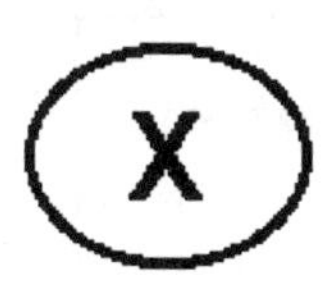

The function F assigns an element y in Y to every element x in X. Familiar examples are functions that map the real numbers to the real numbers such as f(x) = sin(x) and $f(x) = x^2 + 3x - 5$. Here we consider a class of functions called linear functions that have certain properties that are useful when dealing with linear combinations of vectors and matrices. In particular we are concerned with the function defined by a matrix multiply.

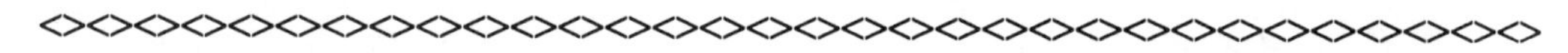

If **A** is an m (rows) by n (columns) matrix, the matrix-vector product defines a function from n-vectors to m-vectors.

<><><><><><><><><><><><><><><><><><><><><><><><><><><><><><><><><><>

Example

$$\mathbf{A} := \begin{pmatrix} 1 & -3 & 7 \\ 0 & 4 & 1 \end{pmatrix} \qquad \mathbf{u} := \begin{pmatrix} x \\ y \\ z \end{pmatrix}$$

$$\begin{pmatrix} 1 & -3 & 7 \\ 0 & 4 & 1 \end{pmatrix} \cdot \begin{pmatrix} x \\ y \\ z \end{pmatrix}$$

<== Evaluate this expression symbolically.

ANSWER:

<><><><><><><><><><><><><><><><><><><><><><><><><><><><><><><><><><>

Let F(**u**) = **A**·**u**, then we see that F is a function from $\mathbf{R}^3$ (3-vectors) to $\mathbf{R}^2$ (2-vectors). We say that $\mathbf{R}^3$ is the **domain** and $\mathbf{R}^2$ is the **range** of F.

In the case where both the range and the domain are $\mathbf{R}^2$ we can experiment with the function defined by the matrix multiply by looking at the graph of the image. Consider a pair of vectors

$$\mathbf{v} := \begin{pmatrix} 2 \\ 0 \end{pmatrix} \qquad \text{and} \qquad \mathbf{w} := \begin{pmatrix} 0 \\ 4 \end{pmatrix}$$

We combine these into a single object that looks like clock hands set a 3 o'clock. For graphing purposes we define the following matrices and index j.

$$\mathbf{C1} := \begin{pmatrix} 0 & 2 \\ 0 & 0 \end{pmatrix} \qquad \mathbf{C2} := \begin{pmatrix} 0 & 0 \\ 0 & 4 \end{pmatrix} \qquad \mathbf{j} := 1..2$$

OBJECT

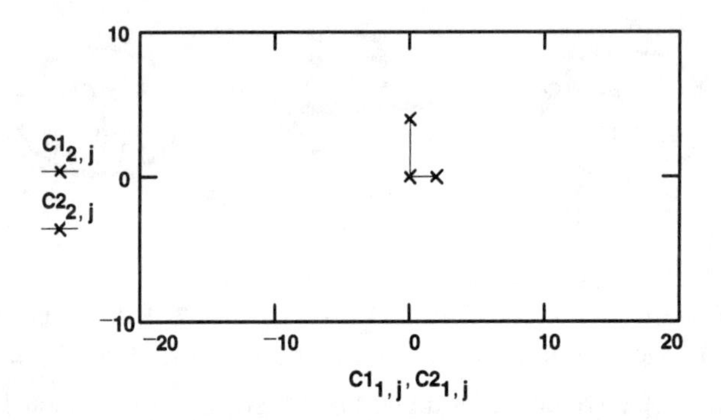

Complete:

The coordinates of the tip of the hour hand are (,).
The coordinates of the tip of the minute hand are (,).

We next define a matrix $\mathbf{E}$ and investigate the function $f(\mathbf{x}) = \mathbf{E}\cdot\mathbf{x}$, where $\mathbf{x}$ is a 2-vector. Perform the following graphics experiments for this function by changing the values of p and q in the matrix $\mathbf{E}$. Here we are mapping vectors in the plane $\mathbf{R}^2$ into other vectors in the plane.

$$\mathbf{E(p,q)} := \begin{pmatrix} p & 0 \\ 0 & q \end{pmatrix} \qquad \mathbf{p} := 8 \qquad \mathbf{q} := -2$$

IMAGE OF THE OBJECT

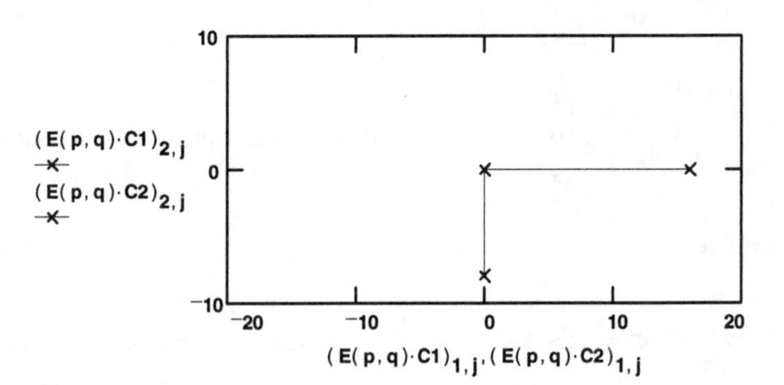

Explain in words what this function ***does*** as a function of p and q in terms of the coordinates of each hand.
ANSWER:

What does the function ***do*** when p = q = 1?
ANSWER:

<><><><><><><><><><><><><><><><><><><><><><><><><><><><><><><><><><>

Examine how the matrix **E(p,q)** maps the unit circle for various values of p and q. Vectors of length 1 from the origin are easily described by vectors of the form $(\cos(t), \sin(t))^T$.

$p := 4 \qquad q := 6$ <== ***Values*** $\qquad j := 0..20$ <== ***Index***

$$U(j) := E(p,q)\cdot\begin{pmatrix}\cos\left(\frac{\pi\cdot j}{10}\right)\\ \sin\left(\frac{\pi\cdot j}{10}\right)\end{pmatrix}$$

IMAGE of the UNIT CIRCLE

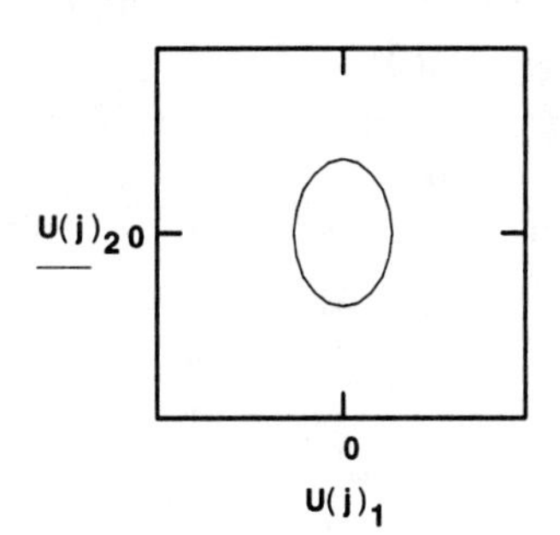

Repeat the experiment for different values of p and q looking for a pattern for the shape of the image.

Give a geometric description of the figure which is the image of the unit circle for the function defined by the matrix **E**.
ANSWER:

How does the value of p affect the image?
ANSWER:

How does the value of q affect the image?
ANSWER:

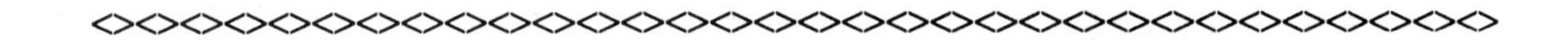

We noted previously that if **A** is a matrix, **u** and **v** are vectors, and c and d are scalars, then

$$\mathbf{A}\cdot(c\cdot\mathbf{u} + d\cdot\mathbf{v}) = c\cdot\mathbf{A}\cdot\mathbf{u} + d\cdot\mathbf{A}\cdot\mathbf{v}$$

A function, f, is called a **LINEAR FUNCTION** if

$$f(c\cdot\mathbf{x} + d\cdot\mathbf{y}) = c\cdot f(\mathbf{x}) + d\cdot f(\mathbf{y})$$

for all vectors **x** and **y** and all scalars c and d.

We conclude that ***a function defined by the MATRIX-VECTOR product is a linear function***. The word ***linear*** has ***line*** as its first four letters. If **v** is any vector in $\mathbf{R}^n$, then as t varies from $-\infty$ to $+\infty$, $t\cdot\mathbf{v}$ sweeps out a line through the origin in n-space. If **A** is a matrix (with m rows and n columns) then $\mathbf{A}\cdot(t\cdot\mathbf{v})$ will sweep out a line in $\mathbf{R}^m$ as t varies from $-\infty$ to $+\infty$. Thus **the linear function defined by the matrix-vector product takes a line to a line**. (Hence the name ***linear***.)

Another way to define a linear function is by the following two properties:

1. $f(\mathbf{u} + \mathbf{v}) = f(\mathbf{u}) + f(\mathbf{v})$
2. $f(c\cdot\mathbf{u}) = c\cdot f(\mathbf{u})$

for all vectors **u** and **v** and scalars c.

Assume that f satisfies the first definition of linear function; i.e., $f(a\cdot\mathbf{u} + b\cdot\mathbf{v}) = a\cdot f(\mathbf{u}) + b\cdot f(\mathbf{v})$. Explain why properties 1 and 2 must hold.
ANSWER:

Assume that properties 1 and 2 hold, explain why this implies that $f(a\cdot\mathbf{u}+b\cdot\mathbf{v}) = a\cdot f(\mathbf{u}) + b\cdot f(\mathbf{v})$.
ANSWER:

Let $f(x) = x + 3$. Is this a linear function? Explain why it is or isn't.
ANSWER:

Let F be a linear function with domain $\mathbf{R}^m$ and range $\mathbf{R}^n$. What is $F(\mathbf{0})$? Explain.
ANSWER:

Let **x** and **y** be vectors and let F be a linear function.

Let $\mathbf{F(x)} = \begin{pmatrix} 1 \\ 0 \\ 5 \end{pmatrix}$ $\quad \mathbf{F(y)} = \begin{pmatrix} 0 \\ 2 \\ -1 \end{pmatrix}$

What is F(3·**x**-2·**y**)?

ANSWER:

Let the linear function G satisfy: $\mathbf{G}\left(\begin{pmatrix} 1 \\ 0 \end{pmatrix}\right) = \begin{pmatrix} 3 \\ 5 \end{pmatrix}$ $\quad \mathbf{G}\left(\begin{pmatrix} 0 \\ 1 \end{pmatrix}\right) = \begin{pmatrix} -2 \\ 1 \end{pmatrix}$

Compute the following and give the answer below. $\mathbf{G}\left(\begin{pmatrix} 10 \\ -5 \end{pmatrix}\right)$ $\quad \mathbf{G}\left(\begin{pmatrix} 2 \\ -3 \end{pmatrix}\right)$

ANSWER:

$\mathbf{G}\left(\begin{pmatrix} 10 \\ -5 \end{pmatrix}\right) = \blacksquare$

$\mathbf{G}\left(\begin{pmatrix} 2 \\ -3 \end{pmatrix}\right) = \blacksquare$

<><><><><><><><><><><><><><><><><><><><><><><><><><><><><><><><>

Examples

Consider the clock hands set at 3 o'clock . $\mathbf{C1} := \begin{pmatrix} 0 & 2 \\ 0 & 0 \end{pmatrix}$ $\quad \mathbf{C2} := \begin{pmatrix} 0 & 0 \\ 0 & 4 \end{pmatrix}$ $\quad \mathbf{i} := 1..2$ $\quad \mathbf{j} := 1..2$

The matrix **T** given below defines a function on the clock hands.

$$\mathbf{T(t)} := \begin{pmatrix} \cos\left(\frac{\pi \cdot t}{180}\right) & 0 \\ \sin\left(\frac{\pi \cdot t}{180}\right) & 1 \end{pmatrix}$$

The matrix **T** depends on the value of t. **Experiment by changing the value of t** (Try t = 45, 90, 135, 180, 225, 270 and -45.) and report back on how this changes the effect of **T** on the clock hands.

In particular, report on how the length of each hand and the angle that each hand makes with the horizontal change.

Change the value of t here. ===> $t := 225$

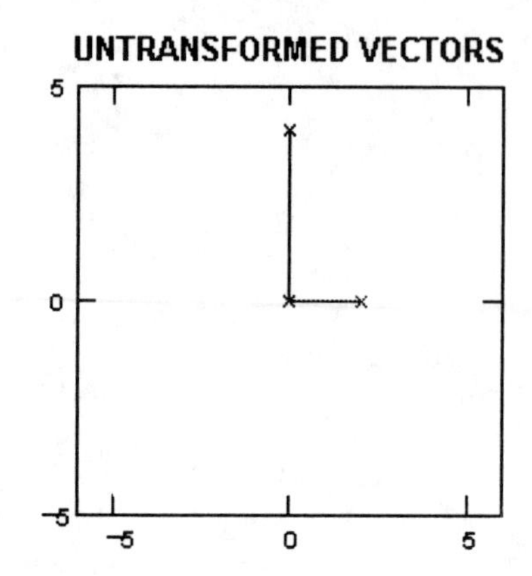

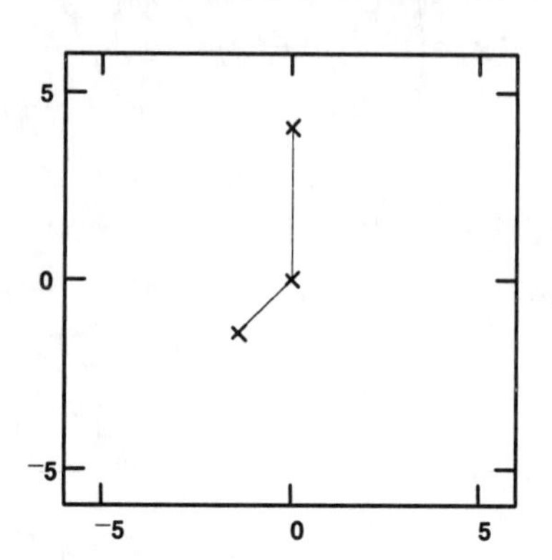

Describe the effect of the function defined by multiplying by matrix **T**(t). How has each hand been affected? Report on how this depends upon the value of t.

ANSWER:

The hour hand is: $h := \begin{pmatrix} 2 \\ 0 \end{pmatrix}$ The minute hand is: $m := \begin{pmatrix} 0 \\ 4 \end{pmatrix}$

To measure the effect of **T**(t) on the length of **h** and **m**, compute: $Lh(t) := \frac{|T(t) \cdot h|}{|h|}$ and $Lm(t) := \frac{|T(t) \cdot m|}{|m|}$

for t = 10, 30, 50, 90, 135, 225, 270. These are the ratios of the transformed lengths to the original lengths. Does this support the answers you gave above?

$t := 10$ $Lh(t) = 1$ $Lm(t) = 1$

ANSWER:

The angle between **h** and **T**(t)·**h** is: $\operatorname{acos}\left[\frac{h \cdot (T(t) \cdot h)}{|h| \cdot |T(t) \cdot h|}\right] = 10 \cdot deg$

The angle between **m** and **T**(t)·**m** is: $\operatorname{acos}\left[\frac{m \cdot (T(t) \cdot m)}{|m| \cdot |T(t) \cdot m|}\right] = 0 \cdot deg$

(Refer to "Converting to degrees" in the Help file, **Index...**, for an explanation of the right hand side of the previous computation)

Compute these angles for t = 10, 30, 50, 90, 135, 225, 270. Does this support the answers you gave above?

ANSWER:

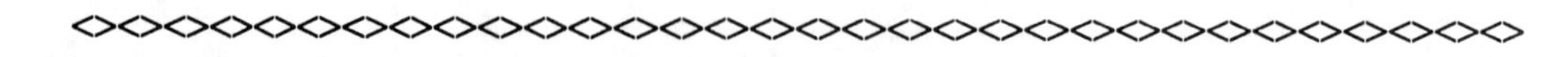

Define a new matrix **U** as follows:

$$U(u) := \begin{pmatrix} 1 & -\sin\left(\frac{\pi \cdot u}{180}\right) \\ 0 & \cos\left(\frac{\pi \cdot u}{180}\right) \end{pmatrix}$$

The matrix **U** depends on the value of u. **Experiment by changing the value of u** (Try u = 45, 90, 135, 180, 225, 270 and -45.) and report back on how this changes the effect of **U** on the clock hands.
In particular, report on how the length of each hand and the angle that each hand makes with the horizontal change.

Change the value of u here. **===>** $u := 135$

UNTRANSFORMED VECTORS

VECTORS TRANSFORMED BY U

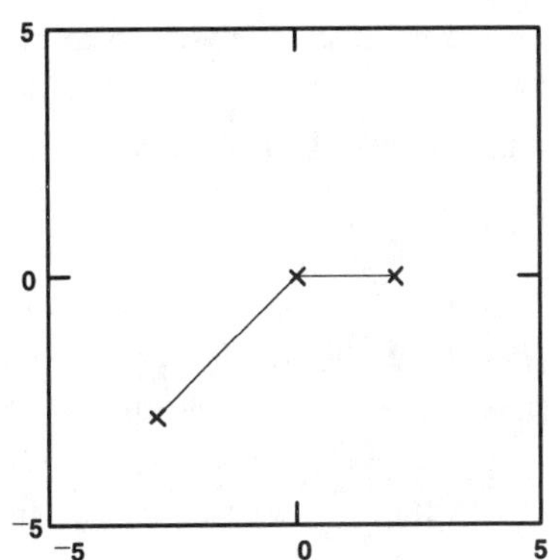

Describe the behavior of the function defined by multiplying by matrix **U**(u). How has each clock hand been affected? Report on how this depends upon the value of u.
ANSWER:

The hour hand is: $h := \begin{pmatrix} 2 \\ 0 \end{pmatrix}$ The minute hand is: $m := \begin{pmatrix} 0 \\ 4 \end{pmatrix}$

To measure the effect of **U**(u) on the length of **h** and **m**, compute

$$Lh(u) := \frac{|U(u) \cdot h|}{|h|} \quad \text{and} \quad Lm(u) := \frac{|U(u) \cdot m|}{|m|}$$

for u = 10, 30, 50, 90, 135, 225, 270. Does this support the answers you gave above?

$u := 137$ $Lh(u) = 1$ $Lh(u) = 1$

ANSWER:

The angle between **h** and **U**(u)·**h** is: $$\mathrm{acos}\left[\frac{h\cdot(U(u)\cdot h)}{|h|\cdot|U(u)\cdot h|}\right] = 0\cdot\mathrm{deg}$$

The angle between **m** and **U**(u)·**m** is: $$\mathrm{acos}\left[\frac{m\cdot(U(u)\cdot m)}{|m|\cdot|U(u)\cdot m|}\right] = 137\cdot\mathrm{deg}$$

Compute these angles for u = 10, 30, 50, 90, 135, 225, 270. Does this support the answers you gave above?

ANSWER:

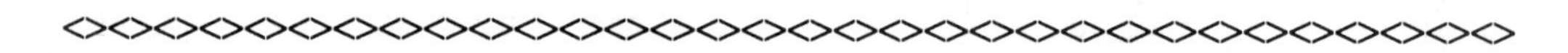

Define a new matrix **R** as follows:

$$R(r) := \begin{pmatrix} \cos\left(\frac{\pi\cdot r}{180}\right) & -\sin\left(\frac{\pi\cdot r}{180}\right) \\ \sin\left(\frac{\pi\cdot r}{180}\right) & \cos\left(\frac{\pi\cdot r}{180}\right) \end{pmatrix}$$

The matrix depends on the value of r. **Experiment by changing the value of r** (Try r = 45, 90, 135, 180, 225, 270 and -45.) and report back on how this changes the effect of **R** on the clock hands. In particular, report on how the length of each hand and the angle that each hand makes with the horizontal change.

Change the value of r here. ===> $r := 90$

UNTRANSFORMED VECTORS

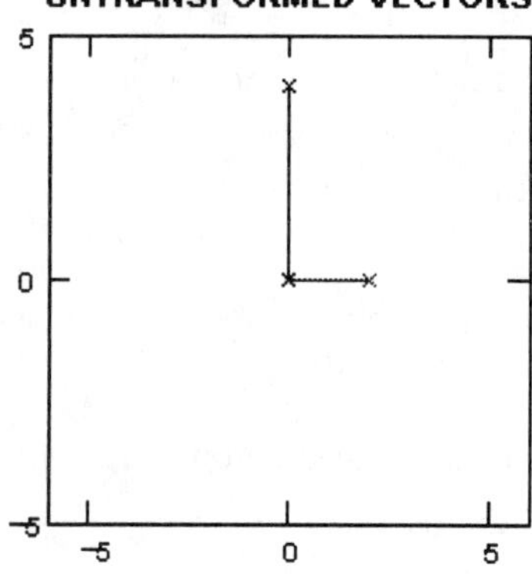

VECTORS TRANSFORMED BY R

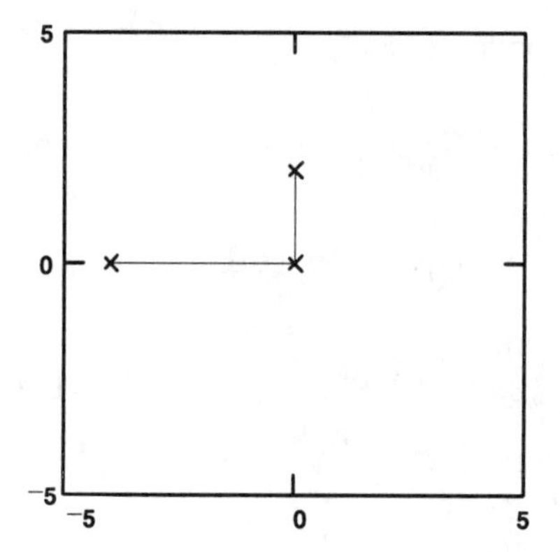

Describe the behavior of the function defined by multiplying by matrix **R**(r). How has each clock hand been affected? Report on how this depends on the value of r.
ANSWER:

The hour hand is: $\mathbf{h} := \begin{pmatrix} 2 \\ 0 \end{pmatrix}$ The minute hand is: $\mathbf{m} := \begin{pmatrix} 0 \\ 4 \end{pmatrix}$

To measure the effect of **R**(r) on the length of **h** and **m**, compute the ratios

$$\mathrm{Lh}(r) := \frac{|\mathbf{R}(r) \cdot \mathbf{h}|}{|\mathbf{h}|} \quad \text{and} \quad \mathrm{Lm}(r) := \frac{|\mathbf{R}(r) \cdot \mathbf{m}|}{|\mathbf{m}|}$$

for r = 10, 30, 50, 90, 135, 225, 270. Does this support the answers you gave above?

$r := 39$ $\quad \mathrm{Lh}(r) = 1$ $\quad \mathrm{Lh}(r) = 1$

ANSWER:

The angle between **h** and **R**(r)·**h** is: $\mathrm{acos}\left[\dfrac{\mathbf{h} \cdot (\mathbf{R}(r) \cdot \mathbf{h})}{|\mathbf{h}| \cdot |\mathbf{R}(r) \cdot \mathbf{h}|}\right] = 39 \cdot \mathrm{deg}$

The angle between **m** and **R**(r)·**m** is: $\mathrm{acos}\left[\dfrac{\mathbf{m} \cdot (\mathbf{R}(r) \cdot \mathbf{m})}{|\mathbf{m}| \cdot |\mathbf{R}(r) \cdot \mathbf{m}|}\right] = 39 \cdot \mathrm{deg}$

Compute these angles for r = 10, 30, 50, 90, 135, 225, 270. Does this support the answers you gave above?

ANSWER:

<><><><><><><><><><><><><><><><><><><><><><><><><><><><><><><><><><><><><><><><><><><>

Exercise

Think of the original vectors given above as hands on a clock (blue = minutes, red = hours) which is set at 3 o'clock. Try to find a combination (i.e., do one and then another) of transformations **T**(t)**,** **U**(u), and **R**(r) for selected values of t, u, and r, so that the resulting image is 7:30. (See the picture.)

Do not change the definitions of the functions **T**(t), **U**(u), or **R**(r). Naturally you must choose values of t, u, and r so that your result is identical to =========>

Hint: Review what each function does, then plan a sequence of *moves. **The same moves are to be applied to both hands.***

ANSWER: (Include your picture.)

DESIRED RESULT

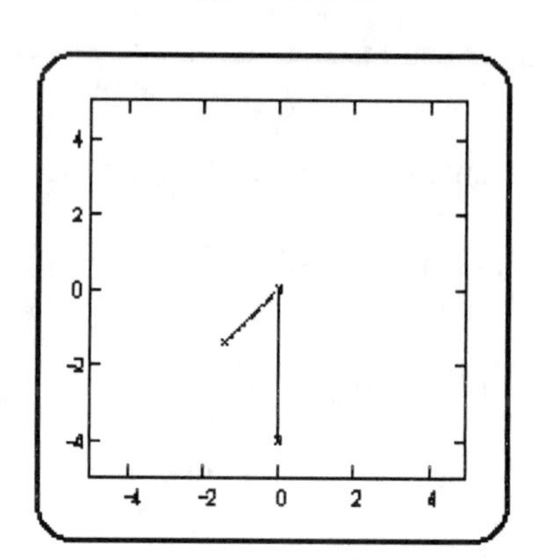

<><><><><><><><><><><><><><><><><><><><><><><><><><><><><><><><>

We showed above that every function defined by the matrix-vector product is a linear function. We now want to show that **every linear function from $\mathbf{R}^n$ to $\mathbf{R}^m$ can be given by the matrix-vector product for an appropriate choice of matrix.**

To make things clearer we examine the special case where n = 4 and m = 3.

First note that if F is a **linear function** and **e1, e2, e3, and e4** are the four vectors given below, then *if we know* F(**e1**), F(**e2**), F(**e3**) and F(**e4**), *we know* F(**v**) for any 4-vector **v**.

$$\mathbf{e1} := \begin{pmatrix} 1 \\ 0 \\ 0 \\ 0 \end{pmatrix} \qquad \mathbf{e2} := \begin{pmatrix} 0 \\ 1 \\ 0 \\ 0 \end{pmatrix} \qquad \mathbf{e3} := \begin{pmatrix} 0 \\ 0 \\ 1 \\ 0 \end{pmatrix} \qquad \mathbf{e4} := \begin{pmatrix} 0 \\ 0 \\ 0 \\ 1 \end{pmatrix}$$

Explain why this is true. (Hint: There are two steps. First show why **v** is a linear combination of the **ej**'s and then show why this means that F(**v**) is determined.)
ANSWER:

<><><><><><><><><><><><><><><><><><><><><><><><><><><><><><><><>

Suppose that

$$F(e1)=\begin{pmatrix}1\\0\\3\end{pmatrix} \qquad F(e2)=\begin{pmatrix}-6\\3\\.5\end{pmatrix} \qquad F(e3)=\begin{pmatrix}7\\2\\2\end{pmatrix} \qquad F(e4)=\begin{pmatrix}0\\2\\0\end{pmatrix}$$

Compute $F\left(\begin{pmatrix}2\\-1\\0\\4\end{pmatrix}\right)$ and $F\left(\begin{pmatrix}2\\1\\-2\\0\end{pmatrix}\right)$

ANSWER:

<><><><><><><><><><><><><><><><><><><><><><><><><><><><><><><>

To determine a matrix, **A**, so that multiplication by **A** will be equal to the action of the linear function F we must have that $\mathbf{A}\cdot\mathbf{v} = \mathbf{F(v)}$. In particular, $\mathbf{A}\cdot\mathbf{ej} = \mathbf{F(ej)}$. In terms of the matrix **A**, what is $\mathbf{A}\cdot(\mathbf{ej})$?

ANSWER:

Explain why if we know that $\mathbf{A}\cdot(\mathbf{ej}) = F(\mathbf{ej})$ for each j this determines **A**.

ANSWER:

<><><><><><><><><><><><><><><><><><><><><><><><><><><><><><><>

Suppose that

$$F(e1)=\begin{pmatrix}1\\0\\3\end{pmatrix} \qquad F(e2)=\begin{pmatrix}-6\\3\\.5\end{pmatrix} \qquad F(e3)=\begin{pmatrix}7\\2\\2\end{pmatrix} \qquad F(e4) := \begin{pmatrix}0\\2\\0\end{pmatrix}$$

Find a matrix **A** such that $\mathbf{A}\cdot\mathbf{v} = \mathbf{F(v)}$ for all **v**.

ANSWER:

Explain in your own words how to find the matrix **A** if you know F(**ej**) for each j.

ANSWER:

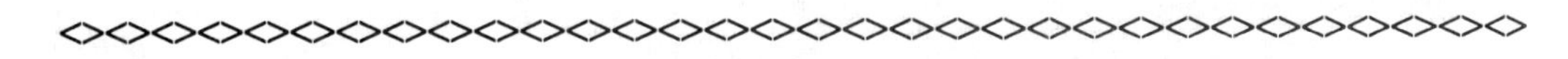

Let Rot(θ) be the map of the plane that rotates any vector through an angle of θ degrees in the counterclockwise direction. Rot(θ) is easily seen to be a linear map since the rotation of a linear combination of two vectors is the corresponding linear combination of the rotated vectors. Explain why this is true by examining what happens under rotation to the parallelogram that determines the sum of two vectors.

ANSWER:

From the discussion above we know that we can determine the matrix that corresponds to this linear map if we compute Rot(θ)·(**e1**) and Rot(θ)·(**e2**) where

$$\mathbf{e1} := \begin{pmatrix} 1 \\ 0 \end{pmatrix} \qquad \text{and} \qquad \mathbf{e2} := \begin{pmatrix} 0 \\ 1 \end{pmatrix}$$

The picture on the right illustrates the rotation of **e1** and **e2** through the angle θ.

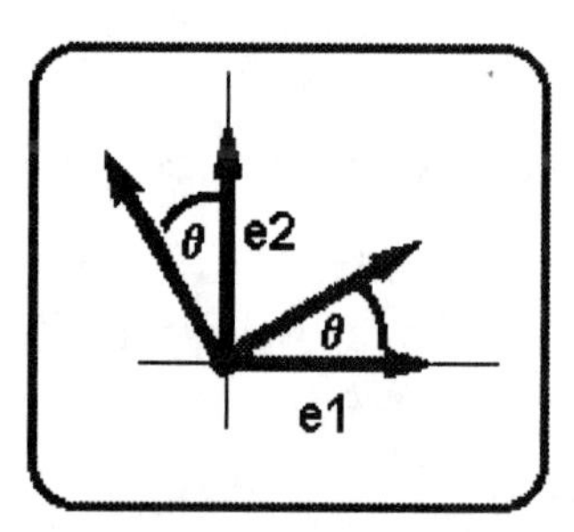

Compute the matrix that corresponds to linear function Rot(θ). Use trigonometry to give the coordinates of Rot(θ)·**e1** and Rot(θ)·**e2**. Since these must be the first and second columns of Rot(θ) they determine Rot(θ)

$$\mathbf{Rot}(\theta) := \begin{pmatrix} \blacksquare & \blacksquare \\ \blacksquare & \blacksquare \end{pmatrix} \qquad \text{Let} \quad \mathbf{v} := \begin{pmatrix} \blacksquare \\ \blacksquare \end{pmatrix} \qquad \text{and} \qquad \theta := \blacksquare$$

Choose several different values for **v** and θ and verify that
(1) Rot(θ) preserves the length of **v** and
(2) the angle between **v** and Rot(θ)·**v** is θ
by computing the following:

$$\frac{|\mathbf{Rot}(\theta)\cdot\mathbf{v}|}{|\mathbf{v}|} = \qquad \qquad \mathbf{acos}\left[\frac{\mathbf{v}\cdot(\mathbf{Rot}(\theta)\cdot\mathbf{v})}{|\mathbf{v}|\cdot|\mathbf{Rot}(\theta)\cdot\mathbf{v}|}\right] =$$

What is the relation between Rot(θ) and the transformation **R**(r) given earlier in this section?

ANSWER:

<><><><><><><><><><><><><><><><><><><><><><><><><><><><><><><><><><><><>

Exercises

1. Compute the matrix **Q** that will transform the clock hands set at 3 o'clock (see above) to 7:30. Follow the steps given below.

Let $\mathbf{Q} := \begin{pmatrix} a & c \\ b & d \end{pmatrix}$ $\quad \mathbf{m} := \begin{pmatrix} 0 \\ 4 \end{pmatrix}$ $\quad \mathbf{h} := \begin{pmatrix} 2 \\ 0 \end{pmatrix}$

m is the minute hand at 3:00 and **h** is the hour hand at 3:00.

Let **hh** be the hour hand at 7:30 and let **mm** be the minute hand at 7:30.

Enter the coordinates of **hh** and **mm**.

$\mathbf{hh} := \begin{pmatrix} \blacksquare \\ \blacksquare \end{pmatrix}$ $\quad \mathbf{mm} := \begin{pmatrix} \blacksquare \\ \blacksquare \end{pmatrix}$

Use the fact that $\mathbf{Q} \cdot \mathbf{h} = \mathbf{hh}$ and $\mathbf{Q} \cdot \mathbf{m} = \mathbf{mm}$ to compute **Q**.

ANSWER:

<><><><><><><><><><><><><><><><><><><><><><><><><><><><><><><><><><><><>

2. In this exercise we consider another derivation for the matrix **Q** of Exercise 1.

Let F be the function that transforms the clock hands at 3 o'clock to the clock hands at 7:30. From the discussion earlier in this section we know that:

$$\mathbf{Q}^{<1>} := F\left(\begin{pmatrix} 1 \\ 0 \end{pmatrix}\right) \quad \text{and} \quad \mathbf{Q}^{<2>} := F\left(\begin{pmatrix} 0 \\ 1 \end{pmatrix}\right)$$

If **m** is minute hand, **m,** at 3:00, then $\mathbf{m} := 4 \cdot \begin{pmatrix} 0 \\ 1 \end{pmatrix}$ or $\begin{pmatrix} 0 \\ 1 \end{pmatrix} = \frac{\mathbf{m}}{4}$

so $\mathbf{Q}^{<2>} := \frac{1}{4} \cdot F(\mathbf{m})$ Thus once we compute the components of the minute hand at 7:30 we will have computed $\mathbf{Q}^{<2>}$.

Explain in a similar way how to compute $\mathbf{Q}^{<1>}$.

ANSWER:

<><><><><><><><><><><><><><><><><><><><><><><><><><><><><><><><><><><><>

3. (Individual) Write a summary of the MATHEMATICS that you learned from this section.

Section 1-8 Matrix Multiplication

Name:_______________

Score:_____________

Overview

Using the previous definition of the matrix-vector product we explore when we can multiply two matrices (matrix-matrix product), the way this product is defined, and the ways that it can be interpreted.

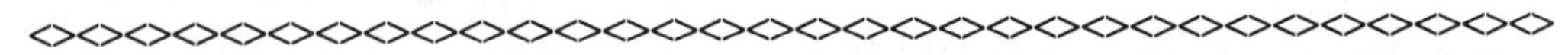

In Section 1-6 we investigated the matrix-vector product. Recall that if matrix **A** is m by n and **v** is a column vector of size n by 1, then the product **A·v** is defined and is a column vector of size m by 1. The entries of **A·v** are the dot products of the rows of **A** with the vector **v**. That is

$$\mathbf{A}\cdot\mathbf{v} = \begin{bmatrix} \text{row}_1(\mathbf{A})\cdot\mathbf{v} \\ \text{row}_2(\mathbf{A})\cdot\mathbf{v} \\ \vdots \\ \text{row}_m(\mathbf{A})\cdot\mathbf{v} \end{bmatrix}$$

The jth element of the product, $(\mathbf{A}\cdot\mathbf{v})_j$, is the dot product of the jth row of A with v.

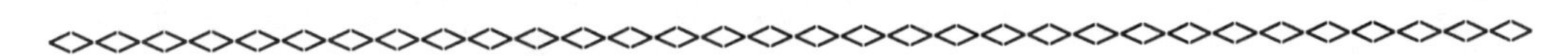

Let **A** and **B** be matrices and let **v** be a (column) vector.

If **v** is an n-vector (i.e., **v** has n rows) then how many columns must **B** have so that **B·v** makes sense?

ANSWER:

Assume that **B** has m rows. Then the product **B·v** is an m vector. For **A·(B·v)** to make sense, how many columns must **A** have?
ANSWER:

One of the nice properties that we require of matrix multiplication is ***associativity***. **Associativity** means that if we multiply ***a*** times ***b*** times ***c*** it doesn't matter whether we multiply (***a*** times ***b***) times ***c*** or ***a*** times (***b*** times ***c***). We have tried to be vague about what ***a***, ***b***, and ***c*** are but you already know that multiplication of real numbers, for example, is associative.

Associativity in the expression **A·(B·v)** above imposes a constraint on the dimensions of **A** and **B**; i.e., if **A·(B·v) = (A·B)·v** is to make sense then the constraints discovered above on the sizes of **A** and **B** must hold.

Complete the following sentence:

If **A** is a p by q matrix and **B** is an m by n matrix then for the product **A·B** to be defined ____________________ and in that case the product is a(n) _____ by _____ matrix.

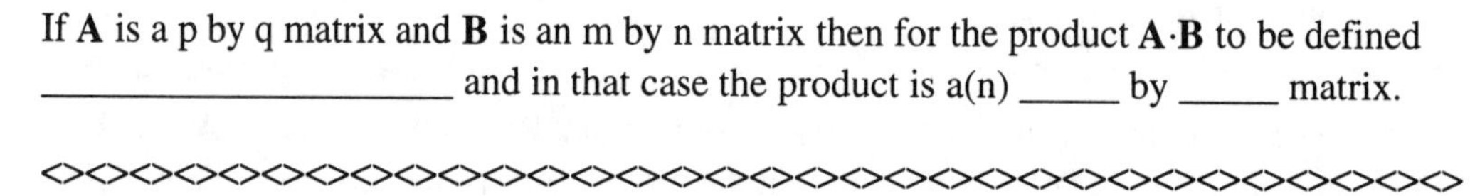

<><><><><><><><><><><><><><><><><><><><><><><><><><><><><>

Example 1

Your response as to the sizes of **A** and **B** for the product **A·B** to make sense is really an educated guess (i.e., a conjecture). To check your response use Mathcad to determine which of the following products make sense. If a product does not make sense to Mathcad write a short explanation of why that is the case.

$$A := \begin{pmatrix} 2 & 0 \\ 6 & 1 \\ 9 & -3 \end{pmatrix} \qquad B := \begin{pmatrix} 5 & 0 & -1 \\ -3 & 5 & 1 \end{pmatrix} \qquad C := \begin{pmatrix} 10 & 7 & 2 \\ -3 & 12 & 3 \\ 4 & -8 & 0 \end{pmatrix}$$

Directions: Click on the dot to box, type =, and then press F9.

$A \cdot B$ $\qquad$ $B \cdot A$ $\qquad$ $C \cdot A$

$B \cdot C$ $\qquad$ $C \cdot B$ $\qquad$ $A \cdot C$

$A \cdot A^T$ $\qquad$ $C \cdot C$ $\qquad$ $B \cdot B$

$A \cdot A$ $\qquad$ $A^T \cdot A$

Based on these experiments check your response to the fill-in-the-blank statement above.

<><><><><><><><><><><><><><><><><><><><><><><><><><><><>

So far you have established that the matrix product **A·B** makes sense only when the sizes of **A** and **B** are ***compatible*** in the sense that you have described above in response to the fill-in-the-blank statement. ***Our next goal is to discover how the entries of the product A·B are computed.***

$e := \text{identity}(4)$

$$e^{\langle 1\rangle} = \begin{bmatrix}1\\0\\0\\0\end{bmatrix} \quad e^{\langle 2\rangle} = \begin{bmatrix}0\\1\\0\\0\end{bmatrix} \quad e^{\langle 3\rangle} = \begin{bmatrix}0\\0\\1\\0\end{bmatrix} \quad e^{\langle 4\rangle} = \begin{bmatrix}0\\0\\0\\1\end{bmatrix}$$

Evaluate the above expressions by pressing F9. Complete the following sentence:
$\mathbf{e}^{\langle j\rangle}$ is the vector whose ith component is ___ unless _____ in which case it is 1 .

Let $\quad M := \begin{pmatrix}-1 & 3 & 0 & 7\\ 0 & .5 & -9 & 0\\ 7 & 2 & 3 & 1\end{pmatrix}$

For j = 1, 2, 3, 4 compute the product $\mathbf{M}\cdot\mathbf{e}^{\langle j\rangle}$.

$$M\cdot e^{\langle 1\rangle} = \begin{pmatrix}-1\\0\\7\end{pmatrix} \qquad M\cdot e^{\langle 2\rangle} = \begin{pmatrix}3\\0.5\\2\end{pmatrix}$$

$$M\cdot e^{\langle 3\rangle} = \begin{pmatrix}0\\-9\\3\end{pmatrix} \qquad M\cdot e^{\langle 4\rangle} = \begin{pmatrix}7\\0\\1\end{pmatrix}$$

In terms of the matrix **M**, complete the following statement: $\mathbf{M}\cdot\mathbf{e}^{\langle j\rangle}$ is the ___________ of **M**.

State a general result about the product $\mathbf{B}\cdot\mathbf{e}^{\langle j\rangle}$ where **B** is an m by n matrix and $\mathbf{e}^{\langle j\rangle}$ is the n-vector described above. What size is the resulting vector?

ANSWER:

Consider $\mathbf{A}\cdot(\mathbf{B}\cdot\mathbf{e}^{<j>})$. Since $\mathbf{B}\cdot\mathbf{e}^{<j>}$ is an m-vector this expression is a matrix-vector product.

If $\mathbf{A}$ is p by q, when does $\mathbf{A}\cdot(\mathbf{B}\cdot\mathbf{e}^{<j>})$ make sense?
ANSWER:

Since $\mathbf{B}\cdot\mathbf{e}^{<j>}$ is an m-vector and $\mathbf{A}\cdot(\mathbf{B}\cdot\mathbf{e}^{<j>})$ is a matrix-vector product it follows that $\mathbf{A}\cdot(\mathbf{B}\cdot\mathbf{e}^{<j>})$ is a p-vector with

$$A\cdot\left(B\cdot e^{<j>}\right) = \begin{bmatrix} row_1(A)\cdot\left(B\cdot e^{<j>}\right) \\ row_2(A)\cdot\left(B\cdot e^{<j>}\right) \\ \dots\dots\dots \\ \dots\dots\dots \\ row_p(A)\cdot\left(B\cdot e^{<j>}\right) \end{bmatrix} = \begin{bmatrix} row_1(A)\cdot col_j(B) \\ row_2(A)\cdot col_j(B) \\ \dots\dots\dots \\ \dots\dots\dots \\ row_p(A)\cdot col_j(B) \end{bmatrix} \tag{1}$$

As discussed above, associativity requires that

$$\mathbf{A}\cdot(\mathbf{B}\cdot\mathbf{e}^{<j>}) = (\mathbf{A}\cdot\mathbf{B})\cdot\mathbf{e}^{<j>}$$

Consider $(\mathbf{A}\cdot\mathbf{B})\cdot\mathbf{e}^{<j>}$ where we think of $\mathbf{A}\cdot\mathbf{B}$ as some matrix. Previously we saw that the following statement was true.

$$(\text{matrix})\cdot\mathbf{e}^{<j>} = \text{column } j \text{ of the matrix}$$

So

$$(\mathbf{A}\cdot\mathbf{B})\cdot\mathbf{e}^{<j>} = \text{col}_j(\mathbf{A}\cdot\mathbf{B}) = (\mathbf{A}\cdot\mathbf{B})^{<j>} \tag{2}$$

Then from (1) and (2) we have

$$col_j(A\cdot B) = (A\cdot B)^{<j>} = \begin{bmatrix} row_1(A)\cdot col_j(B) \\ row_2(A)\cdot col_j(B) \\ \dots\dots\dots \\ \dots\dots\dots \\ row_p(A)\cdot col_j(B) \end{bmatrix} \tag{3}$$

If **A** is p by q, **B** is m by n, and **A·B** makes sense, then what size is **A·B**?

ANSWER:

From (3) state how to compute the (k,j)-entry of **A·B**, $((\mathbf{A}\cdot\mathbf{B})_{k,j})$ when k = 1, 2, ..., p.

ANSWER:

Since j in (3) can be any value from 1 to n, state how to compute the (r,s)-entry of **A·B** where r = 1, 2. ..., p and s = 1, 2, ..., n.

ANSWER:

<><><><><><><><><><><><><><><><><><><><><><><><><><><><><><><><><><><><>

Example 2

Let $\mathbf{A} := \begin{pmatrix} 2 & -1 & 0 \\ 1 & 3 & 4 \end{pmatrix}$ and $\mathbf{B} := \begin{pmatrix} 4 & 0 & -1 \\ 1 & 2 & 1 \\ -3 & 1 & -2 \end{pmatrix}$

From your statement above, what size is **A·B**?

ANSWER:

To check this, execute each of the following statements:

$\mathrm{rows}(\mathbf{A}\cdot\mathbf{B})$ $\qquad$ $\mathrm{cols}(\mathbf{A}\cdot\mathbf{B})$

Calculate each of the following entries of **A·B** using your statement above.

(2,1)-entry $\qquad$ (1,3)-entry $\qquad$ (2,2)-entry

To check your answer evaluate the following variables:

$(\mathbf{A}\cdot\mathbf{B})_{2,1}$ $\qquad$ $(\mathbf{A}\cdot\mathbf{B})_{1,3}$ $\qquad$ $(\mathbf{A}\cdot\mathbf{B})_{2,2}$

Correct your answer to the preceding statement if necessary.

<><><><><><><><><><><><><><><><><><><><><><><><><><><><><><><><><><><><>

Example 3

The following is a symbolic example of a matrix product.

$$\begin{pmatrix} a & b \\ c & d \end{pmatrix} \cdot \begin{pmatrix} 2 & -3 & 5 \\ 3 & 4 & -7 \end{pmatrix}$$ <== ***Evaluate symbolically.***

Note that each entry is the dot product of a row of the first matrix with a column of the second matrix.

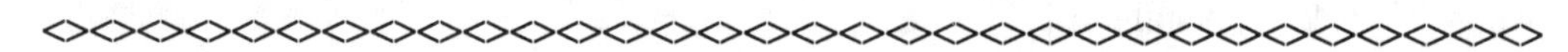

In our exploration of matrix multiplication, $\mathbf{A} \cdot \mathbf{B}$, we required that

$$(\mathbf{A} \cdot \mathbf{B}) \cdot \mathbf{e}^{<j>} = \mathbf{A} \cdot (\mathbf{B} \cdot \mathbf{e}^{<j>})$$

This is certainly the case if we use the definition for $\mathbf{A} \cdot \mathbf{B}$ developed above. But much more is true.

The fact that $(\mathbf{A} \cdot \mathbf{B}) \cdot \mathbf{e}^{<j>} = \mathbf{A} \cdot (\mathbf{B} \cdot \mathbf{e}^{<j>})$ together with the linearity property of the matrix-vector product ensures that for any vector $\mathbf{v}$ we have $(\mathbf{A} \cdot \mathbf{B}) \cdot \mathbf{v} = \mathbf{A} \cdot (\mathbf{B} \cdot \mathbf{v})$. Explain why this is true. (Hint: For any n-vector $\mathbf{v}$, $\mathbf{v} = \mathbf{v}_1 \cdot \mathbf{e}^{<1>} + \mathbf{v}_2 \cdot \mathbf{e}^{<2>} + \cdots + \mathbf{v}_n \cdot \mathbf{e}^{<n>}$.)

ANSWER:

<><><><><><><><><><><><><><><><><><><><><><><><><><><><><><><><><><><><>

Claim: Matrix multiplication is ***ASSOCIATIVE***; i.e., $\mathbf{A} \cdot (\mathbf{B} \cdot \mathbf{C}) = (\mathbf{A} \cdot \mathbf{B}) \cdot \mathbf{C}$.

The objective in the following argument is to show that corresponding columns in the matrices $\mathbf{A} \cdot (\mathbf{B} \cdot \mathbf{C})$ and $(\mathbf{A} \cdot \mathbf{B}) \cdot \mathbf{C}$ are equal.

$$\underbrace{(\mathbf{A} \cdot (\mathbf{B} \cdot \mathbf{C})) \cdot \mathbf{e}^{<j>}}_{\text{column j of } \mathbf{A} \cdot (\mathbf{B} \cdot \mathbf{C})} = \mathbf{A} \cdot ((\mathbf{B} \cdot \mathbf{C}) \cdot \mathbf{e}^{<j>}) = \mathbf{A} \cdot (\mathbf{B} \cdot (\mathbf{C} \cdot \mathbf{e}^{<j>})) = (\mathbf{A} \cdot \mathbf{B}) \cdot (\mathbf{C} \cdot \mathbf{e}^{<j>}) = \underbrace{((\mathbf{A} \cdot \mathbf{B}) \cdot \mathbf{C}) \cdot \mathbf{e}^{<j>}}_{\text{column j of } (\mathbf{A} \cdot \mathbf{B}) \cdot \mathbf{C}}$$

This is true for each column, so the two matrices are equal.

<><><><><><><><><><><><><><><><><><><><><><><><><><><><><><><><><><><><>

Let **e1**, **e2**, **e3** and **e4** be defined as follows:

$e1 := (1\ \ 0\ \ 0\ \ 0)$ $e2 := (0\ \ 1\ \ 0\ \ 0)$

$e3 := (0\ \ 0\ \ 1\ \ 0)$ $e4 := (0\ \ 0\ \ 0\ \ 1)$

Let **A** be the matrix: $\begin{pmatrix} a & b & c \\ d & e & f \\ g & h & i \\ j & k & l \end{pmatrix}$

What is the product **ei·A** in terms of the matrix **A**?

ANSWER:

Generalize the previous result by completing the following:

Let **A** be an m by n matrix and let **ej** be the ____ dimensional row vector whose kth component is 0 if ______ and 1 if ______ then **ej·A** is _________________.

More generally, what is $\mathbf{ei \cdot A \cdot e}^{<j>}$? (In this question, **ei** and $\mathbf{e}^{<j>}$ may have different dimensions.)

ANSWER:

From associativity we can see that **ei·(A·B) = (ei·A)·B**. (4)

Fill in the blanks in the following sentences that explain the above result.

ei·(A·B) is the ith _____ of **A·B** and **(ei·A)·B** is the (ith _____ of **A)·B** . This together with statement (4) implies that the _____________ equals the _____________.

<><><><><><><><><><><><><><><><><><><><><><><><><><><><><><><><><><><><><><><><>

The following diagrams are a good way to remember how matrix multiplication works.

Given an m by n matrix, A, and a n by k matrix B we can define the product matrix A·B to be the m by k matrix whose (i,j)th element is the product of the ith row of A with the jth column of B.

<><><><><><><><><><><><><><><><><><><><><><><><><><><><><><><>

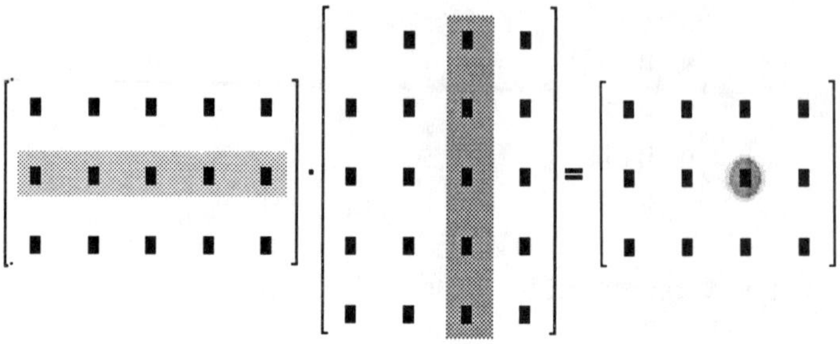

The ith row of the product is the product of the ith row of A with the matrix B.

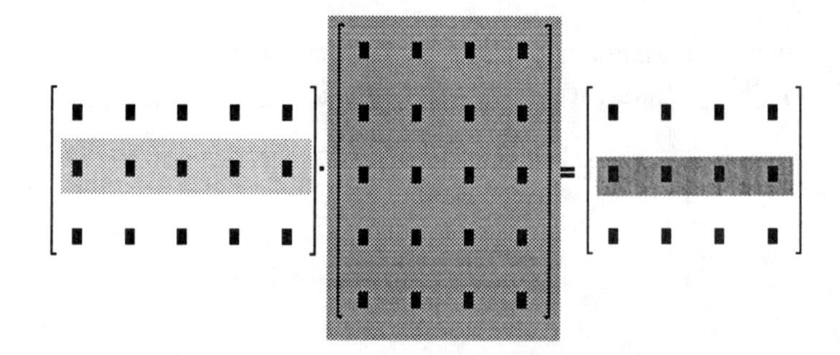

<><><><><><><><><><><><><><><><><><><><><><><><><><><><><><><>

The jth column of the product is the product of A with the jth column of B.

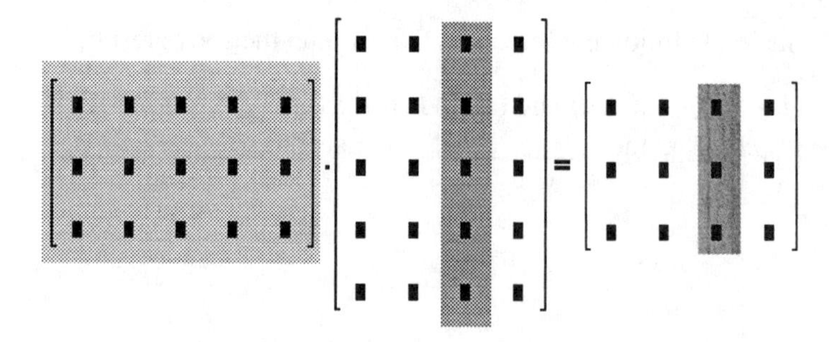

Exercises

1. The product of two matrices **A** and **B** is defined when _________________.

2. Explain in words how to compute the (i,j)th entry of **A·B**.

ANSWER:

3. Explain in words how to compute the jth column of **A·B**. (Do not involve the **e**'s.)

ANSWER:

4. Explain in words how to compute the ith row of **A·B**. (Do not involve the **e**'s.)

ANSWER:

5. Write a general expression that says matrix multiplication is associative. (Use matrices **P**, **Q**, and **R**.)

ANSWER:

6. We began by studying the matrix-vector product and then the vector-matrix product. How are these products related to the matrix-matrix product?

ANSWER:

7. We commented earlier that the product of a row vector and a column vector in Mathcad was not the same as the dot product of the two vectors where we change the row vector into a column vector by transpose. Explain the mathematical justification for this distinction.

ANSWER:

8. We define the outer product of two (column) vectors to be the first times the transpose of the second. (*times* means matrix multiplication.)

$$\mathbf{Outer\_product}(\mathbf{u}, \mathbf{v}) := \mathbf{u} \cdot \mathbf{v}^{\mathbf{T}}$$

How is the ith row of the outer product related to **v**? How is the jth column of the outer product related to **u**? (Hint: Enter two column vectors of symbols and compute their outer product symbolically.)

ANSWER:

9. (Individual) Write a summary of the MATHEMATICS that you learned in this section.

Section 1-9 Matrix Multiplication (continued)

Name:________________

Score:_____________

Overview

In this section we investigate some properties of matrix multiplication: commutativity, the powers of a square matrix, and a special matrix called the identity matrix. We return to the linear functions of Section 1-7 and explore the meaning of matrix multiplication in the context of linear functions. We examine the relationship between the transpose and the matrix product. Finally we come back to the weather model and restudy the problem of determining the stable state of the process by introducing the matrix eigen equation.

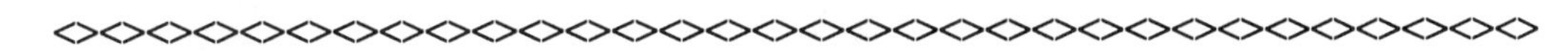

Commutativity

We know that for real numbers *a times b* = *b times a*. This is called **commutativity**.
Is the same property true for matrix multiplication?
In particular

Is matrix multiplication **COMMUTATIVE?**

Is it always true that if matrix #1 is m by n and matrix #2 is n by m then (matrix #1)·(matrix #2) = (matrix #2)·(matrix #1)?

Experiment with the matrices given below, or others you may construct, to aid in answering the questions.

$$\mathbf{A} := \begin{pmatrix} 1 & 2 & 4 \\ 0 & -1 & 3 \end{pmatrix} \qquad \mathbf{B} := \begin{pmatrix} 4 & 5 & -2 \\ 1 & 2 & 0 \\ 0 & 1 & 3 \end{pmatrix} \qquad \mathbf{C} := \begin{pmatrix} 3 & 5 \\ 1 & 0 \\ 2 & 2 \end{pmatrix} \qquad \mathbf{D} := \begin{pmatrix} 5 & 0 & 8 \\ 0 & 1 & 0 \\ 6 & 0 & 3 \end{pmatrix}$$

Let **P** and **Q** be two matrices. For each of the following either give a reason why the statement is true or give an example to show that it is not true.

For all **P** and **Q**, **P·Q = Q·P**.
ANSWER:

If **P·Q** and **Q·P** are the same size, then, **P·Q = Q·P**.
ANSWER:

If **P·Q** and **Q·P** are the same size, what is the relationship between the sizes of **P** and **Q**?
ANSWER:

Construct a pair of 2 by 2 matrices **P** and **Q** (with **P ≠ Q, P ≠ 0**, and **Q ≠ 0**) so that **P·Q = Q·P**.
ANSWER:

Let **P** and **Q** be 2 by 2 matrices. If **P·Q = 0** must either **P = 0** or **Q = 0**?
ANSWER:

Let **P** be a 2 by 2 matrix. If **P·Q = 0** for **all** 2 by 2 matrices **Q** does it follow that **P = 0**? Justify your answer.
ANSWER:

<><><><><><><><><><><><><><><><><><><><><><><><><><><><><><><><><><><><><><><><>

Identity matrix

An identity matrix is a square matrix with ones on the diagonal and zeros elsewhere.

We denote the n by n identity matrix by $\mathbf{I}_n$.

$$I_2=\begin{pmatrix}1 & 0\\ 0 & 1\end{pmatrix} \qquad I_3=\begin{pmatrix}1 & 0 & 0\\ 0 & 1 & 0\\ 0 & 0 & 1\end{pmatrix} \qquad I_4=\begin{pmatrix}1 & 0 & 0 & 0\\ 0 & 1 & 0 & 0\\ 0 & 0 & 1 & 0\\ 0 & 0 & 0 & 1\end{pmatrix}$$

What is the product, $\mathbf{A} \cdot \mathbf{I}_n$ of an m by n matrix **A** and the n by n identity matrix, $\mathbf{I}_n$?

Explain in words why your answer is true. (Hint: What is $\mathbf{A} \cdot \mathbf{I}_n^{\langle j \rangle}$?)

ANSWER:

What is the product, $\mathbf{I}_m \cdot \mathbf{A}$ of an m by n matrix **A** and the m by m identity matrix, $\mathbf{I}_m$?
Explain in words why your answer is true. (Hint: What is the jth row of $\mathbf{I}_m$ times **A**?)

ANSWER:

Based upon the above explain why the n by n identity matrix commutes with every n by n matrix **A** and the result is the matrix **A**. In terms of a matrix expression this is written as

$$\mathbf{A} \cdot \mathbf{I}_n = \mathbf{I}_n \cdot \mathbf{A} = \mathbf{A}$$

ANSWER:

In Mathcad the function **identity(n)** has the n by n identity matrix as its value.

Let $\quad \mathbf{A} := \begin{pmatrix} 9 & 14 \\ -7 & 350 \end{pmatrix}$

Evaluate each of the following expressions.

$\mathbf{identity(2) \cdot A}$

$\mathbf{A \cdot identity(2)}$

<><><><><><><><><><><><><><><><><><><><><><><><><><><><><><><><><><><><>

Matrix powers

Let **A** be an m by n matrix. What restriction must be placed on the size of **A** so that **A·A** is defined?

ANSWER:

If **A·A** is defined so is **A·A·A**. A shorthand for the product of a square matrix with itself is exponent notation; **A·A** is $\mathbf{A}^2$, **A·A·A** is $\mathbf{A}^3$, and so on.

In Mathcad we can compute powers of a square matrix by entering the matrix name followed by either (1) the carat (^) key (Shift 6) or (2) clicking on the power icon on the calculator palette. Then enter the exponent and exit by entering the space bar. Pressing = and F9 will execute the statement.

$$A := \begin{pmatrix} 2 & -1 \\ 1 & 3 \end{pmatrix} \qquad A \cdot A \qquad A^2$$ <== *Execute.*

Compute $\mathbf{A}^3$ and $\mathbf{A}^4$.

ANSWER:

$$B := \begin{pmatrix} 0 & 0 & 0 \\ 5 & 0 & 0 \\ 7 & -3 & 0 \end{pmatrix} \qquad C := \begin{pmatrix} 0 & 0 & 0 & 1 \\ 1 & 0 & 0 & 0 \\ 0 & 1 & 0 & 0 \\ 0 & 0 & 1 & 0 \end{pmatrix}$$

Look for a pattern in $\mathbf{B}^n$ and $\mathbf{C}^n$ for n = 1, 2, 3, 4, 5, 6, ... Report the result of your exploration below.

ANSWER:

<><><><><><><><><><><><><><><><><><><><><><><><><><><><><><><><><><><><>

Linear functions

Let f(**u**) = **A·u** and g(**v**) = **B·v** be given linear functions where **A** and **B** are matrices and **u** and **v** are vectors. Assume that f maps $\mathbf{R}^n \rightarrow \mathbf{R}^m$ and g maps $\mathbf{R}^m \rightarrow \mathbf{R}^p$. What are the dimensions of **A**, **B**, **u**, and **v**?

ANSWER:

The composition (g∘f)(**u**) is defined to be g(f(**u**)).

g∘f means: do f first then g. This is the opposite to the order in which they are written. Be sure you understand which comes first.

Corresponding to the relationships between (f and **A**) and (g and **B**), the matrix expression that corresponds to g(f(**u**)) is **B**·(**A**·**u**). By associativity **B**·(**A**·**u**) = (**B**·**A**)·**u.** Use this relationship to complete the following:

If f maps $\mathbf{R}^n \rightarrow \mathbf{R}^m$ and g maps $\mathbf{R}^m \rightarrow \mathbf{R}^p$ and **A** and **B** are matrices such that f(**u**) = **A**·**u** and g(**v**) = **B**·**v** then **C** = _____ is a matrix such that (g∘f)(**u**) = **C**·**u**.

<><><><><><><><><><><><><><><><><><><><><><><><><><><><><><><><><><><><>

Let's revisit the clock transformations of Section 1-7. Recall that **C1** and **C2** corresponded to the two hands of the clock.

$$C1 := \begin{pmatrix} 0 & 2 \\ 0 & 0 \end{pmatrix} \qquad C2 := \begin{pmatrix} 0 & 0 \\ 0 & 4 \end{pmatrix} \qquad j := 1..2$$

The matrix **R**(r) given below corresponds to a rotation of the clock hands through an angle of r degrees.

$$r := 15$$

$$R(r) := \begin{pmatrix} \cos\left(\frac{\pi \cdot r}{180}\right) & -\sin\left(\frac{\pi \cdot r}{180}\right) \\ \sin\left(\frac{\pi \cdot r}{180}\right) & \cos\left(\frac{\pi \cdot r}{180}\right) \end{pmatrix}$$

Experiment with various values of k to determine the transformation corresponding to the kth power of the matrix.

Set the value of k here. ===> $k := 2$

Vectors Transformed by (R(r))^k

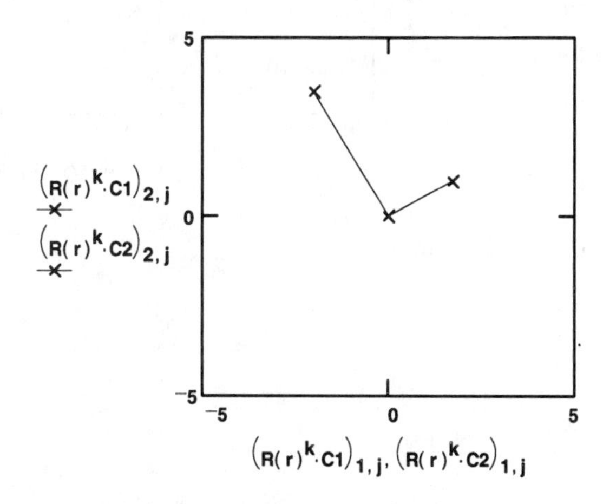

What is the transformation $(\mathbf{R}(r))^k$ for k = 2, 3, 4, 5, ... ? for k = 0? for k = -1, -2, -3?

ANSWER:

Recall that the transformation **T**(t) kept the minute hand fixed and rotated the hour hand through t degrees and the transformation **U**(u) kept the hour hand fixed and rotated the minute hand through u degrees. Do these two transformations commute? Experiment below for various values of t and u.

$$T(t) := \begin{pmatrix} \cos\left(\frac{\pi \cdot t}{180}\right) & 0 \\ \sin\left(\frac{\pi \cdot t}{180}\right) & 1 \end{pmatrix} \qquad U(u) := \begin{pmatrix} 1 & -\sin\left(\frac{\pi \cdot u}{180}\right) \\ 0 & \cos\left(\frac{\pi \cdot u}{180}\right) \end{pmatrix}$$

Set the values of t and u here. ===> $t := 120$ $u := 30$

$T(t) \cdot U(u)$

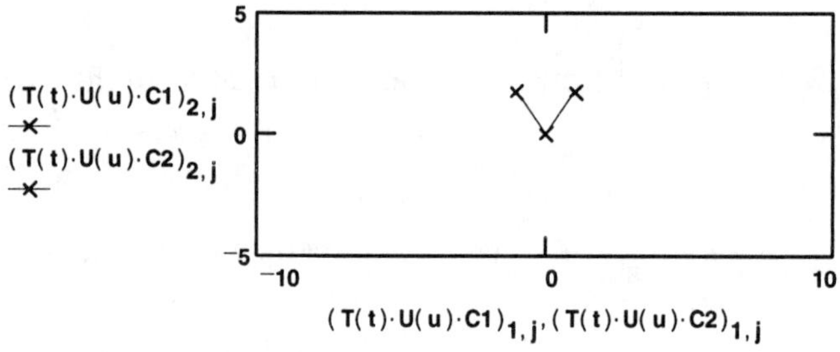

$(T(t) \cdot U(u) \cdot C1)_{2,j}$, $(T(t) \cdot U(u) \cdot C2)_{2,j}$ versus $(T(t) \cdot U(u) \cdot C1)_{1,j}$, $(T(t) \cdot U(u) \cdot C2)_{1,j}$ (axes: −10 to 10, −5 to 5)

$U(u) \cdot T(t)$

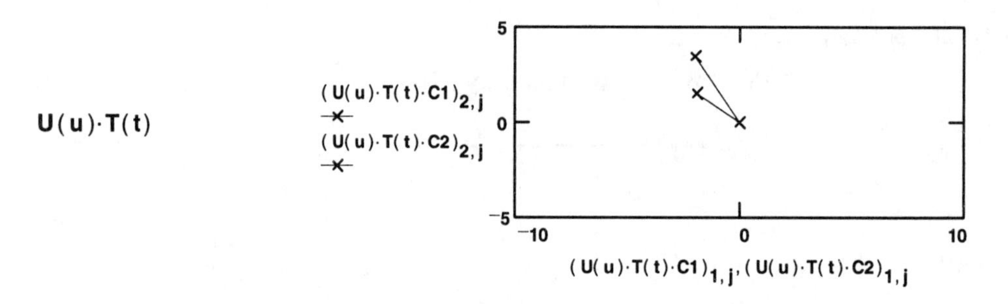

Discuss the results of your experimentation.

ANSWER:

Exercises

1. Transposes. Consider the interaction between matrix products and transposes. In particular what is the relation between $(\mathbf{A}\cdot\mathbf{B})^T$ and $\mathbf{A}^T$ and $\mathbf{B}^T$? Begin by considering the dimensions of matrices involved. Suppose that

$$\mathbf{A} \text{ is m by n and } \mathbf{B} \text{ is n by p}$$

List the dimensions of each of the following:

$\mathbf{A}\cdot\mathbf{B}$	$\mathbf{A}^T$
$(\mathbf{A}\cdot\mathbf{B})^T$	$\mathbf{B}^T$

Based on consideration of dimensions make a conjecture to complete the following expression: $(\mathbf{A}\cdot\mathbf{B})^T =$ _______________.

To check your conjecture, construct a 2 by 3 matrix **A** and a 3 by 4 matrix **B** and compute both sides of the preceding expression. If necessary correct your conjecture.

Express in words the relationship developed above:

ANSWER:

The transpose of a product **A·B** is _________________________ .

<><><><><><><><><><><><><><><><><><><><><><><><><><><><><><><><><><><><><><><><><><><><><>

2. Markov chains: Refer to the Weather Model. (See file weather.mcd.) For the transition matrix **A** and today's probability vector **p** we computed the probability distribution vector for future days by the following scheme:

1 day hence: $\mathbf{A}\cdot\mathbf{p}$
2 days hence: $\mathbf{A}\cdot(\mathbf{A}\cdot\mathbf{p})$
3 days hence: $\mathbf{A}\cdot(\mathbf{A}\cdot(\mathbf{A}\cdot\mathbf{p}))$ and so on.

The corresponding scheme using matrix powers is:

1 day hence: $\mathbf{A}\cdot\mathbf{p}$
2 days hence: $\mathbf{A}^2\cdot\mathbf{p}$
3 days hence: $\mathbf{A}^3\cdot\mathbf{p}$
k days hence: $\mathbf{A}^k\cdot\mathbf{p}$

A new transition matrix has been developed for a city in the midwest. It is the following:

$$\mathbf{A} := \begin{bmatrix} \frac{1}{2} & \frac{1}{8} & \frac{1}{3} \\ \frac{1}{4} & \frac{3}{4} & \frac{1}{3} \\ \frac{1}{4} & \frac{1}{8} & \frac{1}{3} \end{bmatrix}$$

Enter this transition matrix and today's probability vector **p** which has components .3, .6, .1. *Determine the probability distribution for tomorrow, 2 days hence, 3 days hence, and the long term behavior.* Show your work and describe your findings below.

ANSWER:

3. In Exercise 2, we used the iterative procedure for computing the state of the weather for days into the future. Using the same matrix **A** we can compute the long term behavior (the stable state) directly.

For k = 5, 10, 15, 20, 25 compute $\mathbf{A}^k$. Describe the behavior of the columns of the powers of **A**.

ANSWER:

Choose column two of $\mathbf{A}^{25}$ (really any column will do). Name the chosen column **y**. (Hint: Use $\mathbf{y} := (\mathbf{A}^{25})^{\langle 2 \rangle}$.) Show that $\mathbf{A} \cdot \mathbf{y} = \mathbf{y}$.

ANSWER:

<><><><><><><><><><><><><><><><><><><><><><><><><><><><><><><>

4. The long term behavior in the weather model is a probability vector **s** that satisfies the matrix equation $\mathbf{A} \cdot \mathbf{s} = \mathbf{s}$. Such a vector is called an **eigenvector** of the matrix **A** corresponding to the **eigenvalue** 1. Vector **s** is called the **stable state** of Markov Chain for the weather model.

For the matrix **A** given below, use the powers of the matrix **A** to obtain an accurate estimate of the stable state. That is, find an eigenvector of **A** corresponding to eigenvalue 1.

$$\mathbf{A} := \begin{bmatrix} 0 & \frac{1}{3} & \frac{1}{5} & \frac{1}{2} \\ \frac{1}{4} & 0 & \frac{2}{5} & \frac{1}{6} \\ \frac{1}{2} & \frac{1}{3} & \frac{1}{5} & 0 \\ \frac{1}{4} & \frac{1}{3} & \frac{1}{5} & \frac{1}{3} \end{bmatrix}$$

ANSWER:

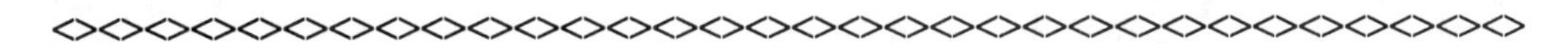

5. Refer back to the Leontief model presented in Section 0-2. This model is saved in the file leontief.mcd. The model was summarized as follows:

	Supplies		Energy		Constr		Transp		consumer demand
Energy	x	=	0.4 x	+	0.2 y	+	0.1 z	+	100
Construction	y	=	0.2 x	+	0.4 y	+	0.1 z	+	50
Transportation	z	=	0.15 x	+	0.2 y	+	0.2 z	+	100

Set this up as a matrix equation as follows:

$\mathbf{s} = \mathbf{A} \cdot \mathbf{s} + \mathbf{c}$ where **c** is the consumer demand.

Enter A here. ==> $\mathbf{A} := \begin{pmatrix} \blacksquare & \blacksquare & \blacksquare \\ \blacksquare & \blacksquare & \blacksquare \\ \blacksquare & \blacksquare & \blacksquare \end{pmatrix}$ $\mathbf{c} := \begin{pmatrix} 100 \\ 50 \\ 100 \end{pmatrix}$

We solve this equation by iteration (a repeated process) as follows:

Set $\quad \mathbf{s}^{<1>} := \mathbf{c} \qquad \mathbf{j} := 1..50$

$$\mathbf{s}^{<j+1>} := \mathbf{A} \cdot \mathbf{s}^{<j>} + \mathbf{c}$$

Evaluate by pressing F9. ===> $\qquad \mathbf{s} =$

Is this the same answer as in Section 0-2?

ANSWER:

Change the coefficients of the matrix **A** consistent with the input constraints (see Section 0-2). Does the process converge?

ANSWER:

6. (Individual) Write a summary of the MATHEMATICS that you learned in this section.

Section 1-10 Graphs

Name:________________

Score:______________

Overview

We develop a way to represent a graph as a matrix. (The matrix is a mathematical model of the graph.) As a result we can investigate properties of the graph by using matrix algebra. In this section we find a way to count the number of paths of a given length between vertices of the graph.

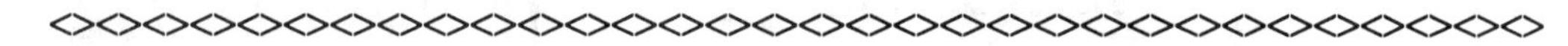

A **directed graph** consists of a number of **points** (or **vertices**) and **directed edges** between these vertices. In the picture below, the vertices are A, B, C, and D. We can label the edges by the vertices that they connect; in this case AB, BA, AD, BC, CB, CA and DC. Note that there are other possible edges (e.g., BD, AC) that do not appear. The arrow indicates the direction of the edge.

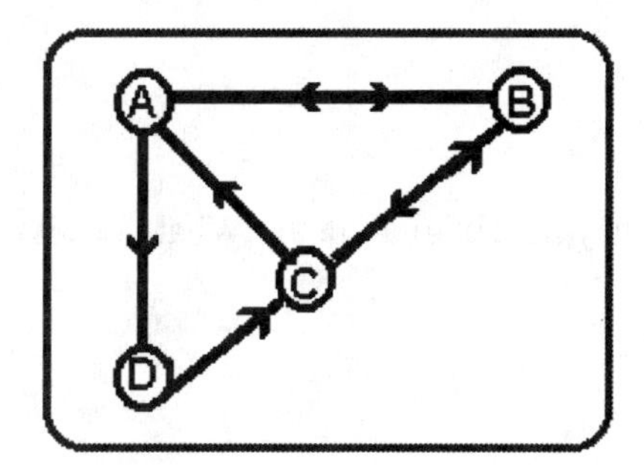

The **adjacency matrix of a graph** 'tells' which vertices are connected.

The (i,j)-entry of the adjacency matrix is 1 if there is an edge between the ith vertex and the jth vertex. The graph above is said to be directed since each edge has a direction.

	A	B	C	D
A	0	1	0	1
B	1	0	1	0
C	1	1	0	0
D	0	0	1	0

==> $Q := \begin{pmatrix} 0 & 1 & 0 & 1 \\ 1 & 0 & 1 & 0 \\ 1 & 1 & 0 & 0 \\ 0 & 0 & 1 & 0 \end{pmatrix}$ is the adjacency matrix of the preceding graph.

<><><><><><><><><><><><><><><><><><><><><><><><><><><><><><><><>

Exercises

1. Construct the adjacency matrix **M** for the graph displayed below.

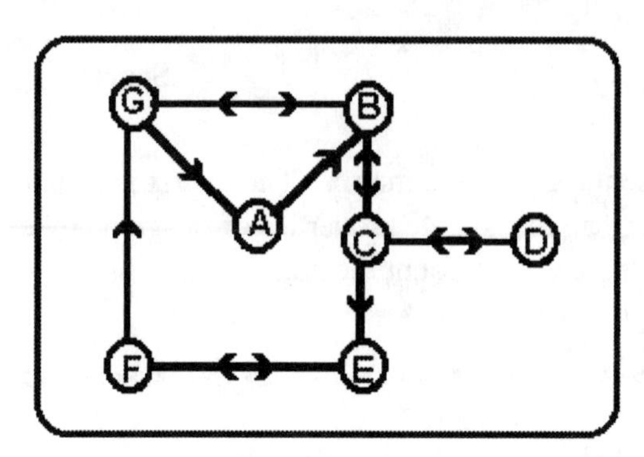

2. For the adjacency matrix below describe the graph. Your description can be in words or use paintbrush or a similar program to draw the graph and paste in its picture.

$$\begin{pmatrix} 0 & 1 & 0 & 1 \\ 1 & 0 & 0 & 1 \\ 0 & 1 & 0 & 1 \\ 1 & 0 & 0 & 0 \end{pmatrix}$$

<== ***Your description.***

3. Suppose that **M** is the adjacency matrix of a graph. What does it mean in terms of the graph if $\mathbf{M}_{2,2} = 1$?

ANSWER:

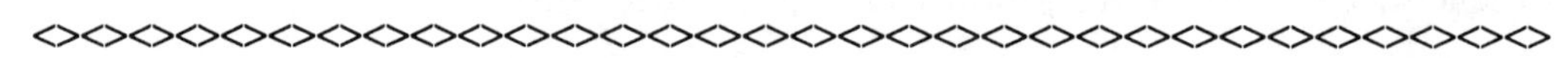

Since the adjacency matrix of a graph is square, its powers are defined. For the graph below we have displayed its adjacency matrix **Q**.

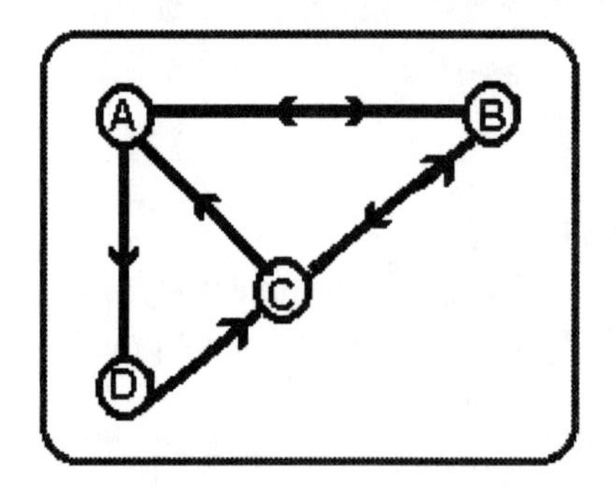

$$\mathbf{Q} := \begin{pmatrix} 0 & 1 & 0 & 1 \\ 1 & 0 & 1 & 0 \\ 1 & 1 & 0 & 0 \\ 0 & 0 & 1 & 0 \end{pmatrix}$$

The (i,j)th-element of $\mathbf{Q}^2$ is the dot product of row_i with col_j:

$$row_i(\mathbf{Q})\cdot col_j(\mathbf{Q}) = q_{i,1}\cdot q_{1,j} + q_{i,2}\cdot q_{2,j} + \ldots + q_{i,n}\cdot q_{n,j}$$

Each member of the sum of terms comprising the dot product above has the form $q_{i,k}\cdot q_{k,j}$. When is the term $q_{i,k}\cdot q_{k,j} = 1$? (Answer in terms of the graph.)

ANSWER:

<><><><><><><><><><><><><><><><><><><><><><><><><><><><><><><><><><><><>

We have the following interpretation:

$$q_{i,k} = \begin{cases} 1 \text{ if there is an edge from the ith vertex to the kth vertex} \\ 0 \text{ if there is no edge from the ith vertex to the kth vertex} \end{cases}$$

$$q_{k,j} = \begin{cases} 1 \text{ if there is an edge from the kth vertex to the jth vertex} \\ 0 \text{ if there is no edge from the kth vertex to the jth vertex} \end{cases}$$

Hence $q_{i,k}\cdot q_{k,j} = 1$ implies that we can start at the _______ vertex, go through the _______ vertex, and reach the _______ vertex. (Fill in the blanks.)

Let the length of a path from one vertex to another be one. What is the length of the path that goes from i to k and then from k to j?

ANSWER:

<><><><><><><><><><><><><><><><><><><><><><><><><><><><><><><><><><><><>

We were discussing the (i,j)th-element of $\mathbf{Q}^2$ which is

$$row_i(\mathbf{Q})\cdot col_j(\mathbf{Q}) = q_{i,1}\cdot q_{1,j} + q_{i,2}\cdot q_{2,j} + \ldots + q_{i,n}\cdot q_{n,j}$$

When this sum is computed each term is either a zero or a one. Explain, **in terms of the graph**, what the sum represents. (Be specific.)

ANSWER:

What do the entries of $\mathbf{Q}^2$ count?

ANSWER:

Extend your answer to $\mathbf{Q}^3, \ldots, \mathbf{Q}^n$.

ANSWER:

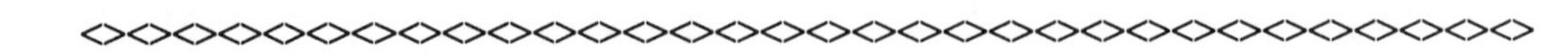

Exercises

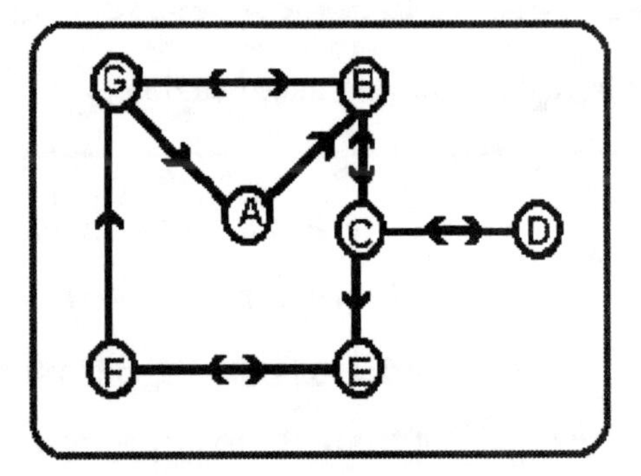

4. Determine the adjacency matrix for the graph on the right. How many paths of length 3 are there from A to D? (Show your work.)

ANSWER:

<><><><><><><><><><><><><><><><><><><><><><><><><><><><><><><><><><><><>

5. A certain graph has adjacency matrix **A**. What do the entries of $\mathbf{A} + \mathbf{A}^2$ represent?

ANSWER:

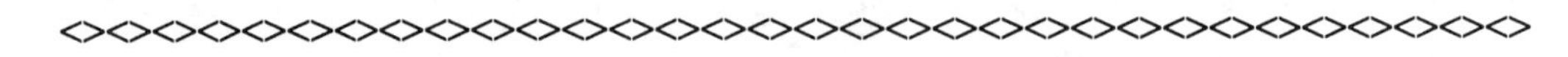

Undirected Graphs

A graph is said to be undirected if each edge is a 'two way' street. That means that we don't distinguish between an edge from A to B and one from B to A. In this case, it must always be true that the (i,j)th-element of the adjacency matrix is equal to the (j,i)th-element. Such matrices are called **symmetric**. This property can be written symbolically as the matrix equation $\mathbf{M}^T = \mathbf{M}$. Symmetric matrices play an important role in a variety of applications.

6. Determine the adjacency matrix for the undirected graph on the right. How many paths of length 4 are there from A to C?

ANSWER:

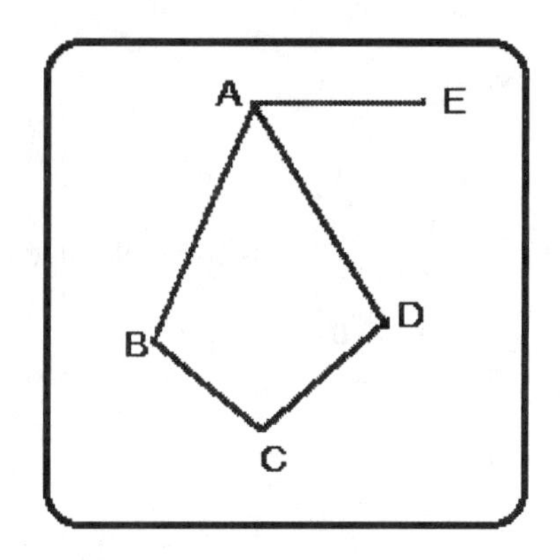

<><><><><><><><><><><><><><><><><><><><><><><><><><><><><><>

The number of paths of length n in a graph

Let **M** be the adjacency matrix of a directed graph. Since $\mathbf{M}^n_{i,j}$ is the number of paths of length n from vertex i to vertex j, the total number of paths of length n is

$$\sum_{(i,j)} (M^n)_{i,j}$$

If the graph has k vertices let **one** be the k-vector all of whose entries are 1.
We can generate this vector as follows. (It is a function of **M** since it must have the same number of rows as **M**.)

$$\mathrm{one}(M) := \overrightarrow{\left(\left(M^{<1>} = M^{<1>}\right)\right)}$$

(The arrow over the previous expression means that the operation is applied component-wise.) For two scalars the value of the operation (a = b) is 1 if a and b are equal and zero otherwise. Since $\mathbf{M}^{<1>}_j = \mathbf{M}^{<1>}_j$ for all values of j the resulting vector has all of its components equal to 1.

Verify the previous statement by evaluating : $\mathrm{one}(\mathrm{identity}(5)) = \begin{bmatrix} 1 \\ 1 \\ 1 \\ 1 \\ 1 \end{bmatrix}$

Explain why the total number of paths of length n can be computed as ==> $\mathrm{one}(M)\cdot(M^n \cdot \mathrm{one}(M))$

(Hint: What does $\mathrm{Row}_i(\mathbf{M})\cdot\mathbf{one}(\mathbf{M})$ compute?)

ANSWER:

We use the dot product in the above expression since we want the answer be a scalar.

<><><><><><><><><><><><><><><><><><><><><><><><><><><><><><>

$$\mathbf{Q} := \begin{pmatrix} 0 & 1 & 0 & 1 \\ 1 & 0 & 1 & 0 \\ 1 & 1 & 0 & 0 \\ 0 & 0 & 1 & 0 \end{pmatrix}$$ is the incidence matrix of

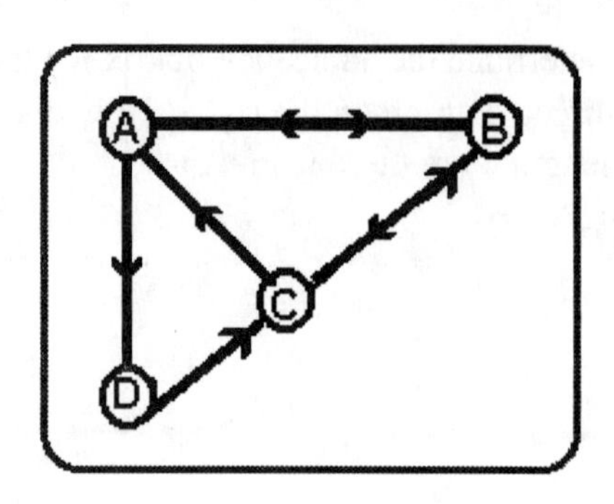

Set $\mathbf{T}(\mathbf{Q},\mathbf{n}) := \mathbf{one}(\mathbf{Q}) \cdot (\mathbf{Q}^{\mathbf{n}} \cdot \mathbf{one}(\mathbf{Q}))$

Graph T(**Q**,n) as a function of n. $\mathbf{n} := \mathbf{1} .. \mathbf{20}$

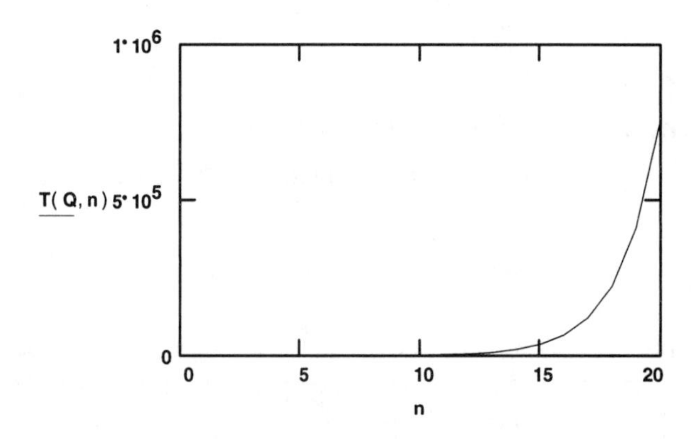

As you can see T(**Q**,n) gets very big very fast. We are interested in knowing how fast T(**Q**,n) gets big. To a mathematician 'how fast' means to compare the current function with other functions such as x^2, x^3, ... or 2^x, 3^x, ... (the function a^x is called the exponential to the base a). Two functions F(x) and G(x) grow at the same rate if the limit of their quotient exists and is a non-zero number. If the function grows like x^k for some k we say that it grows like a polynomial of degree k. If it grows like a^x, we say that it grows exponentially.

The goal of the next set of experiments is to determine from various graphs of a function whether the function grows like a polynomial or grows like an exponential.

Let's examine the graph of a polynomial and the graph of an exponential.

$x := 1 .. 20$ $\quad p(x) := x^{15}$ $\quad e(x) := 3^x$

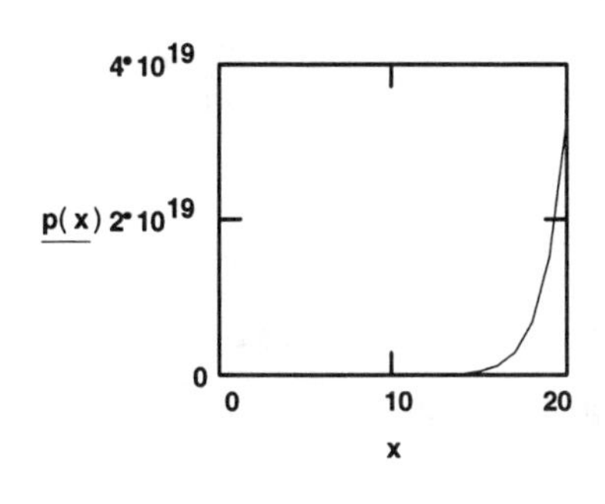

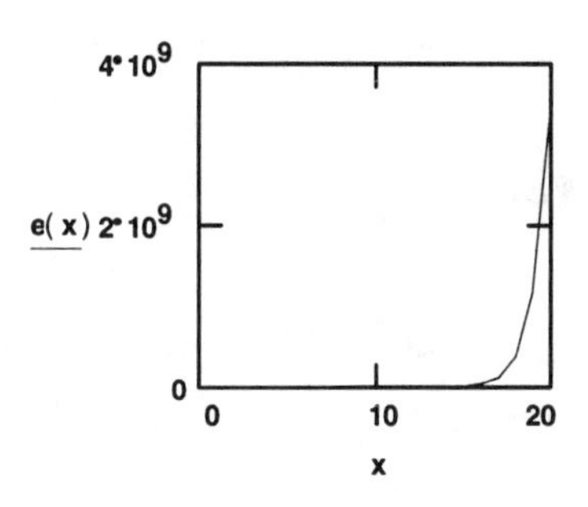

press F9.

As you can see, it is difficult to tell the difference from the graph.

Now double click on the first graph. In the dialogue box click on the choices Log Scale for x and Log Scale for y. Then click on OK. This is called a log-log transformation. The new graph will be log(x) vs. log(y). (**The logs here use base 10.**) Repeat for the second graph.

Describe the graph of x^{15} under the log-log transformation.

ANSWER:

Describe the graph of 3^x under the log-log transformation.

ANSWER:

<><><><><><><><><><><><><><><><><><><><><><><><><><><><><><>

Double click on each graph again. This time choose Log Scale for y but **DO NOT** choose Log Scale for x. Click on OK. This is called a semi-log y transformation. The new graph will be x vs. log(y).

Describe the graph of x^{15} under the semi-log y transformation.

ANSWER:

Describe the graph of 3^x under the semi-log y transformation.

ANSWER:

Suppose the log-log graph is a line with slope k. What was the function? (Hint: This means $\log(y) = k \cdot \log(x) + c$.)

ANSWER:

Suppose the semi-log graph is a line with slope k. What was the function? (Hint: This means $\log(y) = k \cdot x + c$.)

ANSWER:

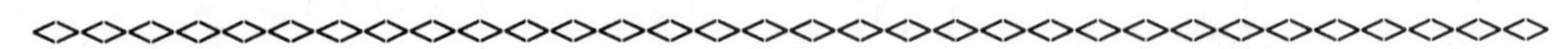

Now examine the function that gives the total number of paths of length n for the adjacency matrix **Q** given above.

$n := 1..20$

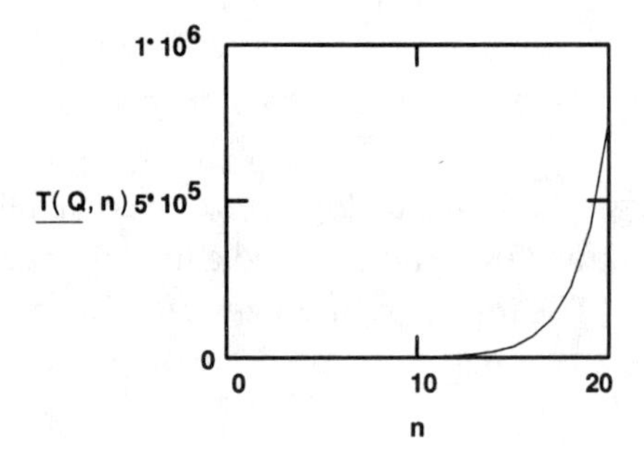

Is the function exponential or polynomial or neither? (Hint: Examine the graphs illustrated above.)

ANSWER:

If it is polynomial, compute the power of the polynomial. If it is exponential, compute the base (the value of a). (Hint: Think about the slope of the line in the appropriate graph. Take into account any transformations. There is some algebra to do here.)

ANSWER:

When we examine the growth rate of a function we are concerned about the asymptotic or long term behavior. This means that we may ignore values of the function for small numbers.

What are your conclusions about the number of paths of length n in the graph with adjacency matrix **Q**?

ANSWER:

What role does 1.839... play in your answer?

ANSWER:

<><><><><><><><><><><><><><><><><><><><><><><><><><><><><><><>

$$\mathbf{M} := \begin{bmatrix} 0 & 1 & 0 & 1 & 0 \\ 1 & 0 & 1 & 0 & 0 \\ 1 & 1 & 0 & 0 & 1 \\ 0 & 0 & 1 & 0 & 1 \\ 1 & 0 & 0 & 0 & 0 \end{bmatrix}$$

M is the adjacency matrix of another example. Examine the growth rate of the number of paths of length n. Construct a polynomial or exponential function which approximates the growth rate, whichever is appropriate.

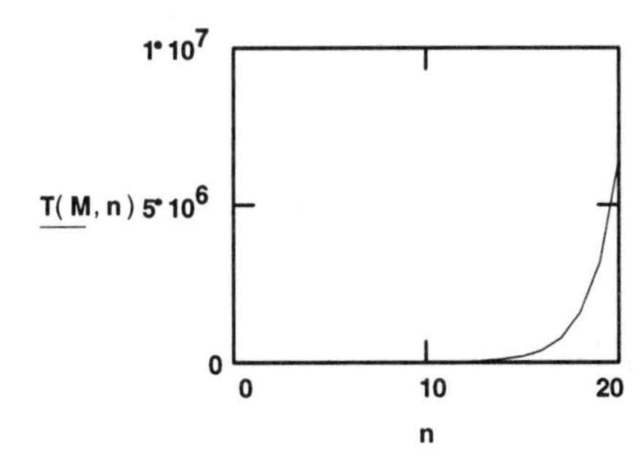

ANSWER:

<><><><><><><><><><><><><><><><><><><><><><><><><><><><><><><>

Let **G** be the adjacency matrix for a graph. Explain **in terms of the graph** what each of the following functions represents. (The Mathcad function max(**A**) returns the largest entry in the matrix or the vector **A**.)

$f(G, n) := \max(G^n)$

ANSWER:

$g(G, n) := \max(G^n \cdot one(G))$

ANSWER:

$h(G, n) := \max\left(one(G)^T \cdot G^n\right)$

ANSWER:

<><><><><><><><><><><><><><><><><><><><><><><><><><><><><><><><><><><><><><><><>

Let f, g, and h be the functions defined above and let **M** be an adjacency matrix. Determine the functions that approximate f(**M**,n), g(**M**,n) and h(**M**,n) as a function of n. How do these compare with the function that approximates T(**M**,n)?

$$M := \begin{bmatrix} 0 & 1 & 0 & 1 & 0 \\ 1 & 0 & 1 & 0 & 0 \\ 1 & 1 & 0 & 0 & 1 \\ 0 & 0 & 1 & 0 & 1 \\ 1 & 0 & 0 & 0 & 0 \end{bmatrix}$$

ANSWER:

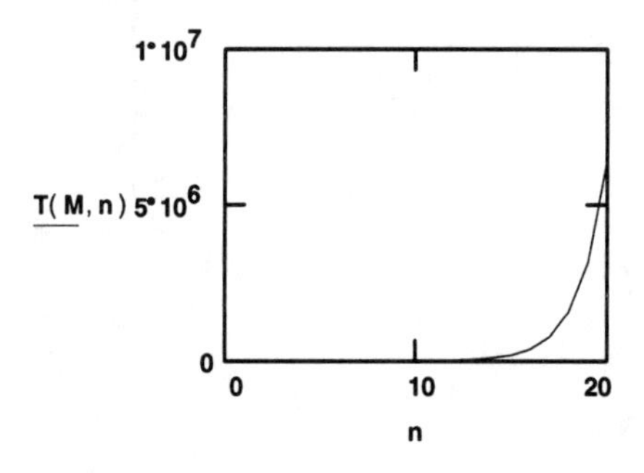

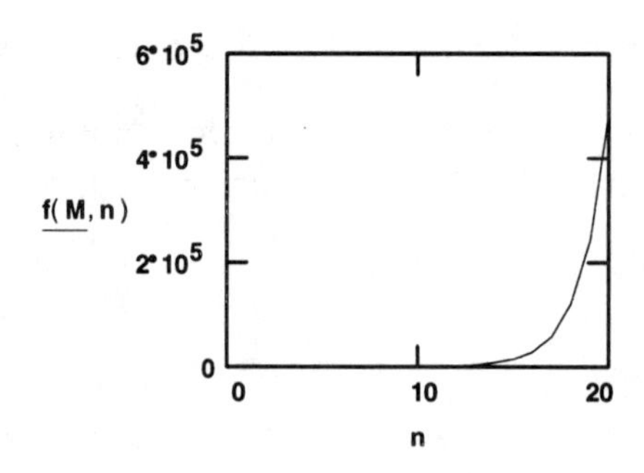

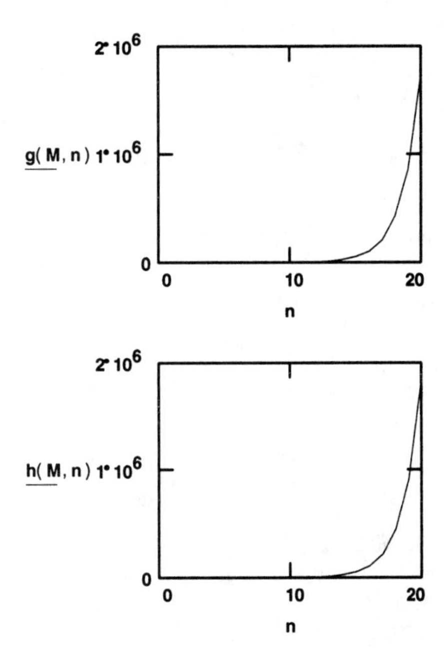

<><><><><><><><><><><><><><><><><><><><><><><><><><><><><><><><>

Exercise: (Individual) Write a summary of the MATHEMATICS that you learned from this section.

CHAPTER 2

SYSTEMS OF LINEAR EQUATIONS AND THEIR SOLUTION SETS

$$2x + y - 3z = -3$$

$$x + 3y + x = 6$$

$$3x + 2y - z = 3$$

Section 2-1 Systems of Linear Equations

Name:________________

Score:_____________

Overview

We examine systems of linear equations and their solutions. Our goal is to replace the algebraic manipulations that we learned in high school with an algorithm using matrices. This algorithm will be more efficient and will solve large systems quickly. We explore solutions to systems of linear equations with the goal of classifying them.

<><><><><><><><><><><><><><><><><><><><><><><><><><><><><><><><><><><><><>

Systems of linear equations are used to model applications in engineering and in the natural and social sciences. We introduced such a model in Section 0-2, the Refinery Model.

The description of the Refinery Model has been pasted on the page to the right in the electronic book. Scroll to that page using the slide bar on the bottom of this page and review the model. When you have completed reviewing the model return to the top of this page.

<><><><><><><><><><><><><><><><><><><><><><><><><><><><><><><><><><><><><>

Before we try to solve a system of linear equations, let's look at a pair of linear equations in two unknowns, x and y.

$$a \cdot x + b \cdot y = s$$
$$c \cdot x + d \cdot y = t$$

The set of points (x,y) that satisfy an equation of the form $ax + by = s$ form a line in the plane $\mathbf{R}^2$. What does it mean *geometrically* for a point (x,y) to satisfy both the first equation and the second equation?

ANSWER:

We call the set of all pairs of numbers (x,y) that satisfy both equations **the solution set of the system of equations**. What does it mean to say that (x,y) *satisfies the equations?*

ANSWER:

Based on the answer to the preceding question, **describe geometrically all possible solution sets** to two equations in two unknowns.
ANSWER:

What would the possible solution sets be if there were three equations in two unknowns?
ANSWER:

Another way to classify the solution sets is by the number of elements in the set. Fill in the blanks below with the number of elements in the geometric objects you have described above.

A solution set for a system of equations in two unknowns must have either ________ or _________ or _________ points.

Given a linear system of m equations in n unknowns. What do we mean by the solution set? (What kind of objects are in the solution set?) Does this help explain why linear algebra is used to study the solution set of a linear equation?
ANSWER:

State a conjecture about the number of vectors in the solution set to a system of m equations in n unknowns.
ANSWER:

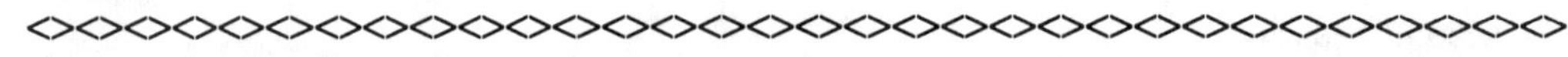

Solving Linear Systems Algebraically

Consider the system of linear equations:

$$3\cdot x + 2\cdot y - z = 3$$
$$x + 3\cdot y + z = 6$$
$$2\cdot x + y - 3\cdot z = -3$$

You learned how to solve such systems in high school algebra. There you learned to solve these systems by adding and subtracting equations to eliminate variables. The objective is to obtain an equation in one variable that can be solved directly. Back substitution is then used to solve for the other variables.

Take a piece of paper and solve the system given above.

Describe the process that you used to solve the system. By the way, the solution set is the 3-vector (x, y, z) where x = 1, y = 1, z = 2.
ANSWER:

<><><><><><><><><><><><><><><><><><><><><><><><><><><><><><><><><><><><><><>

The algebraic process applied to solve the preceding system of equations implicitly assumed that the operations performed in no way altered the solution set. In effect the manipulations produced a new, but related, linear system with the same solution set. We say that two linear systems are **equivalent** if their solution sets are the same.

If you solved the system above in a systematic fashion you made use of several basic facts. These include:

1. Multiplying both sides of the equation by the same nonzero constant doesn't change the solution set.

2. The solution set is not affected by the order in which you wrote the equations.

3. Adding a multiple of one equation to another doesn't change the solution set.

(1) and (2) are totally obvious. To show that (3) is true, let Eq1 and Eq2 be two linear equations. To simplify our calculations we will restrict ourselves to linear equations in four unknowns. Our discussion is easily generalized to equations with more variables.

$$\begin{aligned} &\textbf{Eq1}:\quad a_1 \cdot x_1 + a_2 \cdot x_2 + a_3 \cdot x_3 + a_4 \cdot x_4 = c_1 \\ &\textbf{Eq2}:\quad b_1 \cdot x_1 + b_2 \cdot x_2 + b_3 \cdot x_3 + b_4 \cdot x_4 = c_2 \end{aligned} \qquad (1)$$

We form a new system of equations by replacing Eq2 with Eq2 + k·Eq1.

$$\begin{aligned} &a_1 \cdot x_1 + a_2 \cdot x_2 + a_3 \cdot x_3 + a_4 \cdot x_4 = c_1 \\ &(k \cdot a_1 + b_1) \cdot x_1 + (k \cdot a_2 + b_2) \cdot x_2 + (k \cdot a_3 + b_3) \cdot x_3 + (k \cdot a_4 + b_4) \cdot x_4 = k \cdot c_1 + c_2 \end{aligned} \qquad (2)$$

To show that (3) is true we must show two things.

1. If (s1, ..., s4) is a solution of system (1) then it is also a solution to system (2).

2. If (t1, ..., t4) is a solution of system (2) then it is also a solution to system (1).

Explain why statement 1 is true.

ANSWER:

It is easy to see that statement 2 is true. Let's renumber the systems.

$$a_1 \cdot x_1 + a_2 \cdot x_2 + a_3 \cdot x_3 + a_4 \cdot x_4 = c_1 \qquad (1)$$
$$(k \cdot a_1 + b_1) \cdot x_1 + (k \cdot a_2 + b_2) \cdot x_2 + (k \cdot a_3 + b_3) \cdot x_3 + (k \cdot a_4 + b_4) \cdot x_4 = k \cdot c_1 + c_2$$

$$a_1 \cdot x_1 + a_2 \cdot x_2 + a_3 \cdot x_3 + a_4 \cdot x_4 = c_1 \qquad (2)$$
$$b_1 \cdot x_1 + b_2 \cdot x_2 + b_3 \cdot x_3 + b_4 \cdot x_4 = c_2$$

Then system 2 is derived from system 1 by adding -k times the first equation to the second but the proof that you just gave shows that these two sets have the same solutions.

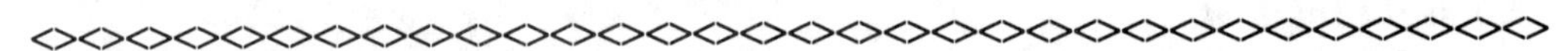

To solve the linear system above you may have proceeded as follows:

Original system:

$$3 \cdot x + 2 \cdot y - z = 3 \qquad \text{eq 1}$$
$$x + 3 \cdot y + z = 6 \qquad \text{eq 2}$$
$$2 \cdot x + y - 3 \cdot z = -3 \qquad \text{eq 3}$$

Step 1:

$$x + 3 \cdot y + z = 6$$
$$3 \cdot x + 2 \cdot y - z = 3$$
$$2 \cdot x + y - 3 \cdot z = -3$$

Interchange eq 1 and eq 2.

Step 2:

$$x + 3 \cdot y + z = 6$$
$$-7 \cdot y - 4 \cdot z = -15 \qquad -3 \cdot \text{eq 1} + \text{eq 2}$$
$$-5 \cdot y - 5 \cdot z = -15 \qquad -2 \cdot \text{eq 1} + \text{eq 3}$$

Eliminates x from eq 2 and eq 3.

Step 3:

$x + 3\cdot y + z = 6$	interchange eq 2 & eq 3	
$y + z = 3$	-(1/5)·eq 2	*This gives new eq 2 with y as the first term.*
$7\cdot y + 4\cdot z = 15$	- 1·eq 3	

<><><><><><><><><><><><><><><><><><><><><><><><><><><><><><><>

Step 4:

$x + 3\cdot y + z = 6$		
$y + z = 3$		
$-3\cdot z = -6$	- 7·eq 2 + eq 3	*Eliminates y from eq 3.*

<><><><><><><><><><><><><><><><><><><><><><><><><><><><><><><>

Step 5:

$z = 2$	from equation 3	
$y = 1$	from equation 2	*Solving for the variables by back substitution.*
$x = 1$	from equation 1	

<><><><><><><><><><><><><><><><><><><><><><><><><><><><><><><>

Our goal is to translate this algebraic solution process into a matrix algebra process. To do this it is important to keep in mind that the symbols x, y, and z play no inherent role in the calculation. They are just placeholders. If we used 🖫 ⌘ ♦ for the variable names they would work just as well.

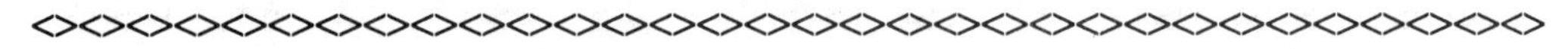

Return to the linear system

$$3\cdot x + 2\cdot y - z = 3$$
$$x + 3\cdot y + z = 6 \qquad (3)$$
$$2\cdot x + y - 3\cdot z = -3$$

Arrange the coefficients into a matrix **A**. This is called the **coefficient matrix**. The (i,j)th entry is the coefficient of the jth variable in the ith equation. The jth column corresponds to the jth variable; the ith row to the ith equation.

$$A := \begin{pmatrix} 3 & 2 & -1 \\ 1 & 3 & 1 \\ 2 & 1 & -3 \end{pmatrix}$$

Arrange the unknowns x, y, and z into a vector **u** and the constants on the right side into a vector **b**.

$$u := \begin{pmatrix} x \\ y \\ z \end{pmatrix} \qquad b := \begin{pmatrix} 3 \\ 6 \\ -3 \end{pmatrix}$$

If you are using Mathcad 5.0, click on the Maple Leaf icon **on the toolbar** to load the Symbolic Processor. This is not necessary in Mathcad 6.0.

Display the product **A·u**: $\begin{pmatrix} 3 & 2 & -1 \\ 1 & 3 & 1 \\ 2 & 1 & -3 \end{pmatrix} \cdot \begin{pmatrix} x \\ y \\ z \end{pmatrix}$ <== ***Click on the dot, evaluate symbolically (shift F9).***

The result is a vector whose components are the equations on the left side of (3).

An important observation about this matrix-vector product is that the system of equations in (3) can be written as the **matrix equation A·u = b**. (We call this the ***matrix form*** of the linear system.) We say that a vector **v** is a solution of the matrix equation **A·u = b** if **A·v = b.**

$$\begin{aligned} 3\cdot x + 2\cdot y - z &= 3 \\ x + 3\cdot y + z &= 6 \\ 2\cdot x + y - 3\cdot z &= -3 \end{aligned} \quad \Longleftrightarrow \quad \begin{pmatrix} 3 & 2 & -1 \\ 1 & 3 & 1 \\ 2 & 1 & -3 \end{pmatrix} \cdot \begin{pmatrix} x \\ y \\ z \end{pmatrix} = \begin{pmatrix} 3 \\ 6 \\ -3 \end{pmatrix}$$

<><><><><><><><><><><><><><><><><><><><><><><><><><><><><><><><>

For computational purposes we form a new matrix representation of the system consisting of coefficient matrix **A** and right-hand side **b**. This new matrix is called the **augmented coefficient matrix** (or ***augmented matrix*** for short) and is formed by adjoining **A** and **b**. For the system in (3) the augmented matrix is

$$\begin{pmatrix} 3 & 2 & -1 & 3 \\ 1 & 3 & 1 & 6 \\ 2 & 1 & -3 & -3 \end{pmatrix}$$

Before developing our matrix algorithm we have two comments:

1. In some texts an augmented matrix often appears with a vertical bar separating the coefficient matrix **A** and the right-hand side **b**, as follows

$$[\mathbf{A} \mid \mathbf{b}] \quad \text{or} \quad \left(\begin{array}{ccc|c} 3 & 2 & -1 & 3 \\ 1 & 3 & 1 & 6 \\ 2 & 1 & -3 & -3 \end{array}\right)$$

In Mathcad no such bar appears. Mathcad has a command to form an augmented matrix. Since we previously defined the coefficient matrix as **A** and the right-hand side as **b**, we obtain the augmented matrix as

$$\mathbf{Aug} := \mathbf{augment}(\mathbf{A}, \mathbf{b})$$

Checking we see $\mathbf{Aug} = \begin{pmatrix} 3 & 2 & -1 & 3 \\ 1 & 3 & 1 & 6 \\ 2 & 1 & -3 & -3 \end{pmatrix}$ <== *Press F9.*

(We named the augmented matrix **Aug**; we could have used any name.)

2. In matrix form we write a linear system as **A·u** = **b** to be consistent with Mathcad. In most books this is shortened to **Au** = **b**, and the matrix-vector product · is implied.

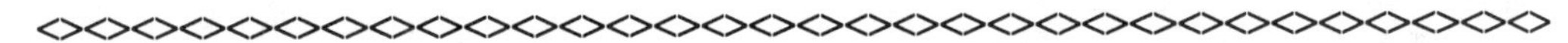

In our algebraic manipulation of the system of equations, whenever we performed the same actions on the left and right hand sides of the = sign the solution set remained the same. In the matrix form this means we can act on the matrix **A** and the vector **b** simultaneously. The augmented matrix allows us to do this.

Any manipulations done on equations can be performed by corresponding operations on the rows of the augmented matrix.

Algebraically we used three operations to change the form of the system of equations. As we saw above, the operations produce equivalent systems.

<><><><><><><><><><><><><><><><><><><><><><><><><><><><><><><><><><><><>

We are so familiar with the fact that various systems are equivalent that we sometimes fail to recognize that

$$\begin{matrix} x + y = 5 \\ 2\cdot x - 3\cdot y = 7 \end{matrix} \qquad \text{and} \qquad \begin{matrix} 2\cdot x - 3\cdot y = 7 \\ x + y = 5 \end{matrix}$$

or

$$x - 5\cdot y = 2 \qquad \text{and} \qquad 10\cdot x - 50\cdot y = 20$$

are indeed different systems because we *know* that they are *not really different*. **Equivalent** is another way of saying that they are *not really different*.

<><><><><><><><><><><><><><><><><><><><><><><><><><><><><><><><><><><><>

Elementary Row Operations

The three algebraic operations used to produce equivalent systems are.

1. Interchange a pair of equations.

2. Multiply an equation by a nonzero number.

3. Add a multiple of one equation to another equation.

The operations on equations correspond to operations on the rows of the corresponding augmented matrix [**A** | **b**] where they are called **ELEMENTARY ROW OPERATIONS.** The symbol on the right is a convenient way to indicate the row operation.

1. Interchange a pair of rows. $kR_i \rightarrow R_i$

2. Multiply a row by a nonzero number.

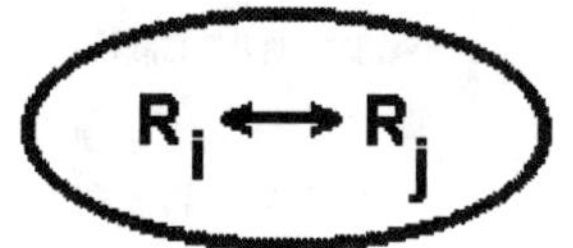

3. Add a multiple of one row to another row.

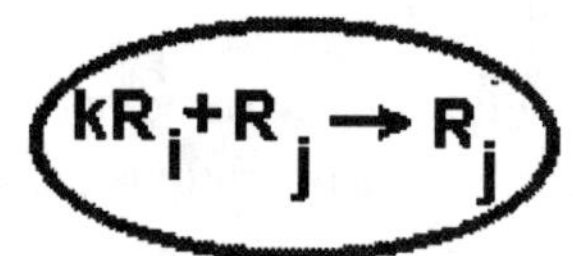

We use row operations on the augmented matrix, just as in the algebraic method, to obtain an equivalent system from which we can easily compute the solution set. We use the term **row-reduce the matrix** to indicate this process.

We row-reduce the matrix first by using some Mathcad functions that have been written for this text. (In the next section we will do the reduction directly using matrix multiplication.) We show the result of the reduction on the equations and the row-reduction steps side by side.

The Row-reduction Process

Let $\mathbf{M} := \mathbf{augment(A, b)}$ $\qquad \mathbf{M} = \begin{pmatrix} 3 & 2 & -1 & 3 \\ 1 & 3 & 1 & 6 \\ 2 & 1 & -3 & -3 \end{pmatrix}$ $\qquad$ <== *Display* **M**.

The following functions perform the elementary row operations on the augmented matrix **M**.

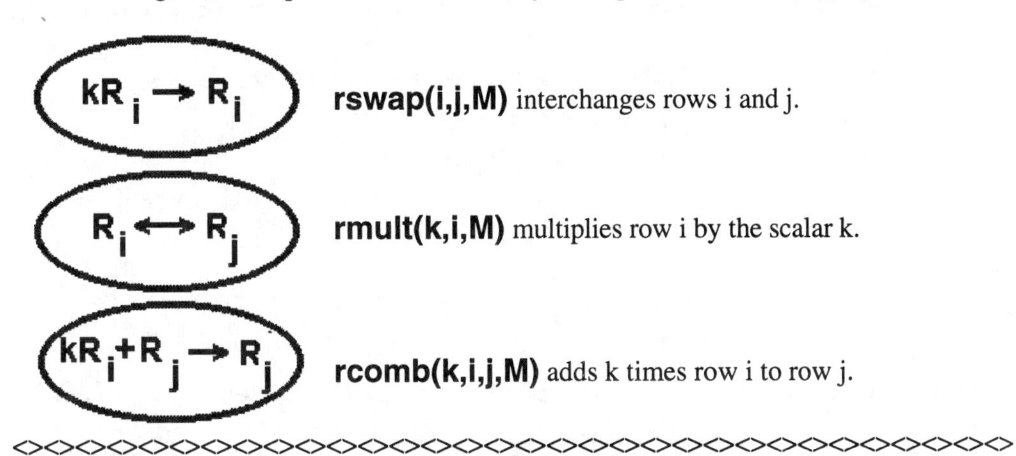

rswap(i,j,M) interchanges rows i and j.

rmult(k,i,M) multiplies row i by the scalar k.

rcomb(k,i,j,M) adds k times row i to row j.

<><><><><><><><><><><><><><><><><><><><><><><><><><><><><><><><><><><><><><><><><><><><><><><>

The Mathcad functions, **rswap**, **rmult**, and **rcomb** are contained in the file rowops.dll that should be in the efi directory. That file was on the disks that were distributed with this text. If for some reason you do not have that file you may define these functions by inserting the file rowops.mcd in the text. If you access the functions in that way, you will have to insert the functions in each worksheet in which they are used.

To insert a file, put the cursor below, then click on **Insert** on the **File** menu, select the file you wish to insert and then click on **OK**.

<======== ***put cursor here***

<><><><><><><><><><><><><><><><><><><><><><><><><><><><><><><><><><><><><><><><><><><><><><><>

$$3 \cdot x + 2 \cdot y - z = 3$$
$$x + 3 \cdot y + z = 6$$
$$2 \cdot x + y - 3 \cdot z = -3$$

$$\mathbf{M} = \begin{pmatrix} 3 & 2 & -1 & 3 \\ 1 & 3 & 1 & 6 \\ 2 & 1 & -3 & -3 \end{pmatrix}$$

<== ***Box, press =***
then press F9.

<><><><><><><><><><><><><><><><><><><><><><><><><><><><><><><><><><><><><><><><><><><><><><><>

$\mathbf{M1} := \mathbf{rswap(1, 2, M)}$ $\qquad$ $\mathbf{R_1 <==> R_2}$

$$x + 3 \cdot y + z = 6$$
$$3 \cdot x + 2 \cdot y - z = 3$$
$$2 \cdot x + y - 3 \cdot z = -3$$

$$\mathbf{M1} = \begin{pmatrix} 1 & 3 & 1 & 6 \\ 3 & 2 & -1 & 3 \\ 2 & 1 & -3 & -3 \end{pmatrix}$$

<== ***Box, press =***
then press F9.

<><><><><><><><><><><><><><><><><><><><><><><><><><><><><><><><><><><><><><><><><><><><><><><>

$x + 3 \cdot y + z = 6$

$-7 \cdot y - 4 \cdot z = -15$

$-5 \cdot y - 5 \cdot z = -15$

M2 := rcomb(- 3, 1, 2, M1) $\qquad -3 \cdot \mathbf{R}_1 + \mathbf{R}_2 \Longrightarrow \mathbf{R}_2$

M3 := rcomb(- 2, 1, 3, M2) $\qquad -2 \cdot \mathbf{R}_1 + \mathbf{R}_3 \Longrightarrow \mathbf{R}_3$

$$M3 = \begin{pmatrix} 1 & 3 & 1 & 6 \\ 0 & -7 & -4 & -15 \\ 0 & -5 & -5 & -15 \end{pmatrix}$$

<== Box, press = then press F9.

<><><><><><><><><><><><><><><><><><><><><><><><><><><><><><><>

$x + 3 \cdot y + z = 6$

$y + z = 3$

$7 \cdot y + 4 \cdot z = 15$

M4 := rswap(2, 3, M3) $\qquad \mathbf{R}_2 \Longleftrightarrow \mathbf{R}_3$

M5 := rmult(- .2, 2, M4) $\qquad -(1/5) \cdot \mathbf{R}_2 \Longrightarrow \mathbf{R}_2$

M6 := rmult(- 1, 3, M5) $\qquad -1 \cdot \mathbf{R}_3 \Longrightarrow \mathbf{R}_3$

$$M6 = \begin{pmatrix} 1 & 3 & 1 & 6 \\ 0 & 1 & 1 & 3 \\ 0 & 7 & 4 & 15 \end{pmatrix}$$

<== Box, press = then press F9.

<><><><><><><><><><><><><><><><><><><><><><><><><><><><><><><>

$x + 3 \cdot y + z = 6$

$y + z = 3$

$-3 \cdot z = -6$

M7 := rcomb(- 7, 2, 3, M6) $\qquad -7 \cdot \mathbf{R}_2 + \mathbf{R}_3 \Longrightarrow \mathbf{R}_3$

$$M7 = \begin{pmatrix} 1 & 3 & 1 & 6 \\ 0 & 1 & 1 & 3 \\ 0 & 0 & -3 & -6 \end{pmatrix}$$

<== Box, press = then press F9.

<><><><><><><><><><><><><><><><><><><><><><><><><><><><><><><>

$x + 3 \cdot y + z = 6$

$y + z = 3$

$z = 2$

$M8 := \mathrm{rmult}\left(-\frac{1}{3}, 3, M7\right)$ $\qquad -(1/3) \cdot \mathbf{R}_3 \Longrightarrow \mathbf{R}_3$

$$M8 = \begin{pmatrix} 1 & 3 & 1 & 6 \\ 0 & 1 & 1 & 3 \\ 0 & 0 & 1 & 2 \end{pmatrix}$$

<== Box, press = then press F9.

<><><><><><><><><><><><><><><><><><><><><><><><><><><><><><><>

There is nothing special about the names **M1, M2**, ... above. We could have called all the successive matrices **M** or anything we wished.

Use the equations above to find the solution set for the system.

ANSWER:

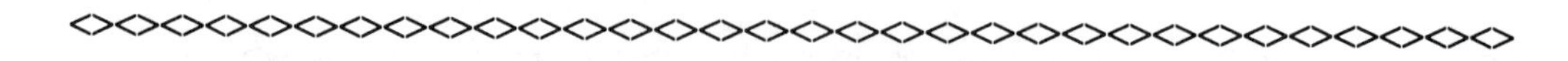

We have used the functions defined above to systematically transform our original systems to an equivalent system that can be solved more easily. The important facts about the new system are the following:

Each equation in the new system can be solved for the '*first*' variable that appears with a nonzero coefficient in terms of the succeeding variables. We refer to this variable as the *first* variable below.

No variable is the '*first*' for more than one equation.

Once a variable is used as a *first* variable it does not appear in any of the equations that follow.

The equations are listed in the same order as their *first* variable.

For example:

System 1

$x + 3\cdot y - z = 0$ <== ***First variable** = x.*

$y + 2\cdot z = 5$ <== ***First variable** = y.*

System 2

$y + 2\cdot z = 5$ <== ***First variable** = y.*

$x + 3\cdot y - z = 0$ <== ***First variable** = x.*

System 3

$x + 3\cdot y - z = 0$ <== ***First variable** = x.*

$2\cdot z = 8$ <== ***First variable** = z.*

System 4

$x + 3\cdot y - z = 0$ <== ***First variable** = x.*

$x + 2\cdot z = 5$ <== ***First variable** = x.*

System 5

$2\cdot z = 8$ <== ***First variable** = z.*

$x + 3\cdot y = 0$ <== ***First variable** = x.*

System 6

$x + 3\cdot y - z = 0$ <== ***First variable** = x.*

Systems 1, 3 and 6 are in reduced form; the others are not.

The augmented matrices that correspond to a reduced system are called **row-reduced**. The first nonzero element in each row is called the **pivot**. The requirements for a reduced system then become:

No column can contain more than one pivot.

The entries below a pivot (in the same column) are all zero.

Each pivot lies to the left of the pivot in the following row.

Any zero rows come last.

A matrix that is row-reduced has a descending staircase whose corners are the pivots. All entries below the staircase are 0. The following is a generic example.

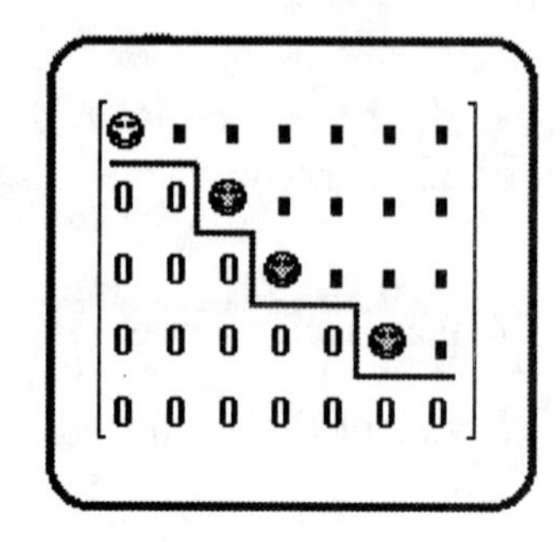

The faces are the pivots and must be nonzero. The placeholders, ■ , can contain any values.

Why is it easy to find the solution set of the row-reduced system?

ANSWER:

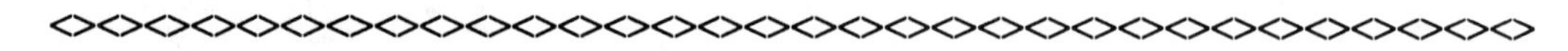

The system of equations: $x + y = 3$

$x + y = 0$

has no solutions since the sum of x and y can not simultaneously be both 3 and 0. A system of equations with no solutions is called **inconsistent**. If a system of equations has at least one solution it is called **consistent**.

Explain how you can tell that a system is inconsistent after the augmented matrix has been row-reduced.

ANSWER:

Explain how, after the augmented matrix has been row-reduced, you can tell that a system has an infinite number of solutions.

ANSWER:

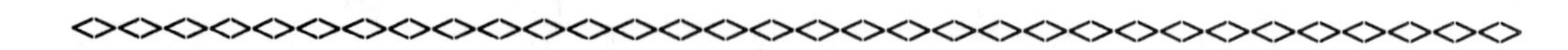

Exercises

1. Solve the following linear system by using row operations to row-reduce the system and then applying back substitution either in Mathcad or by hand. Record your result below. Carefully follow the directions.

$$\begin{aligned} 2x + y + 3z &= 8 \\ x + 3y - z &= -1 \\ -4x + z &= -11 \end{aligned}$$

***Create the coefficient matrix* A *here.* ==>**

***Create the right-hand side* b *here.* ==>**

***Create the augmented matrix* M *here.* ==>**

Use row operations below. Display each step of the row-reduction.

Insert more lines as you need them by entering Control F9 or by using the Edit Menu.

Solution: $x =$ _______ $y =$ _______ $z =$ ________

<><><><><><><><><><><><><><><><><><><><><><><><><><><><><><><>

2. The REFINERY MODEL linear system is:

$$\begin{aligned} 4x_1 + 2x_2 + 2x_3 &= 600 \\ 2x_1 + 5x_2 + 2x_3 &= 800 \\ x_1 + 2.5\,x_2 + 5x_3 &= 1000 \end{aligned}$$

Solve this linear system by using row operations to reduce the system to row-reduced form and then applying back substitution either in Mathcad or by hand. Record your work below.

Solution: $x_1 =$ _________ $x_2 =$ _________ $x_3 =$ _________

<><><><><><><><><><><><><><><><><><><><><><><><><><><><><><><>

3. Each of the following augmented matrices represents a system that has been row-reduced. On the line below each system indicate whether there are no solutions, a unique solution (that is only one solution), or infinitely many solutions.

$$\left(\begin{array}{cccc|c} 1 & -1 & 2 & 1 & -1 \\ 0 & 0 & 1 & 1 & 2 \\ 0 & 0 & 0 & -1 & 1 \\ 0 & 0 & 0 & 0 & 3 \end{array}\right) \qquad \left(\begin{array}{cccc|c} 1 & 2 & 2 & 3 & -1 \\ 0 & 0 & 1 & 2 & 0 \\ 0 & 0 & 0 & 1 & 3 \end{array}\right) \qquad \left(\begin{array}{cccc|c} 1 & -1 & 2 & 1 & -1 \\ 0 & 0 & 1 & 1 & 2 \\ 0 & 0 & 0 & -1 & 1 \\ 0 & 0 & 0 & 0 & 0 \end{array}\right)$$

______________ ______________ ______________

<><><><><><><><><><><><><><><><><><><><><><><><><><><><><><><><><><>

4. The REFINERY MODEL has been modified so that no gasoline is to be produced. The resulting linear system is

$$\begin{aligned} 4x_1 + \quad 2x_2 + 2x_3 &= 600 \\ 2x_1 + \quad 5x_2 + 2x_3 &= 800 \\ x_1 + 2.5\,x_2 + 5x_3 &= 0 \end{aligned}$$

Solve this linear system by using row operations to row-reduce the system and then applying back substitution either in Mathcad or by hand. Record your work below.

ANSWER:

If the system is consistent, does the computed solution make sense? Explain.

ANSWER:

<><><><><><><><><><><><><><><><><><><><><><><><><><><><><><><><><><>

5. The following linear systems have the same coefficient matrices but different right-hand sides. We can solve them at the same time by creating a matrix **M** with two augmented columns and row-reducing it.

System #1	**System #2**
$x - y + 3z = 5$	$x - y + 3z = 5$
$2x + y - z = 3$	$2x + y - z = 3$
$x - 7y + 17z = 19$	$x - 7y + 17z = 1$

Coefficient matrix ==> $\mathbf{A} := \begin{pmatrix} 1 & -1 & 3 \\ 2 & 1 & -1 \\ 1 & -7 & 17 \end{pmatrix}$

Right sides ==> $\mathbf{b} := \begin{pmatrix} 5 \\ 3 \\ 19 \end{pmatrix}$ $\quad \mathbf{c} := \begin{pmatrix} 5 \\ 3 \\ 1 \end{pmatrix}$

Special Augmented Matrix ==> $\mathbf{M} := \mathbf{augment}(\mathbf{augment}(\mathbf{A}, \mathbf{b}), \mathbf{c})$

Display M here. ==>

Solve these systems simultaneously. Show your work and the solution to each system.

ANSWER:

Check your work by calculating **A** times your answer.

<><><><><><><><><><><><><><><><><><><><><><><><><><><><><><><><><><><><><><>

6. The city traffic commission has collected data on the traffic flow for the street network shown below. The direction of the traffic flow between intersections is indicated by arrows. The intersections are labeled A through D and the average number of vehicles per hour on portions of the streets is indicated by x_1 through x_5. The average number of vehicles that enter or exit a street appears near the street. For instance, at intersection B, 200 vehicles per hour exit the network, and at intersection D, 300 vehicles per hour enter from the left side and 200 per hour exit on the diagonal street. The vehicles entering an intersection must also exit the intersection, so we can construct an **equilibrium** or **input-equals-output equation** for each intersection.

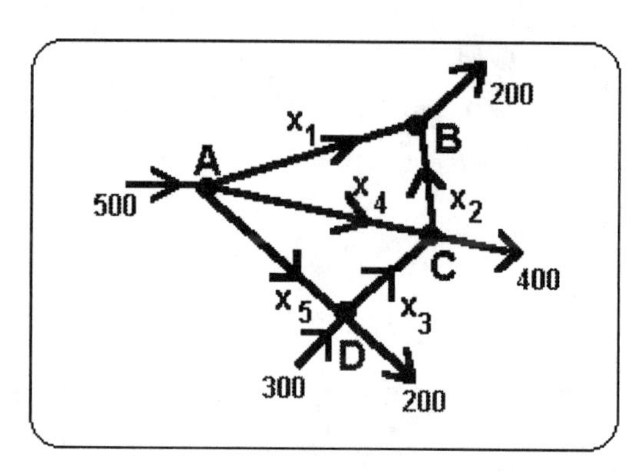

Write an equilibrium equation below for each intersection. **Do not solve the system.**

Equation for intersection A:

Equation for intersection B:

Equation for intersection C:

Equation for intersection D:

Show that each of the following vectors is a solution of the traffic flow system. (Entries are listed in the order x_1 through x_5.)

$u := (100\ \ 100\ \ 200\ \ 300\ \ 100)^T$ $\qquad$ $v := (300\ \ -100\ \ 200\ \ 100\ \ 100)^T$

How many solutions are there to system of four equations given for the traffic flow problem?

ANSWER:

Does the solution in vector **v** make sense? Explain.

ANSWER:

<><><><><><><><><><><><><><><><><><><><><><><><><><><><><><><><><><><><><><><>

7. (Individual) Write a summary of the MATHEMATICS you learned in this section.

Section 2-2 Elementary Matrices

Name:________________

Score:_____________

Overview

We have seen that the row operations

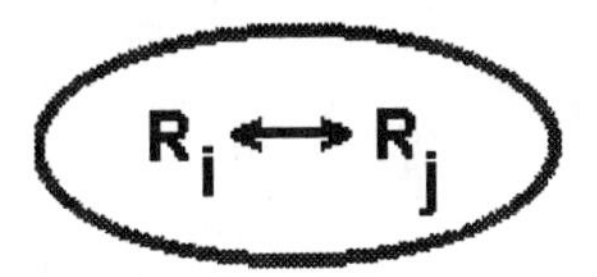

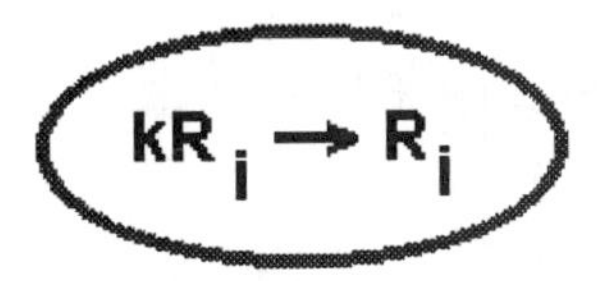

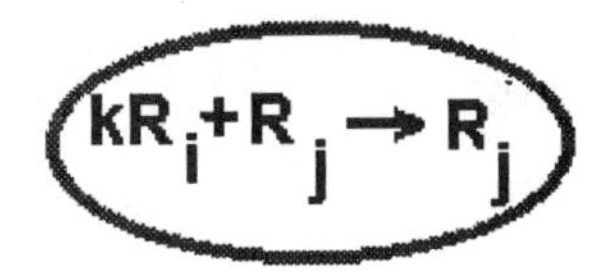

can be used to transform a system of linear equations into row-reduced form from which back substitution gives the solution set. It is convenient for certain purposes to be able to view the reduction process as a **matrix equation**. That is, given the augmented matrix **M** = [**A** | **b**] there is a matrix **T** such that **T·M** is an equivalent system in row-reduced form.

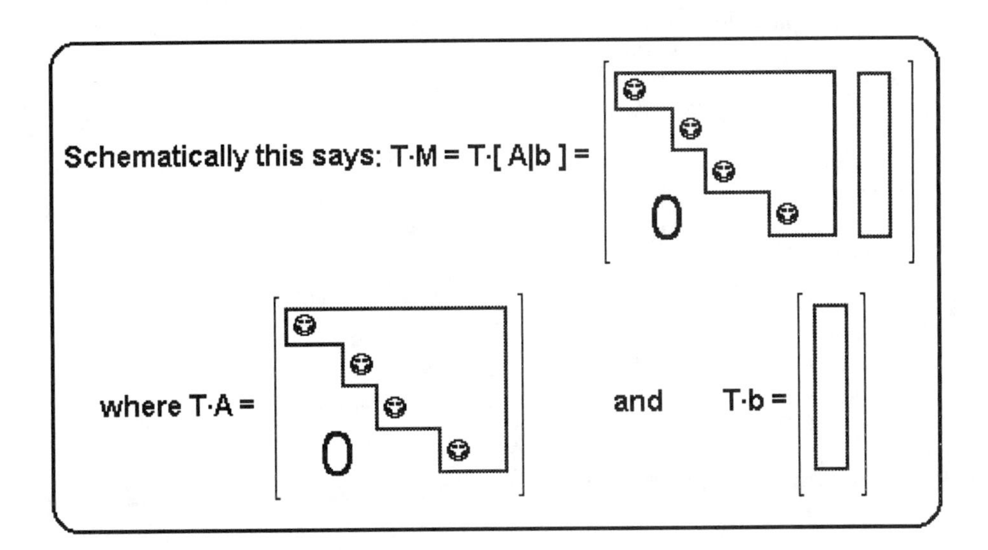

The matrix **T** performs the transformation to row-reduced form.

Since a sequence of row operations is performed to get to row-reduced form, it is reasonable to imagine that the matrix **T** is 'composed' of a sequence of 'pieces' corresponding to the row operations. We discover both the pieces and the way they are 'composed' in this section.

<><><><><><><><><><><><><><><><><><><><><><><><><><><><><><>

Example 1

Let the matrix **M** correspond to a system of linear equations.

$$M := \begin{pmatrix} 2 & 31 & -25 & 7 \\ 11 & 25 & 9 & 8 \\ 60 & 6 & 0 & 17 \end{pmatrix}$$

Multiply **M** by each of the following matrices and describe the relationship between the product **Eij·M** and the original matrix **M**. (Similarly for **Fi** and **Gj**.)

$$E12 := \begin{pmatrix} 0 & 1 & 0 \\ 1 & 0 & 0 \\ 0 & 0 & 1 \end{pmatrix} \qquad E23 := \begin{pmatrix} 1 & 0 & 0 \\ 0 & 0 & 1 \\ 0 & 1 & 0 \end{pmatrix} \qquad F1 := \begin{pmatrix} -2 & 0 & 0 \\ 0 & 1 & 0 \\ 0 & 0 & 1 \end{pmatrix}$$

$$F2 := \begin{pmatrix} 1 & 0 & 0 \\ 0 & 1 & 0 \\ 0 & 0 & 3 \end{pmatrix} \qquad G1 := \begin{pmatrix} 1 & 0 & 0 \\ 0 & 1 & 0 \\ -30 & 0 & 1 \end{pmatrix} \qquad G2 := \begin{pmatrix} 1 & 0 & 0 \\ -5.5 & 1 & 0 \\ 0 & 0 & 1 \end{pmatrix}$$

(Hint: Is there a row operation that will produce the same result?)
Answer on the line provided.

a) $E12 \cdot M$

____

b) $E23 \cdot M$

____

M is repeated for easy reference.

$$M = \begin{pmatrix} 2 & 31 & -25 & 7 \\ 11 & 25 & 9 & 8 \\ 60 & 6 & 0 & 17 \end{pmatrix}$$

c) $F1 \cdot M$

____

d) $F2 \cdot M$

____

e) $G1 \cdot M$

____

f) $G2 \cdot M$

____

On each line above you should have mentioned a row operation. Check your answers by computing the following products symbolically.

a) $\begin{pmatrix} 0 & 1 & 0 \\ 1 & 0 & 0 \\ 0 & 0 & 1 \end{pmatrix} \cdot \begin{pmatrix} a & d & g & j \\ b & e & h & k \\ c & f & i & l \end{pmatrix}$

b) $\begin{pmatrix} 1 & 0 & 0 \\ 0 & 0 & 1 \\ 0 & 1 & 0 \end{pmatrix} \cdot \begin{pmatrix} a & d & g & j \\ b & e & h & k \\ c & f & i & l \end{pmatrix}$

c) $\begin{pmatrix} -2 & 0 & 0 \\ 0 & 1 & 0 \\ 0 & 0 & 1 \end{pmatrix} \cdot \begin{pmatrix} a & d & g & j \\ b & e & h & k \\ c & f & i & l \end{pmatrix}$

d) $\begin{pmatrix} 1 & 0 & 0 \\ 0 & 1 & 0 \\ 0 & 0 & 3 \end{pmatrix} \cdot \begin{pmatrix} a & d & g & j \\ b & e & h & k \\ c & f & i & l \end{pmatrix}$

e) $\begin{pmatrix} 1 & 0 & 0 \\ 0 & 1 & 0 \\ -30 & 0 & 1 \end{pmatrix} \cdot \begin{pmatrix} a & d & g & j \\ b & e & h & k \\ c & f & i & l \end{pmatrix}$

f) $\begin{pmatrix} 1 & 0 & 0 \\ -5.5 & 1 & 0 \\ 0 & 0 & 1 \end{pmatrix} \cdot \begin{pmatrix} a & d & g & j \\ b & e & h & k \\ c & f & i & l \end{pmatrix}$

For each of the matrices **Eij**, **Fi**, and **Gj** in Example 1 describe how it can be created from the 3 by 3 identity matrix using a **single** row operation. The matrices are repeated below for convenience.

$E12 := \begin{pmatrix} 0 & 1 & 0 \\ 1 & 0 & 0 \\ 0 & 0 & 1 \end{pmatrix}$ $E23 := \begin{pmatrix} 1 & 0 & 0 \\ 0 & 0 & 1 \\ 0 & 1 & 0 \end{pmatrix}$ $F1 := \begin{pmatrix} -2 & 0 & 0 \\ 0 & 1 & 0 \\ 0 & 0 & 1 \end{pmatrix}$

$F2 := \begin{pmatrix} 1 & 0 & 0 \\ 0 & 1 & 0 \\ 0 & 0 & 3 \end{pmatrix}$ $G1 := \begin{pmatrix} 1 & 0 & 0 \\ 0 & 1 & 0 \\ -30 & 0 & 1 \end{pmatrix}$ $G2 := \begin{pmatrix} 1 & 0 & 0 \\ -5.5 & 1 & 0 \\ 0 & 0 & 1 \end{pmatrix}$

ANSWERS:

E12: **E23**: **F1**:

F2: **G1**: **G2**:

<><><><><><><><><><><><><><><><><><><><><><><><><><><><><><><><><>

Compare the results of the last two exercises and complete the following sentence:

To perform a row operation on a matrix **M**, multiply **M** on the left by a matrix **E** where **E** is the matrix created by ______________________ on the appropriate size identity matrix.

This leads us to the concept of an elementary matrix.

> **An *elementary matrix* is an identity matrix on which a *single* row operation has been performed.**

<><><><><><><><><><><><><><><><><><><><><><><><><><><><><><><><><><><><><><><><>

Summary

Given a system of linear equations $\mathbf{A}\cdot\mathbf{x} = \mathbf{b}$, we transform the augmented matrix $\mathbf{M} = [\mathbf{A} \mid \mathbf{b}]$ to row-reduced form by multiplying **M** (on the left) by a sequence of elementary matrices. Let $\mathbf{E}_i$ denote the elementary matrix corresponding to the ith row operation and suppose s row operations give us the row-reduced form; then

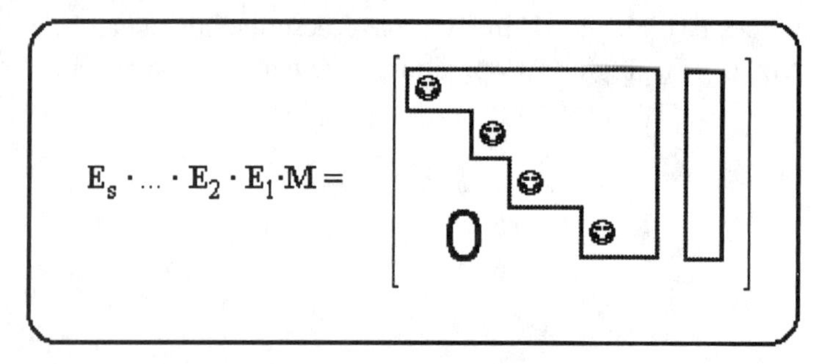

If we set $\mathbf{E}_s \cdot \ldots \cdot \mathbf{E}_2 \cdot \mathbf{E}_1 = \mathbf{T}$, then **T** transforms **M** to row-reduced form. The matrix **T** is not unique. Many different sequences of row operations can perform the task of generating row-reduced form.

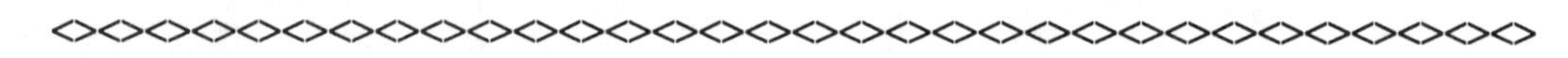

Example 2

Transform the linear system

$$\begin{aligned} 2y + z &= 0 \\ 3x + y - 2z &= 9 \\ -6x \quad + 5z &= -16 \end{aligned}$$

to row-reduced form using elementary matrices.

Step 1: Form the augmented matrix **M** from the coefficient matrix **A** and the right-hand side **b**.

$$\mathbf{A} := \begin{pmatrix} 0 & 2 & 1 \\ 3 & 1 & -2 \\ -6 & 0 & 5 \end{pmatrix} \quad \mathbf{b} := \begin{pmatrix} 0 \\ 9 \\ -16 \end{pmatrix} \quad \mathbf{M} := \mathbf{augment}(\mathbf{A},\mathbf{b}) \quad \mathbf{M} = \begin{pmatrix} 0 & 2 & 1 & 0 \\ 3 & 1 & -2 & 9 \\ -6 & 0 & 5 & -16 \end{pmatrix}$$

$$\mathbf{origM} := \mathbf{M}$$

<><><><><><><><><><><><><><><><><><><><><><><><><><><><><><><><><><><><><><>

Step 2: Interchange rows to get a nonzero value into the (1,1) position.

$\mathbf{R}_1 \iff \mathbf{R}_2$; corresponds to the elementary matrix

$$\mathbf{E1} := \begin{pmatrix} 0 & 1 & 0 \\ 1 & 0 & 0 \\ 0 & 0 & 1 \end{pmatrix}$$

Compute $\mathbf{M} := \mathbf{E1} \cdot \mathbf{M}$

$$\mathbf{M} = \begin{pmatrix} 3 & 1 & -2 & 9 \\ 0 & 2 & 1 & 0 \\ -6 & 0 & 5 & -16 \end{pmatrix}$$

<><><><><><><><><><><><><><><><><><><><><><><><><><><><><><><><><><><><><><>

Step 3: Add 2 times row 1 to row 3 to get a zero in the (3,1) position.

$2\mathbf{R}_1 + \mathbf{R}_3 \implies \mathbf{R}_3$; corresponds to the elementary matrix

$$\mathbf{E2} := \begin{pmatrix} 1 & 0 & 0 \\ 0 & 1 & 0 \\ 2 & 0 & 1 \end{pmatrix}$$

Compute $\mathbf{M} := \mathbf{E2} \cdot \mathbf{M}$

$$\mathbf{M} = \begin{pmatrix} 3 & 1 & -2 & 9 \\ 0 & 2 & 1 & 0 \\ 0 & 2 & 1 & 2 \end{pmatrix}$$

<><><><><><><><><><><><><><><><><><><><><><><><><><><><><><><><><><><><><><>

Step 4: Subtract row 2 from row 3 to get a zero in the (3,2) position.

$-1\mathbf{R}_2 + \mathbf{R}_3 \implies \mathbf{R}_3$; corresponds to the elementary matrix

$$\mathbf{E3} := \begin{pmatrix} 1 & 0 & 0 \\ 0 & 1 & 0 \\ 0 & -1 & 1 \end{pmatrix}$$

Compute $\mathbf{M} := \mathbf{E3} \cdot \mathbf{M}$

$$\mathbf{M} = \begin{pmatrix} 3 & 1 & -2 & 9 \\ 0 & 2 & 1 & 0 \\ 0 & 0 & 0 & 2 \end{pmatrix}$$

<><><><><><><><><><><><><><><><><><><><><><><><><><><><><><><><><><><><><><>

We see that this system has no solution (is inconsistent). ***WHY***?

ANSWER:

Set $\mathbf{T} := \mathbf{E3} \cdot \mathbf{E2} \cdot \mathbf{E1}$

Compute $\mathbf{T} \cdot \mathbf{origM} = \begin{pmatrix} 3 & 1 & -2 & 9 \\ 0 & 2 & 1 & 0 \\ 0 & 0 & 0 & 2 \end{pmatrix}$ <== ***Display this product.***

This product is the row-reduced matrix we computed above. The matrix **T** is a product of elementary matrices and is

$$\mathbf{T} = \begin{pmatrix} 0 & 1 & 0 \\ 1 & 0 & 0 \\ -1 & 2 & 1 \end{pmatrix}$$ <== ***Display* T.**

It is very difficult to construct **T** without doing it step-by-step in terms of the elementary matrices.

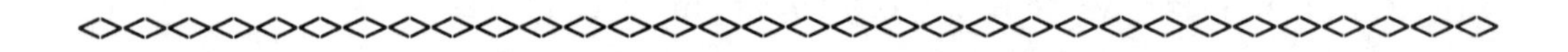

Let **T** be the matrix constructed in Example 2. Explain how we can use **T** to row-reduce the linear system

$$\begin{aligned} 2y + z &= 0 \\ 3x + y - 2z &= 6.5 \\ -6x \quad + 5z &= -13 \end{aligned}$$

ANSWER:

Row-reduce this system using **T**.

ANSWER:

<><><><><><><><><><><><><><><><><><><><><><><><><><><><><><><><><><><><><>

Exercises

1. Given the system

$$\begin{aligned} 2x + y + 3z &= 1 \\ -6x - 3y + 2z &= -14 \\ 4x \quad\quad + z &= 3 \end{aligned}$$

Find a sequence of elementary matrices that reduce $\mathbf{M} = [\mathbf{A} \mid \mathbf{b}]$ to row-reduced form. Call the elementary matrices $\mathbf{E}_1$, $\mathbf{E}_2$, ... etc. as in Example 2. Compute the product of the elementary matrices as in Example 2 and call it **T**. Display **T** below.

ANSWER:

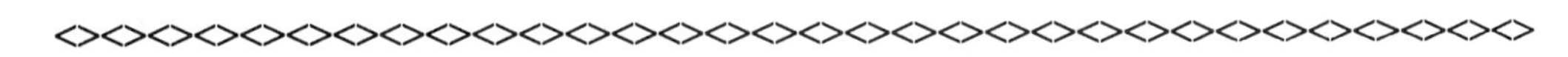

2. In Example 1 we multiplied the matrix, **M**, by the elementary matrices, **E12** and **E23**, given below.

$$\mathbf{E12} := \begin{pmatrix} 0 & 1 & 0 \\ 1 & 0 & 0 \\ 0 & 0 & 1 \end{pmatrix} \qquad \mathbf{E23} := \begin{pmatrix} 1 & 0 & 0 \\ 0 & 0 & 1 \\ 0 & 1 & 0 \end{pmatrix} \qquad \mathbf{M} := \begin{pmatrix} 2 & 31 & -25 & 7 \\ 11 & 25 & 9 & 8 \\ 60 & 6 & 0 & 17 \end{pmatrix}$$

a) Compute $(\mathbf{E12})^2$

ANSWER:

Describe the result of $\mathbf{E12 \cdot E12 \cdot M} = (\mathbf{E12})^2 \cdot \mathbf{M}$ below.

ANSWER:

b) Compute $(\mathbf{E23})^2$

ANSWER:

Describe the result of $\mathbf{E23 \cdot E23 \cdot M} = (\mathbf{E23})^2 \cdot \mathbf{M}$ below.

ANSWER:

c) Let **Eij** be the elementary matrix that corresponds to row operation $\mathbf{R_i} \Longleftrightarrow \mathbf{R_j}$. Form a conjecture about the behavior of $(\mathbf{Eij})^2$ when multiplied times **M**. Explain your answer.

ANSWER:

<><><><><><><><><><><><><><><><><><><><><><><><><><><><><><>

3. In Example 1 we multiplied the matrix, **M**, by the elementary matrices, **F1** and **F2**, given below.

$$\mathbf{F1} := \begin{pmatrix} -2 & 0 & 0 \\ 0 & 1 & 0 \\ 0 & 0 & 1 \end{pmatrix} \qquad \mathbf{F2} := \begin{pmatrix} 1 & 0 & 0 \\ 0 & 1 & 0 \\ 0 & 0 & 3 \end{pmatrix} \qquad \mathbf{M} := \begin{pmatrix} 2 & 31 & -25 & 7 \\ 11 & 25 & 9 & 8 \\ 60 & 6 & 0 & 17 \end{pmatrix}$$

a) Suppose you have computed **F1·M**, but you want to reverse the effect of **F1**; that is, undo the row operation. Construct an elementary matrix **F3** to accomplish this. (Find a matrix **F3** so that **F3·F1·M = M**.)

ANSWER:

b) Repeat part a) using **F2** in place of **F1**.

ANSWER:

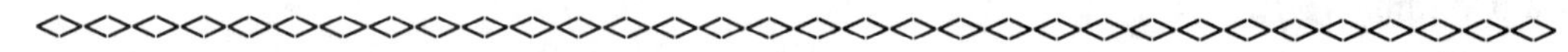

4. In Example 1 we multiplied the matrix, **M**, by the elementary matrices, **G1** and **G2**, given below.

$$\mathbf{G1} := \begin{pmatrix} 1 & 0 & 0 \\ 0 & 1 & 0 \\ -30 & 0 & 1 \end{pmatrix} \qquad \mathbf{G2} := \begin{pmatrix} 1 & 0 & 0 \\ -5.5 & 1 & 0 \\ 0 & 0 & 1 \end{pmatrix} \qquad \mathbf{M} := \begin{pmatrix} 2 & 31 & -25 & 7 \\ 11 & 25 & 9 & 8 \\ 60 & 6 & 0 & 17 \end{pmatrix}$$

a) Suppose you have computed **G1·M**, but you want to reverse the effect of **G1**; that is, undo the row operation. Construct an elementary matrix **G3** to accomplish this. (Find a matrix **G3** so that **G3·G1·M = M**.)

ANSWER:

b) Repeat part a) using **G2** in place of **G1**.

ANSWER:

<><><><><><><><><><><><><><><><><><><><><><><><><><><><><><><>

5. Exercises 2 - 4 showed that each row operation (elementary matrix) is reversible by using another row operation (elementary matrix) of the same kind. State the reversing matrix for each of the following elementary matrices.

a) $\mathbf{E1} := \begin{pmatrix} 1 & 0 & 0 \\ 0 & 0 & 1 \\ 0 & 1 & 0 \end{pmatrix}$

Define its 'reverse' here and call it **E2** ==>

Compute the products **E2·E1** and **E1·E2** below.

Do the products make sense in light of the reversal of the product order? Explain.
ANSWER:

b) $\mathbf{E3} := \begin{pmatrix} 1 & 0 & 0 \\ 0 & -.5 & 0 \\ 0 & 0 & 1 \end{pmatrix}$

Define its 'reverse' here and call it **E4** ==>

Compute the products **E4·E3** and **E3·E4**

Do the products make sense in light of the reversal of the product order? Explain.
ANSWER:

c) $\quad \mathbf{E5} := \begin{pmatrix} 1 & 0 & 0 \\ -2 & 1 & 0 \\ 0 & 0 & 1 \end{pmatrix}$

Define its 'reverse' here and call it **E6** ==>

Compute the products **E6·E5** and **E5·E6**

Do the products make sense in light of the reversal of the product order? Explain.

ANSWER:

Let **Q** be a matrix. If there is another matrix **N** such that **N** 'reverses' or 'undoes' **Q** then **N** is called the **inverse of Q**. In particular, this means that **N·Q** is an appropriate size identity matrix. Pull down the **Symbolic** menu and note that there is an entry called **Invert Matrix**. Box each of the following matrices and use that operation to invert the matrix. Are the results the same as your results above?

$$\begin{pmatrix} 0 & 1 & 0 \\ 1 & 0 & 0 \\ 0 & 0 & 1 \end{pmatrix} \qquad \begin{pmatrix} 1 & 0 & 0 \\ 0 & 0 & 1 \\ 0 & 1 & 0 \end{pmatrix} \qquad \begin{pmatrix} -2 & 0 & 0 \\ 0 & 1 & 0 \\ 0 & 0 & 1 \end{pmatrix}$$

$$\begin{pmatrix} 1 & 0 & 0 \\ 0 & 1 & 0 \\ 0 & 0 & 3 \end{pmatrix} \qquad \begin{pmatrix} 1 & 0 & 0 \\ 0 & 1 & 0 \\ -30 & 0 & 1 \end{pmatrix} \qquad \begin{pmatrix} 1 & 0 & 0 \\ -5.5 & 1 & 0 \\ 0 & 0 & 1 \end{pmatrix}$$

A **permutation** of a set is an arrangement of the set in a specified order. For example, here are the permutations of a set with three elements.

☺😐☹ ☺☹😐 😐☺☹ 😐☹☺ ☹☺😐 ☹😐☺

The word **permutation** is also used to mean a rearrangement of the set from one order to another.

<><><><><><><><><><><><><><><><><><><><><><><><><><><><><><><><><><><><><>

6. A square matrix **P** which has exactly one 1 in each row and exactly one 1 in each column with all other entries 0 is called a **permutation matrix**. Describe in words the effect of each of the following permutation matrices **P** as it multiplies matrix **A** on the left.

a) $\mathbf{P} := \begin{pmatrix} 0 & 1 & 0 \\ 0 & 0 & 1 \\ 1 & 0 & 0 \end{pmatrix}$ $\qquad$ $\mathbf{A} := \begin{pmatrix} 1 & 2 & 3 \\ 4 & 5 & 6 \\ 7 & 8 & 9 \end{pmatrix}$

ANSWER:

b) $\mathbf{P} := \begin{pmatrix} 0 & 0 & 1 \\ 1 & 0 & 0 \\ 0 & 1 & 0 \end{pmatrix}$

ANSWER:

c) Explain why such matrices are called permutation matrices?

<><><><><><><><><><><><><><><><><><><><><><><><><><><><><><><><>

7. (Individual) Write a summary about the MATHEMATICS you learned in this section.

Section 2-3 Row-reduction and Solution Sets

Name:_______________

Score:_____________

Overview

In Sections 2-1 and 2-2 we described the process called **row-reduction**. We modify this slightly in this section to require that the first element in each nonzero row of the row-reduced matrix is 1. We will refer to the modified process as row-reduction or **Gaussian elimination**. It is a systematic process or **algorithm** for eliminating variables from a system of linear equations. The solution set can easily be computed from the row-reduced system by back substitution. The goal of this section is to **classify** the set of possible outcomes of row-reduction and to describe *qualitatively* the corresponding solution sets.

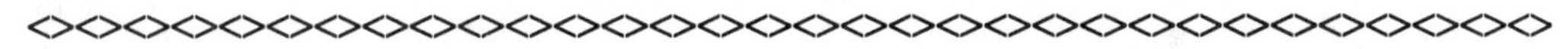

If the first element in each nonzero row of a row-reduced matrix is 1 we say that the matrix is in row echelon form. Such a 1 is called a **leading 1**. For the remainder of this text we shall use **row-reduced matrix** to mean that the matrix is in **row echelon form (ref)**. We summarize the definition of **ref** below.

A matrix is in Row Echelon Form (REF) provided

i) The first nonzero entry in each row is a 1. (This is called a leading one or a pivot.)

ii) Any rows of all zeros appear at the bottom of the matrix.

iii) The leading 1's are arranged in a staircase downward form. That is, for i > 1 the leading 1 in Row_i is to the right and below the leading 1's in previous rows.

Computing solutions symbolically

In Sections 2-1 and 2-2 we explored two procedures for row-reducing a matrix: row operations and multiplication by elementary matrices. Multiplication by elementary matrices has two computational advantages: (1) It can easily be modified so that we can do several operations simultaneously and (2) by using symbolic operations we can get fractions as answers. There is nothing inherently better about fractions than decimal numbers except that if many reduction steps are required the lack of accuracy that results from rounding off the fraction sometimes tends to accumulate. (This topic is discussed further in Section 2-9.)

Example 1

Determine the solution set of the linear system

$$w + 7\cdot x + 5\cdot y + 2\cdot z = 0$$
$$2\cdot w + 14\cdot x + 6\cdot y + z = 2$$
$$7\cdot y = 1$$

We use elementary matrices to reduce the system represented by the augmented matrix **M**.

$$M := \begin{pmatrix} 1 & 7 & 5 & 2 & 0 \\ 2 & 14 & 6 & 1 & 2 \\ 0 & 0 & 7 & 0 & 1 \end{pmatrix}$$

Write in symbols the row operation that corresponds to the matrix multiplication in each case below. Then compute the product.

$$\begin{pmatrix} 1 & 0 & 0 \\ -2 & 1 & 0 \\ 0 & 0 & 1 \end{pmatrix} \cdot \begin{pmatrix} 1 & 7 & 5 & 2 & 0 \\ 2 & 14 & 6 & 1 & 2 \\ 0 & 0 & 7 & 0 & 1 \end{pmatrix}$$

ANSWER:

$$\begin{pmatrix} 1 & 0 & 0 \\ 0 & 1 & 0 \\ 0 & \frac{7}{4} & 1 \end{pmatrix} \cdot \begin{pmatrix} 1 & 7 & 5 & 2 & 0 \\ 0 & 0 & -4 & -3 & 2 \\ 0 & 0 & 7 & 0 & 1 \end{pmatrix}$$

ANSWER:

$$\begin{bmatrix} 1 & 0 & 0 \\ 0 & -\frac{1}{4} & 0 \\ 0 & 0 & -\frac{4}{21} \end{bmatrix} \cdot \begin{pmatrix} 1 & 7 & 5 & 2 & 0 \\ 0 & 0 & -4 & -3 & 2 \\ 0 & 0 & 0 & \frac{-21}{4} & \frac{9}{2} \end{pmatrix}$$

ANSWER:

$$\begin{bmatrix} 1 & 7 & 5 & 2 & 0 \\ 0 & 0 & 1 & \frac{3}{4} & -\frac{1}{2} \\ 0 & 0 & 0 & 1 & -\frac{6}{7} \end{bmatrix}$$

The matrix on the left is in row echelon form.

We can write the set of equations corresponding to the augmented matrix as follows. (We do this in reverse order to facilitate back substitution.)

$$z = -\frac{6}{7}$$

$$y = -\frac{1}{2} - \frac{3}{4} \cdot z$$

$$w + 7 \cdot x + 5 \cdot y + 2 \cdot z = 0$$

Use Mathcad to back substitute as follows:

1. **Box just the value for z and copy it to the clipboard.**
2. **Place the cursor next to z in the second equation and click on Substitute for Variable in the Symbolic menu.**
3. **Place the cursor next to z in the third equation and Substitute for Variable.**
4. **Copy just the resulting value for y to the clipboard and substitute for y in the bottom equation.**
5. **Now put your cursor next to w in the last equation and click on Solve for Variable in the Symbolic Menu.**

Record the expression for w. w =

What does it mean that there is an x in the expression for w? Explain.

ANSWER:

What characteristic of the original system could have predicted that there might be more than one solution to the system?

ANSWER:

In this example we used elementary matrices to perform the row-reduction. We could have used the functions **rswap**, **rmult**, and **rcomb** instead.

<><><><><><><><><><><><><><><><><><><><><><><><><><><><><><><><><><><><><><><><><>

Example 2

Consider $\mathbf{A} := \begin{pmatrix} 2 & -6 & 1 \\ 3 & 7 & 5 \\ 4 & 2 & 7 \end{pmatrix}$

We begin to row-reduce this matrix by performing the following operations:

a) multiply the first row by 1/2

b) add -3 times the new first row to the second row

c) add -4 times the new first row to the third row

These operations correspond to left multiplication of **A** by the following three matrices:

$$\begin{pmatrix} 1 & 0 & 0 \\ 0 & 1 & 0 \\ -4 & 0 & 1 \end{pmatrix} \cdot \begin{pmatrix} 1 & 0 & 0 \\ -3 & 1 & 0 \\ 0 & 0 & 1 \end{pmatrix} \cdot \begin{pmatrix} \frac{1}{2} & 0 & 0 \\ 0 & 1 & 0 \\ 0 & 0 & 1 \end{pmatrix} = \begin{bmatrix} \frac{1}{2} & 0 & 0 \\ \frac{-3}{2} & 1 & 0 \\ -\frac{4}{2} & 0 & 1 \end{bmatrix}$$

We can do these three operations simultaneously by using the product matrix. We have used symbolic evaluation to see the relation between the product and the individual operations. Numerical evaluation would have also worked.

Let $\quad \mathbf{Q} := \begin{bmatrix} \frac{1}{2} & 0 & 0 \\ \frac{-3}{2} & 1 & 0 \\ -\frac{4}{2} & 0 & 1 \end{bmatrix}$. Compute the result of the three row operations on **A** by calculating **Q·A**.

ANSWER:

<><><><><><><><><><><><><><><><><><><><><><><><><><><><><><><>

Assume that *a* is not zero. Write the 3 by 3 matrix that will row-reduce the first column of the symbolic matrix given below.

$$\begin{pmatrix} a & d & g \\ b & e & h \\ c & f & i \end{pmatrix}$$

Verify that your answer is correct by showing that it causes the desired actions. (Use symbolic calculations.)

ANSWER:

Assume that we have row-reduced part of a matrix and the result of this reduction is the matrix on the right. Describe in words how you would find a matrix to row-reduce the next column (this is the column containing a, b, c, d). Under what circumstances would you need to begin by permuting the rows?

$$\begin{bmatrix} 1 & s & s & s & s & s & s & s & s \\ 0 & 1 & s & s & s & s & s & s & s \\ 0 & 0 & 0 & 1 & s & s & s & s & s \\ 0 & 0 & 0 & 0 & 1 & s & s & s & s \\ 0 & 0 & 0 & 0 & 0 & 1 & s & s & s \\ 0 & 0 & 0 & 0 & 0 & 0 & a & s & s \\ 0 & 0 & 0 & 0 & 0 & 0 & b & s & s \\ 0 & 0 & 0 & 0 & 0 & 0 & c & s & s \\ 0 & 0 & 0 & 0 & 0 & 0 & d & s & s \end{bmatrix}$$

s = stuff

ANSWER:

<><><><><><><><><><><><><><><><><><><><><><><><><><><><><><><>

Row-reduce each of the following non-augmented matrices using symbolic operations. Do your work in a scrap document and copy the answers below.

To work on a scrap document
(1) copy the matrices to the clipboard
(2) open a new document
(3) paste the matrix
Copy your answers back by copying to the clipboard and pasting below.

a) $\begin{pmatrix} 3 & 2 & 1 \\ 0 & 5 & 1 \\ 1 & 0 & 1 \end{pmatrix}$ b) $\begin{pmatrix} 0 & 0 & 17 \\ 0 & 0 & 17 \\ 0 & 0 & 17 \end{pmatrix}$ c) $\begin{pmatrix} 1 & 2 & -7 \\ 1 & 2 & -7 \\ 1 & 2 & -7 \end{pmatrix}$ d) $\begin{pmatrix} 1 & 0 & 3 \\ 2 & -2 & -1 \\ 3 & -2 & 2 \end{pmatrix}$

e) $\begin{pmatrix} 0 & 2 & -9 \\ 0 & 2 & -9 \\ 0 & 2 & -9 \end{pmatrix}$ f) $\begin{pmatrix} 0 & 5 & 1 \\ 0 & 2 & 3 \\ 0 & 3 & -2 \end{pmatrix}$ g) $\begin{pmatrix} 1 & 3 & -2 \\ 1 & 3 & -2 \\ 0 & 0 & 3 \end{pmatrix}$

ANSWERS:

a) b) c) d)

e) f) g)

Consider the following symbolically row-reduced 3 by 3 matrices.

1) $\begin{pmatrix} 1 & \blacksquare & \blacksquare \\ 0 & 1 & \blacksquare \\ 0 & 0 & 1 \end{pmatrix}$ 2) $\begin{pmatrix} 1 & \blacksquare & \blacksquare \\ 0 & 1 & \blacksquare \\ 0 & 0 & 0 \end{pmatrix}$ 3) $\begin{pmatrix} 1 & \blacksquare & \blacksquare \\ 0 & 0 & 1 \\ 0 & 0 & 0 \end{pmatrix}$ 4) $\begin{pmatrix} 1 & \blacksquare & \blacksquare \\ 0 & 0 & 0 \\ 0 & 0 & 0 \end{pmatrix}$

5) $\begin{pmatrix} 0 & 1 & \blacksquare \\ 0 & 0 & 1 \\ 0 & 0 & 0 \end{pmatrix}$ 6) $\begin{pmatrix} 0 & 1 & \blacksquare \\ 0 & 0 & 0 \\ 0 & 0 & 0 \end{pmatrix}$ 7) $\begin{pmatrix} 0 & 0 & 1 \\ 0 & 0 & 0 \\ 0 & 0 & 0 \end{pmatrix}$ $\blacksquare$ = some number

Explain how these matrix forms relate to the row-reductions that you have done above. (Label them with the letter that corresponds to the row-reductions above.)

ANSWER:

Explain in words what it means for an m by n matrix to be in row-reduced form. (Hint: Look at the patterns of entries in the display above.)

ANSWER:

We can label the symbolic matrices as follows:

$$\underset{1,2,3}{\begin{pmatrix} 1 & \blacksquare & \blacksquare \\ 0 & 1 & \blacksquare \\ 0 & 0 & 1 \end{pmatrix}} \quad \underset{(b)}{\begin{pmatrix} 1 & \blacksquare & \blacksquare \\ 0 & 1 & \blacksquare \\ 0 & 0 & 0 \end{pmatrix}} \quad \underset{1,3}{\begin{pmatrix} 1 & \blacksquare & \blacksquare \\ 0 & 0 & 1 \\ 0 & 0 & 0 \end{pmatrix}} \quad \underset{1}{\begin{pmatrix} 1 & \blacksquare & \blacksquare \\ 0 & 0 & 0 \\ 0 & 0 & 0 \end{pmatrix}}$$

$$\underset{(e)}{\begin{pmatrix} 0 & 1 & \blacksquare \\ 0 & 0 & 1 \\ 0 & 0 & 0 \end{pmatrix}} \quad \underset{2}{\begin{pmatrix} 0 & 1 & \blacksquare \\ 0 & 0 & 0 \\ 0 & 0 & 0 \end{pmatrix}} \quad \underset{(g)}{\begin{pmatrix} 0 & 0 & 1 \\ 0 & 0 & 0 \\ 0 & 0 & 0 \end{pmatrix}} \qquad \blacksquare = \text{some number}$$

There is a pattern to the labeling. Figure it out then determine the labels in terms of the number pattern for each of the following.

(b) (e) (g)

Explain in words the labeling pattern for the row-reduced matrices given above.

ANSWER:

Generalize this labeling pattern to all possible types of row-reduced 3 by 4 nonzero matrices. Give a list of the labels. (There are 14. For example, label 123 implies there are leading 1's in columns 1, 2, and 3.)

ANSWER:

We could go on to seek a formula for the number of types of row-reduced m by n matrices. It is clear that this corresponds to certain increasing sequences of the numbers less than n; however, that is not our goal here. The branch of mathematics that studies counting problems of this type is called **combinatorics**.

<><><><><><><><><><><><><><><><><><><><><><><><><><><><><><><><>

The corner entries (the ones) in a row-reduced matrix are called ***pivots*** and our labeling scheme tells us which columns contain pivots. We return to our original goal which is to describe the solution set of a system of linear equations. ***For the remainder of this section we shall work with row-reduced systems; i.e., matrices in row echelon form.*** Corresponding to each of the row-reduced coefficient matrices above there are a number of row-reduced augmented matrices. For example, corresponding to the row-reduced matrix labeled, (1,2) we could have either of the augmented matrices shown below. The critical issue here is whether the (3,4) entry is zero or nonzero.

$$\begin{pmatrix} 1 & \blacksquare & \blacksquare \\ 0 & 1 & \blacksquare \\ 0 & 0 & 0 \end{pmatrix} \qquad \left(\begin{array}{ccc|c} 1 & \blacksquare & \blacksquare & \blacksquare \\ 0 & 1 & \blacksquare & \blacksquare \\ 0 & 0 & 0 & 0 \end{array}\right) \qquad \left(\begin{array}{ccc|c} 1 & \blacksquare & \blacksquare & \blacksquare \\ 0 & 1 & \blacksquare & \blacksquare \\ 0 & 0 & 0 & 1 \end{array}\right)$$

1,2

What is the equation that corresponds to the third row of the augmented matrix on the right?

ANSWER:

What is the solution set to the equation that you have given? What does this imply about the solution set to the original system that led to the row-reduced system?

ANSWER:

Complete the following sentence to summarize the row-reduced form

$$\left(\begin{array}{ccc|c} 1 & \blacksquare & \blacksquare & \blacksquare \\ 0 & 1 & \blacksquare & \blacksquare \\ 0 & 0 & 0 & 1 \end{array}\right)$$

If some row of a row-reduced augmented matrix has the form

____

then the original system of equations has no solution (inconsistent).

<><><><><><><><><><><><><><><><><><><><><><><><><><><><><><><><><><>

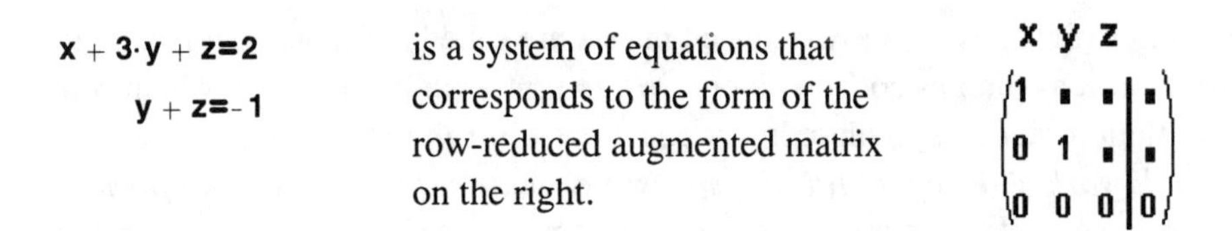

$x + 3 \cdot y + z = 2$

$y + z = -1$

is a system of equations that corresponds to the form of the row-reduced augmented matrix on the right.

$$\begin{array}{c} x \; y \; z \\ \left(\begin{array}{ccc|c} 1 & \blacksquare & \blacksquare & \blacksquare \\ 0 & 1 & \blacksquare & \blacksquare \\ 0 & 0 & 0 & 0 \end{array}\right) \end{array}$$

In each equation we can solve for the variable corresponding to the pivots in terms of the variables that 'come later' in the system. In our case this means that:

$y = -z - 1$

$x = 2 - 3 \cdot y - z$

Copy the value of y to the clipboard and substitute for variable in the second equation.

This results in the following :

$x = 5 + 2 \cdot z$

$y = -z - 1$

This says that for any value of unknown z the vector on the right is a solution to the original system of equations.

$$\begin{pmatrix} 5 + 2 \cdot z \\ -z - 1 \\ z \end{pmatrix}$$

This is a description of the set of all solutions to the system. It is, of course, an infinite set; however, it is a particular infinite set. We can rewrite this solution as

$$\begin{pmatrix} 5 \\ -1 \\ 0 \end{pmatrix} + z \cdot \begin{pmatrix} 2 \\ -1 \\ 1 \end{pmatrix}$$

Notice first that z is simply an unknown value. We could use t or s or any symbol we want as the name of this unknown.

The second summand represents all multiples of a given vector. It is a line in 3-space. Adding the first summand translates the line through the origin to a parallel line through the given point (5, -1, 0). Thus the solution set can be described as a line in 3-space. Every point along this line is a solution of the system. We discuss lines and planes in Chapter 4.

The previous example is a prototype for much of our work.
Be sure that you understand it!

<><><><><><><><><><><><><><><><><><><><><><><><><><><><><><><><><><><><><><><>

We have described an algorithm for finding solution sets of systems of linear equations. In this algorithm it is our practice to solve for the pivot variables in terms of the remaining variables. The remaining variables are called ***free*** since they can assume any value.

Complete the following sentence:

The number of free variables in a solution set for a consistent system of linear equations is the number of columns in the original matrix minus ______________________ in the associated row-reduced augmented matrix.

There will be an infinite number of solutions whenever there are free variables. As a result there will be a unique solution only when there are no free variables.

Describe the row echelon form of an augmented system that has a unique solution.

ANSWER:

Write a system of equations that corresponds to the augmented matrix

$$\left(\begin{array}{ccccc|c} 1 & 0 & 3 & -2 & 1 & 3 \\ 0 & 0 & 1 & 5 & 1 & 0 \\ 0 & 0 & 0 & 1 & 1 & 2 \end{array}\right)$$

ANSWER:

Solve for the pivot variables and back substitute using the symbolic functions. Write the resulting solution set in the most general form as a column.

ANSWER:

How many variables are there? How many pivots? How many free variables?

ANSWER:

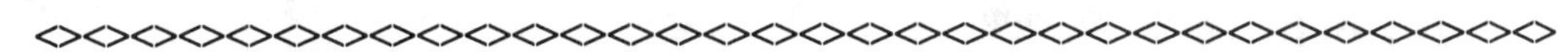

Summary of the algorithm for solving a system of linear equations

Step 1: Create the associated augmented matrix.

Step 2: Use one of the methods described above to reduce the augmented matrix to Row Echelon Form.

Step 3: Back substitute to find all solutions.

The general solution of the system $\mathbf{A}\cdot\mathbf{x} = \mathbf{b}$ where $\mathbf{A}$ is m by n can be written as the set of linear combinations:

$$\mathbf{c} + t_1 \cdot \mathbf{v}_1 + t_2 \cdot \mathbf{v}_2 + \ldots + t_j \cdot \mathbf{v}_j \qquad -\infty < t_i < \infty \qquad i = 1, \ldots, j$$

where $\mathbf{c}$ and $\mathbf{v}_1, \ldots, \mathbf{v}_j$ are vectors and the t_i can be any real number.

How many rows do the vectors have? How is the value of j determined?

ANSWER:

Complete each of the following in terms of the REF, pivots, or free variables.

If ____ then the original system has no solutions.

If the original system is consistent and if ______________________________ then the original system has a unique solution.

<><><><><><><><><><><><><><><><><><><><><><><><><><><><><><><>

We refer to the expression $\{c + t_1 \cdot v_1 + t_2 \cdot v_2 + \ldots + t_j \cdot v_j\}$

as the **general solution** to the system. The vector **c** is called a **particular solution** to the system. Symbolically we denote the members of the solution set as the following linear combination:

$$u := \begin{bmatrix}5\\0\\7\\0\\2\end{bmatrix} + t_1 \begin{bmatrix}\cdot\\\cdot\\\cdot\\\cdot\\\cdot\end{bmatrix} + t_2 \begin{bmatrix}\cdot\\\cdot\\\cdot\\\cdot\\\cdot\end{bmatrix}$$

General Solution

Particular Solution

vector v_1

vector v_2

Our goal is to investigate the connection between the general solutions to

a) $\begin{pmatrix}1 & -2 & 0 & 3\\0 & 0 & 1 & 4\\0 & 0 & 0 & 0\end{pmatrix} \cdot \begin{pmatrix}w\\x\\y\\z\end{pmatrix} = \begin{pmatrix}2\\5\\0\end{pmatrix}$ and b) $\begin{pmatrix}1 & -2 & 0 & 3\\0 & 0 & 1 & 4\\0 & 0 & 0 & 0\end{pmatrix} \cdot \begin{pmatrix}w\\x\\y\\z\end{pmatrix} = \begin{pmatrix}-7\\9\\0\end{pmatrix}$

The corresponding augmented matrices are:

a) $\begin{pmatrix}1 & -2 & 0 & 3 & 2\\0 & 0 & 1 & 4 & 5\\0 & 0 & 0 & 0 & 0\end{pmatrix}$ b) $\begin{pmatrix}1 & -2 & 0 & 3 & -7\\0 & 0 & 1 & 4 & 9\\0 & 0 & 0 & 0 & 0\end{pmatrix}$

Construct the general solutions as follows:

Solve for w and y and complete the following:

w=■
x=x
y=■
z=z

Write the solution as a vector

$$\begin{pmatrix} \blacksquare \\ \blacksquare \\ \blacksquare \\ \blacksquare \end{pmatrix}$$

Now factor and rewrite as:

$$\begin{pmatrix} \blacksquare \\ \blacksquare \\ \blacksquare \\ \blacksquare \end{pmatrix} + x \cdot \begin{pmatrix} \blacksquare \\ \blacksquare \\ \blacksquare \\ \blacksquare \end{pmatrix} + z \cdot \begin{pmatrix} \blacksquare \\ \blacksquare \\ \blacksquare \\ \blacksquare \end{pmatrix}$$

w=■
x=x
y=■
z=z

Write the solution as a vector

$$\begin{pmatrix} \blacksquare \\ \blacksquare \\ \blacksquare \\ \blacksquare \end{pmatrix}$$

Now factor and rewrite as:

$$\begin{pmatrix} \blacksquare \\ \blacksquare \\ \blacksquare \\ \blacksquare \end{pmatrix} + x \cdot \begin{pmatrix} \blacksquare \\ \blacksquare \\ \blacksquare \\ \blacksquare \end{pmatrix} + z \cdot \begin{pmatrix} \blacksquare \\ \blacksquare \\ \blacksquare \\ \blacksquare \end{pmatrix}$$

Describe the relationship between the two general solutions.

ANSWER:

If we changed the right hand side of the original system how would this affect the vectors $\mathbf{v}_1, \ldots, \mathbf{v}_j$?

ANSWER:

<><><><><><><><><><><><><><><><><><><><><><><><><><><><><><>

System of equations	Coefficient Matrix	Augmented Matrix
$x - 2 \cdot y + z = 4$ $x + y = 1$	$A := \begin{pmatrix} 1 & -2 & 1 \\ 1 & 1 & 0 \end{pmatrix}$	$M := \begin{pmatrix} 1 & -2 & 1 & 4 \\ 1 & 1 & 0 & 1 \end{pmatrix}$

Row Echelon Form	General Solution	Particular Solution
$\begin{pmatrix} 1 & -2 & 1 & 4 \\ 0 & 1 & \frac{-1}{3} & -1 \end{pmatrix}$	$\begin{pmatrix} 2 \\ -1 \\ 0 \end{pmatrix} + t \cdot \begin{bmatrix} -\frac{1}{3} \\ \frac{1}{3} \\ 1 \end{bmatrix}$	$\begin{pmatrix} 2 \\ -1 \\ 0 \end{pmatrix}$

Check these results by evaluating the following expressions symbolically.

A·general solution	**A**·particular solution	**A**·(general - particular)
$\begin{pmatrix} 1 & -2 & 1 \\ 1 & 1 & 0 \end{pmatrix} \cdot \left[\begin{pmatrix} 2 \\ -1 \\ 0 \end{pmatrix} + t \cdot \begin{bmatrix} -\frac{1}{3} \\ \frac{1}{3} \\ 1 \end{bmatrix} \right]$	$\begin{pmatrix} 1 & -2 & 1 \\ 1 & 1 & 0 \end{pmatrix} \cdot \begin{pmatrix} 2 \\ -1 \\ 0 \end{pmatrix}$	$\begin{pmatrix} 1 & -2 & 1 \\ 1 & 1 & 0 \end{pmatrix} \cdot \left[t \cdot \begin{bmatrix} -\frac{1}{3} \\ \frac{1}{3} \\ 1 \end{bmatrix} \right]$

What did you discover about the (general solution - particular solution)? Is this always true? Explain your answer.

ANSWER:

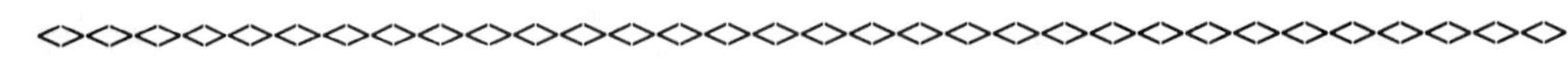

Exercises

1. Let **A·x** = **b** be a system of linear equations and let **sol1** and **sol2** be two different solutions to the system.

a) Write the mathematical equations corresponding to the statement that **sol1** and **sol2** are solutions to the system.

ANSWER:

b) Construct an infinite family of solutions. (Hint: Use a special linear combination of **sol1** and **sol2.**)

ANSWER:

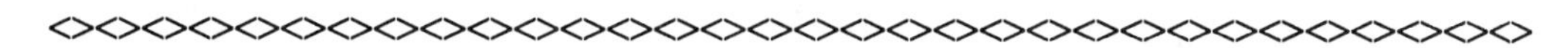

2. In Exercise 4 of Section 1-9 you examined the long term behavior of the weather model. You concluded that

(1) there is a vector **x,** called the long range behavior or stable state of the Markov process, such that $\mathbf{A}\cdot\mathbf{x} = \mathbf{x}$; and

(2) if there is a stable state, it can be computed by taking the limit of $\mathbf{A}^n\cdot\mathbf{p}$ as n gets large, where **p** is any probability vector.

Any vector **x** such that $\mathbf{A}\cdot\mathbf{x} = \mathbf{x}$ is called an **eigenvector of the matrix A corresponding to the eigenvalue** 1.

Let **A** be the matrix given to the right. To find a vector **x** such that $\mathbf{A}\cdot\mathbf{x} = \mathbf{x}$, rewrite the equation as $\mathbf{A}\cdot\mathbf{x} = \mathbf{I}\cdot\mathbf{x}$ where **I** is the identity matrix. This is the same as the system of equations, $(\mathbf{A} - \mathbf{I})\cdot\mathbf{x} = \mathbf{0}$. Use the methods of this chapter to find all solutions for this equation.

ANSWER:

$$\mathbf{A} := \begin{bmatrix} 0 & \frac{1}{3} & \frac{1}{5} & \frac{1}{2} \\ \frac{1}{4} & 0 & \frac{2}{5} & \frac{1}{6} \\ \frac{1}{2} & \frac{1}{3} & \frac{1}{5} & 0 \\ \frac{1}{4} & \frac{1}{3} & \frac{1}{5} & \frac{1}{3} \end{bmatrix}$$

<><><><><><><><><><><><><><><><><><><><><><><><><><><><><><><><><><><>

3. If you computed the answer to the previous exercise correctly you should have seen that there are an infinite number of vectors **x** such that $\mathbf{A}\cdot\mathbf{x} = \mathbf{x}$. Only one of these can be the stable state of the corresponding Markov process. What is that vector and what equation must be added to the system $\mathbf{A}\cdot\mathbf{x} = \mathbf{x}$ to ensure that there is a unique solution?

ANSWER:

<><><><><><><><><><><><><><><><><><><><><><><><><><><><><><><><><><><>

4. (Individual) Write a summary of the MATHEMATICS that you learned in this section.

Section 2-4 Reduced Row Echelon Form (RREF)

Name:________________

Score:______________

Overview

In Sections 2-1 to 2-3 we developed an algorithm for solving a system of linear equations. This algorithm (Gaussian Elimination) led to the row echelon form (REF) of the associated augmented matrix from which we could back substitute to get the general solution to the system. Our goal in this section is to replace back substitution in the algorithm with matrix operations. The result is called ***Gauss-Jordan Elimination*** and leads to the Reduced Row Echelon Form (RREF). We will be able to 'read' the general solution directly from the RREF of the augmented matrix.

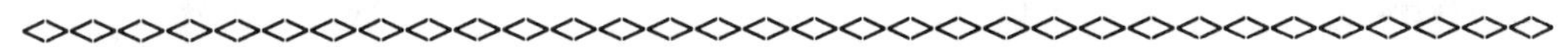

Row Echelon Augmented Matrix

$$\begin{pmatrix} 1 & 2 & 3 & -4 & 7 \\ 0 & 0 & 1 & 5 & 11 \\ 0 & 0 & 0 & 1 & 3 \end{pmatrix}$$

Equations

$$w + 2\cdot x + 3\cdot y - 4\cdot z = 7$$
$$y + 5\cdot z = 11$$
$$z = 3$$

Back substituting (step 1)

$$w + 2\cdot x + 3\cdot y = 19$$
$$y = -4$$
$$z = 3$$

Augmented Matrix for this System

$$\begin{pmatrix} 1 & 2 & 3 & 0 & 19 \\ 0 & 0 & 1 & 0 & -4 \\ 0 & 0 & 0 & 1 & 3 \end{pmatrix}$$

We see that back substituting for z is equivalent to 'cleaning out' the column above the pivot corresponding to z. We can do this in the matrix by adding multiples of row 3 to row 1 and row 2 respectively. This can be done either using row operations or by using the product of elementary matrices as follows:

Step 1

$$\begin{pmatrix} 1 & 0 & 4 \\ 0 & 1 & -5 \\ 0 & 0 & 1 \end{pmatrix} \cdot \begin{pmatrix} 1 & 2 & 3 & -4 & 7 \\ 0 & 0 & 1 & 5 & 11 \\ 0 & 0 & 0 & 1 & 3 \end{pmatrix}$$

<==== Evaluate this symbolically. These row operations clear out above the pivot corresponding to z which is equivalent to back substitution for z.

The new system

$$w + 2\cdot x + 3\cdot y = 19$$
$$y = -4$$
$$z = 3$$

Augmented Matrix for this System

$$\begin{pmatrix} 1 & 2 & 3 & 0 & 19 \\ 0 & 0 & 1 & 0 & -4 \\ 0 & 0 & 0 & 1 & 3 \end{pmatrix}$$

Step 2

Evaluate the product on the right symbolically and explain what was done in this step.

$$\begin{pmatrix} 1 & -3 & 0 \\ 0 & 1 & 0 \\ 0 & 0 & 1 \end{pmatrix} \cdot \begin{pmatrix} 1 & 2 & 3 & 0 & 19 \\ 0 & 0 & 1 & 0 & -4 \\ 0 & 0 & 0 & 1 & 3 \end{pmatrix}$$

ANSWER:

The general solution of the system:

$$w = 31 - 2 \cdot x$$
$$x = x$$
$$y = -4$$
$$z = 3$$

in vector form ==>

$$\begin{pmatrix} 31 \\ 0 \\ -4 \\ 3 \end{pmatrix} + x \cdot \begin{pmatrix} -2 \\ 1 \\ 0 \\ 0 \end{pmatrix}$$

Explain why free variables appear only in the equations that correspond to pivots that come before them in our procedure.

ANSWER:

Why did we need to include the second equation above?

ANSWER:

A matrix is in Reduced Row Echelon Form (RREF) if it is in Row Echelon Form and each column containing a pivot has all of its other entries zero.

Reduction to RREF can be done at the same time as reduction to REF by using a matrix to 'clean out' the given column both above and below the pivot and convert the pivot to a leading 1.

Evaluate the product on the right symbolically and explain the result.

$$\begin{bmatrix} 1 & -\frac{a}{b} & 0 \\ 0 & \frac{1}{b} & 0 \\ 0 & -\frac{c}{b} & 1 \end{bmatrix} \cdot \begin{pmatrix} 1 & a & 7 & 1 \\ 0 & b & 9 & 5 \\ 0 & c & 2 & 21 \end{pmatrix}$$

ANSWER:

We assumed that *b* was not zero. Several row operations were done simultaneously here. They could be done one at a time giving the same result.

<><><><><><><><><><><><><><><><><><><><><><><><><><><><><><><><><><><><><><>

Assume that we have row-reduced a matrix to the following point.

$$\begin{bmatrix} 1 & 0 & 0 & 0 & b & \blacksquare & \blacksquare & \blacksquare & \blacksquare \\ 0 & 1 & a & 0 & c & \blacksquare & \blacksquare & \blacksquare & \blacksquare \\ 0 & 0 & 0 & 1 & d & \blacksquare & \blacksquare & \blacksquare & \blacksquare \\ 0 & 0 & 0 & 0 & e & \blacksquare & \blacksquare & \blacksquare & \blacksquare \\ 0 & 0 & 0 & 0 & f & \blacksquare & \blacksquare & \blacksquare & \blacksquare \\ 0 & 0 & 0 & 0 & g & \blacksquare & \blacksquare & \blacksquare & \blacksquare \end{bmatrix}$$

Assume that $e \neq 0$. Explain what the next step is in the algorithm to obtain the RREF.

ANSWER:

What would the next step be if $e = 0$? if $e = f = 0$?

ANSWER:

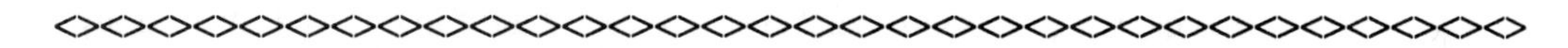

Determine the RREF for each of the following matrices.

$$\begin{pmatrix} 0 & 1 & 0 & -2 \\ 1 & 3 & 4 & 7 \\ 0 & 1 & -1 & 3 \\ 0 & 0 & 0 & 0 \end{pmatrix} \qquad \begin{pmatrix} 1 & 0 & 0 \\ 0 & 1 & 0 \\ 0 & 0 & 1 \end{pmatrix} \qquad \begin{pmatrix} 2 & -4 \\ 3 & 1 \\ 4 & 0 \end{pmatrix}$$

ANSWER:

<><><><><><><><><><><><><><><><><><><><><><><><><><><><><><><><><><><><><>

For a linear system $\mathbf{A} \cdot \mathbf{x} = \mathbf{b}$, the RREF of $\mathbf{M} = [\mathbf{A} \mid \mathbf{b}]$ gives us a quick way to decide if there is one solution, no solutions, or infinitely many solutions to the system. Complete the following using information that would be available from the RREF. Be sure that your answers take the following examples into account.

$$\left(\begin{array}{cc|c} 1 & 0 & 5 \\ 0 & 1 & 2 \\ 0 & 0 & 1 \end{array}\right) \quad \left[\begin{array}{cc|c} 1 & 0 & 5 \\ 0 & 1 & 2 \\ 0 & 0 & 1 \\ 0 & 0 & 0 \end{array}\right] \quad \left(\begin{array}{ccc|c} 1 & 2 & 0 & 5 \\ 0 & 0 & 1 & 2 \\ 0 & 0 & 0 & 1 \end{array}\right) \quad \left[\begin{array}{ccc|c} 1 & 2 & 0 & 5 \\ 0 & 0 & 1 & 2 \\ 0 & 0 & 0 & 1 \\ 0 & 0 & 0 & 0 \end{array}\right]$$

(1) If **RREF(M)** ____

____, then there is a

unique solution. (There are no free variables.)

(2) If **RREF(M)** ____________________________________,

then there is ***no solution***.

(3) If **RREF(M)** ____________________________________

____

then there are ***infinitely many solutions***. (There will be at least one free variable.)

<><><><><><><><><><><><><><><><><><><><><><><><><><><><><><>

Exercises

1. Each of the following matrices represents an augmented matrix that is in RREF. Analyze the nature of the solution set. If the system is consistent, express the general solution in terms of a set of vectors.

(a)
$$\begin{pmatrix} 1 & 0 & 3 & 0 & 4 \\ 0 & 1 & -1 & 0 & 0 \\ 0 & 0 & 0 & 1 & 2 \end{pmatrix}$$

ANSWER:

<><><><><><><><><><><><><><><><><><><><><><><><><><><><><><>

(b)
$$\begin{pmatrix} 1 & 0 & 0 & -4 \\ 0 & 1 & 0 & 3 \\ 0 & 0 & 1 & 5 \end{pmatrix}$$

ANSWER:

<><><><><><><><><><><><><><><><><><><><><><><><><><><><><><>

(c) $\begin{pmatrix} 1 & 0 & 3 & 1 & 4 & 5 \\ 0 & 1 & 0 & -2 & -3 & 6 \\ 0 & 0 & 0 & 0 & 0 & 1 \end{pmatrix}$

ANSWER:

<><><><><><><><><><><><><><><><><><><><><><><><><><><><><><><><><><><><><><><><>

The rref function in Mathcad

Mathcad Plus contains a function that makes row-reduction much easier. This function is the **rref** function. The function takes as its argument a matrix and returns the reduced row echelon form of the matrix. This function is provided for Mathcad users on the disk that contains this text. It is installed when the handbook is 'unzipped.' If you are using a copy of Mathcad and have not installed the text on your computer from a zipped file, the function is not available for your use.

rref examples: (Press F9 to compute the values.)

$$\text{rref}\left(\begin{pmatrix} 6 & 0 & 9 \\ -8 & 2 & 6 \\ 1 & 5 & 3 \end{pmatrix}\right) = \begin{pmatrix} 1 & 0 & 0 \\ 0 & 1 & 0 \\ 0 & 0 & 1 \end{pmatrix} \qquad \text{rref}\left(\begin{pmatrix} 3 & 0 & 3 \\ 1 & 5 & -4 \\ 7 & 5 & 2 \end{pmatrix}\right) = \begin{pmatrix} 1 & 0 & 1 \\ 0 & 1 & -1 \\ 0 & 0 & 0 \end{pmatrix}$$

$$\text{rref}\left(\begin{pmatrix} 3 & 5 & 7 & 9 & 1 \\ 1 & 2 & 3 & 0 & 5 \\ 2 & 3 & 0 & 0 & 2 \end{pmatrix}\right) = \begin{pmatrix} 1 & 0 & 0 & 20.25 & -24.5 \\ 0 & 1 & 0 & -13.5 & 17 \\ 0 & 0 & 1 & 2.25 & -1.5 \end{pmatrix}$$

$$\text{rref}\left(\begin{pmatrix} 1 & 3 & -5 & 12 & 31 \\ 1 & 3 & 2 & 17 & -17 \\ 1 & 3 & 0 & 0 & 4 \end{pmatrix}\right) = \begin{pmatrix} 1 & 3 & 0 & 0 & 4 \\ 0 & 0 & 1 & 0 & -6.523 \\ 0 & 0 & 0 & 1 & -0.468 \end{pmatrix}$$

Because **rref** is a Mathcad numerical procedure it is not always exact as the following example shows.

$$\text{rref}\begin{pmatrix} 3 & 1 & 0 & 7 \\ 5 & 3 & 1 & 5 \\ 1 & -1 & -1 & 9 \end{pmatrix} = \begin{pmatrix} 1 & 0 & -0.25 & 4 \\ 0 & 1 & 0.75 & -5 \\ 0 & 0 & 0 & -1.776\cdot 10^{-15} \end{pmatrix}$$

If the matrix on the left is the matrix of an augmented system of linear equations then you must use your good judgment to determine if the system is consistent. We will see later that we can determine if the last term in the fourth row is zero by calculating the determinant of $\mathbf{A}^T \cdot \mathbf{A}$. In this case that is zero and the term should be zero.

$$\left| \begin{pmatrix} 3 & 1 & 0 & 7 \\ 5 & 3 & 1 & 5 \\ 1 & -1 & -1 & 9 \end{pmatrix}^T \cdot \begin{pmatrix} 3 & 1 & 0 & 7 \\ 5 & 3 & 1 & 5 \\ 1 & -1 & -1 & 9 \end{pmatrix} \right| = 0$$

In general it may be necessary to perform an ancillary computation to verify the accuracy of the **rref** function. We discuss this in Chapter 3. Until that time you should use the **rref** function and check your answers.

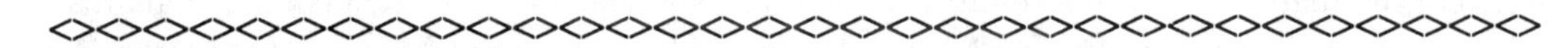

2. Each of the following is the augmented matrix of a system of linear equations. Reduce it to RREF and give the general form of the solution. Verify that your solution is correct by forming the product of the coefficient matrix with the general solution.

Coefficient Matrix **Augmented Matrix**

a) $\begin{pmatrix} 0 & 3 & 0 & 1 & 0 \\ 2 & 5 & 2 & 1 & 2 \\ 1 & 7 & 1 & 0 & 1 \end{pmatrix}$ $\begin{pmatrix} 0 & 3 & 0 & 1 & 0 & 6 \\ 2 & 5 & 2 & 1 & 2 & 7 \\ 1 & 7 & 1 & 0 & 1 & 1 \end{pmatrix}$

ANSWER:

Coefficient Matrix **Augmented Matrix**

b) $\begin{pmatrix} 3 & 5 & 0 \\ 5 & 1 & 2 \\ 1 & 9 & -2 \end{pmatrix}$ $\begin{pmatrix} 3 & 5 & 0 & 7 \\ 5 & 1 & 2 & 5 \\ 1 & 9 & -2 & 9 \end{pmatrix}$

ANSWER:

<><><><><><><><><><><><><><><><><><><><><><><><><><><><><><><><><><><><><><>

3. Dietician problem: The staff dietician at Hill House has to make a meal with 600 calories, 20 grams of protein, and 200 milligrams of vitamin C. There are three food types to choose from: rubbery jello, dried fish sticks and mystery meat. They have the following nutritional content per ounce.

▪	Jello	Fish_Sticks	Mystery_Meat
Calories	10	50	200
Protein	1	3	.2
Vitamin_C	30	10	0

Set up and solve the system of equations to determine how much of each food should be used.

ANSWER:

<><><><><><><><><><><><><><><><><><><><><><><><><><><><><><><><><><><><><>

4. Now suppose that cottage cheese is added to the menu.

▪	Jello	Fish_Sticks	Mystery_Meat	Cottage_Cheese
Calories	10	50	200	150
Protein	1	3	.2	1.3
Vitamin_C	30	10	0	0

What is the largest quantity of cottage cheese that can be used in the menu and still exactly satisfy the nutritional requirements? (Note that if we use more than four units of cottage cheese we will have too many calories. We already know that we can use no cottage cheese and solve the problem. Thus the answer will be some number between 0 and 4.)

ANSWER:

<><><><><><><><><><><><><><><><><><><><><><><><><><><><><><><><><><><><>

5. Verify that $\begin{bmatrix} \frac{6}{7} \\ \frac{2}{7} \\ 0 \end{bmatrix} + t\cdot\begin{bmatrix} -\frac{2}{7} \\ \frac{4}{7} \\ 1 \end{bmatrix}$ is the general solution to

$$\begin{aligned} 2\cdot x + \quad y \qquad &= 2 \\ 3\cdot x + 5\cdot y - 2\cdot z &= 4 \\ 7\cdot y - 4\cdot z &= 2 \end{aligned}$$

ANSWER:

6. Verify that

$$\begin{bmatrix} \frac{6}{7} \\ \frac{2}{7} \\ 0 \end{bmatrix} + t\cdot\begin{pmatrix} -2 \\ 4 \\ 7 \end{pmatrix} \quad \text{and} \quad \begin{pmatrix} 0 \\ 2 \\ 3 \end{pmatrix} + t\cdot\begin{bmatrix} -\frac{2}{7} \\ \frac{4}{7} \\ 1 \end{bmatrix} \quad \text{are also solutions}$$

of the system in Exercise 5 and are therefore equivalent to the solution given in 5.

ANSWER:

<><><><><><><><><><><><><><><><><><><><><><><><><><><><><><><><><><><><>

7. The augmented matrix $\begin{pmatrix} 1 & 2 & -1 & 4 \end{pmatrix}$ corresponding to the system $x + 2y - z = 4$ is in RREF form. What is the general solution?

ANSWER:

<><><><><><><><><><><><><><><><><><><><><><><><><><><><><><><><><><><><>

8. The Leontief model was introduced in Section 0-2. In Section 0-2 you computed the equilibrium of this system using Mathcad's 'solve block.' In Section 1-9, you computed the equilibrium iteratively. Recall the system of equations:

	Supplies	Energy	Constr	Transp	consumer demand
Energy	x	= 0.4 x	+ 0.2 y	+ 0.1 z	+ 100
Construction	y	= 0.2 x	+ 0.4 y	+ 0.1 z	+ 50
Transportation	z	= 0.15 x	+ 0.2 y	+ 0.2 z	+ 100

Enter the augmented matrix for this system of equations and compute the equilibrium by finding the RREF of the matrix. Check your answer.

ANSWER:

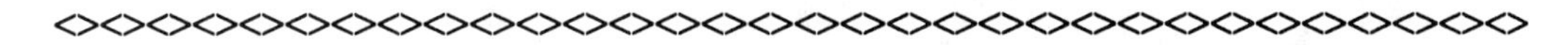

9. (Individual) Write a summary of the MATHEMATICS that you learned in this section.

Section 2-5 Sets in R^n

Name:________________

Score:______________

This section does not require use of the computer but you may use Mathcad if you wish.

Overview

This section contains a discussion of subsets of $\mathbf{R}^n$ and gives the background necessary to discuss the set of solutions to a system of linear equations.

R^n is the set of all n-tuples of real numbers.

<><><><><><><><><><><><><><><><><><><><><><><><><><><><><><><><><><><><><><><><><><><><><>

We call $\mathbf{R}^n$ the **space of all n-vectors** or an **n-dimensional real vector space**.

An **element** of $\mathbf{R}^n$ is called a **real n-tuple**, or a **real n-vector** or **a point in real n-space**.

These are three different interpretations of the same objects. For some applications one description is most useful while for other applications another description is more useful.

We sometimes denote the points that belong to $\mathbf{R}^n$ by symbols such as **x** and **y**, at other times we denote them as n-tuples such as $(x_1, x_2, x_3, \ldots, x_n)$ and sometimes as column vectors such as

$$\begin{bmatrix} x_1 \\ x_2 \\ \cdots \\ \cdots \\ x_n \end{bmatrix}$$

The real numbers in the n-tuples or in the vectors are called the **components** or **entries**.

A **subset** of $\mathbf{R}^n$ is a collection of vectors in $\mathbf{R}^n$. Such a collection can be given in terms of a list of vectors, in terms of a property that the components of the vectors must satisfy, or in terms of a property that the vectors themselves must satisfy. Here are some examples.

a) All vectors $\mathbf{x}$ such that $|\mathbf{x}| \le 1$. (i.e., all vectors with length ≤ 1). This is read 'all vectors of length less than or equal to 1' and is called the **unit ball** in n-space.

Which of the following 3-vectors are in the unit ball in 3-space?

i) $\begin{pmatrix} -1 \\ 0 \\ .5 \end{pmatrix}$ ii) $\begin{pmatrix} .5 \\ .5 \\ .5 \end{pmatrix}$ iii) $\begin{pmatrix} 0 \\ 0 \\ 0 \end{pmatrix}$ iv) $\begin{pmatrix} 0 \\ .7 \\ .8 \end{pmatrix}$

ANSWER:

<><><><><><><><><><><><><><><><><><><><><><><><><><><><><><><>

b) Let $\mathcal{A}$ be the subset of all 4-vectors, (x_1, x_2, x_3, x_4), such that $x_1 + x_2 - x_3 - x_4 = 1$.

Which of the following 4-vectors are in $\mathcal{A}$?

i) (1,2,3,0) ii) (1,0,1,1) iii) (178, -15, 93, -69)

ANSWER:

<><><><><><><><><><><><><><><><><><><><><><><><><><><><><><><>

c) Let $\mathcal{B}$ be the subset of all 3-vectors $\begin{pmatrix} x \\ y \\ z \end{pmatrix}$ such that $\begin{pmatrix} 1 & 5 & 3 \\ -5 & 1 & 0 \end{pmatrix} \cdot \begin{pmatrix} x \\ y \\ z \end{pmatrix} = \begin{pmatrix} 0 \\ 0 \end{pmatrix}$

Which of the following 3-vectors are in $\mathcal{B}$?

i) $\begin{pmatrix} -3 \\ 15 \\ 26 \end{pmatrix}$ ii) $\begin{pmatrix} 0 \\ 0 \\ 0 \end{pmatrix}$ iii) $\begin{bmatrix} \frac{-3}{26} \\ \frac{15}{26} \\ 1 \end{bmatrix}$

ANSWER:

<><><><><><><><><><><><><><><><><><><><><><><><><><><><><><><>

d) Let $\mathcal{C}$ be the subset of all 5-vectors such that the product of their components is zero.

Give an example of two vectors that are in $\mathcal{C}$ and two that are not.

ANSWER:

<><><><><><><><><><><><><><><><><><><><><><><><><><><><><><><>

e) Let $\mathcal{S}$ be the subset of all 3-tuples (x, y, z) such that

$$2 \cdot x + y - z = 0$$
$$-x + 5 \cdot y + 12 \cdot z = 0$$
$$11 \cdot x - 3 \cdot y + 5 \cdot z = 0$$

Give an example of a 3-vector in $\mathcal{S}$ and one that is not in $\mathcal{S}$.

ANSWER:

<><><><><><><><><><><><><><><><><><><><><><><><><><><><><><><>

f) Let $\mathcal{P}$ be the subset of all pairs (x, y) such that

$$3 \cdot x - 3 \cdot y = 1$$
$$x - y = 1$$

Give another description of the vectors in $\mathcal{P}$.

ANSWER:

<><><><><><><><><><><><><><><><><><><><><><><><><><><><><><><>

g) Let $\mathcal{L}$ be the subset of vectors $t \cdot \begin{pmatrix} 1 \\ 0 \\ -5 \end{pmatrix}$ where *t* is any real number.

Give an example of a 3-vector in $\mathcal{L}$ and one that is not in $\mathcal{L}$. This subset is called a line in 3-space.

ANSWER:

<><><><><><><><><><><><><><><><><><><><><><><><><><><><><><><>

h) Let $\mathcal{LC}$ be the subset of vectors $t \cdot \begin{pmatrix} 1 \\ 0 \\ -5 \end{pmatrix} + s \cdot \begin{pmatrix} 0 \\ 1 \\ 3 \end{pmatrix}$

where *s* and *t* are any scalars.

This subset is called **all linear combinations** of $\begin{pmatrix} 1 \\ 0 \\ -5 \end{pmatrix}$ and $\begin{pmatrix} 0 \\ 1 \\ 3 \end{pmatrix}$

Find a 3-vector not in $\mathcal{LC}$ and explain why your example is not in the set.

ANSWER:

<><><><><><><><><><><><><><><><><><><><><><><><><><><><><><><><>

i) Let $\mathcal{G}$ be the subset of all 2-vectors of the form $(x, x^2 + 1)$. These points are called the **graph of the function** $f(x) = x^2 + 1$.

Determine two points that belong to this subset.

ANSWER:

<><><><><><><><><><><><><><><><><><><><><><><><><><><><><><><><>

$\mathbf{R}^n$ is more than just a collection of elements. It is a set with two important operations: **addition** and **scalar multiplication**. We can look for subsets of $\mathbf{R}^n$ that have special properties with respect to these operations. In particular, we can look at subsets that are **closed** under either or both of these operations.

A subset, $\mathcal{S}$, is said to be **closed under addition** if whenever **x** and **y** are vectors (also called members) in $\mathcal{S}$, then $\mathbf{x} + \mathbf{y}$ is also a vector in $\mathcal{S}$.

A subset, $\mathcal{S}$, is said to be **closed under scalar multiplication** if whenever **x** is a vector in $\mathcal{S}$, and c is a real number then $c \cdot \mathbf{x}$ is a vector in $\mathcal{S}$.

<><><><><><><><><><><><><><><><><><><><><><><><><><><><><><><><>

For each of the subsets described above determine if it is closed under addition and if it is closed under scalar multiplication. To show that it is not closed under the operation you *must give an example*. To show that it is closed you must carefully explain why the property holds for all vectors in the subset. We illustrate this below in examples a) and c).

a) All vectors $\mathbf{x}$ in $\mathbf{R}^5$ such that $|\mathbf{x}| \leq 1$.

ANSWER: **Addition**: The vectors (1,0,0,0,0) and (0,1,0,0,0) are both members of the subset but their sum has length sqrt(2) and is not a member of the subset.

Scalar multiplication: (1,0,0,0,0) is a member of the subset but 3·(1,0,0,0,0) is not since it has length 3.

<><><><><><><><><><><><><><><><><><><><><><><><><><><><><><><>

b) All 4-tuples, (x_1, x_2, x_3, x_4), such that $x_1 + x_2 - x_3 - x_4 = 1$.

ANSWER: **Addition**:

Scalar multiplication:

<><><><><><><><><><><><><><><><><><><><><><><><><><><><><><><>

c) All vectors $\begin{pmatrix} x \\ y \\ z \end{pmatrix}$ such that $\begin{pmatrix} 1 & 5 & 3 \\ -5 & 1 & 0 \end{pmatrix} \cdot \begin{pmatrix} x \\ y \\ z \end{pmatrix} = \begin{pmatrix} 0 \\ 0 \end{pmatrix}$.

ANSWER: **Addition**: Call the matrix $\mathbf{A}$ and suppose that $\mathbf{u}$ and $\mathbf{v}$ are two vectors such that $\mathbf{A}\cdot\mathbf{u} = \mathbf{A}\cdot\mathbf{v} = \mathbf{0}$. Then $\mathbf{A}\cdot(\mathbf{u} + \mathbf{v}) = \mathbf{A}\cdot\mathbf{u} + \mathbf{A}\cdot\mathbf{v} = \mathbf{0} + \mathbf{0} = \mathbf{0}$. So it is closed under addition.

Scalar multiplication: If $\mathbf{A}\cdot\mathbf{u} = \mathbf{0}$, then $\mathbf{A}(\mathbf{c}\cdot\mathbf{u}) = \mathbf{c}\cdot\mathbf{A}(\mathbf{u}) = \mathbf{c}\cdot\mathbf{0} = \mathbf{0}$. So it is closed under scalar multiplication as well.

<><><><><><><><><><><><><><><><><><><><><><><><><><><><><><><>

d) The subset of all vectors in 5-space such that the product of the components is zero.

ANSWER: **Addition**:

Scalar multiplication:

<><><><><><><><><><><><><><><><><><><><><><><><><><><><><><><>

e) The subset of 3-vectors (x, y, z) such that

$$2\cdot x + y - z = 0$$
$$-x + 5\cdot y + 12\cdot z = 0$$
$$11\cdot x - 3\cdot y + 5\cdot z = 0$$

ANSWER: **Addition**:

Scalar multiplication:

<><><><><><><><><><><><><><><><><><><><><><><><><><><><><><>

f) The subset of pairs (x,y) such that

$$3\cdot x - 3\cdot y = 1$$
$$x - y = 1$$

ANSWER: **Addition**:

Scalar multiplication:

<><><><><><><><><><><><><><><><><><><><><><><><><><><><><><>

g) The subset of vectors $t\cdot\begin{pmatrix}1\\0\\-5\end{pmatrix}$ where t is any scalar.

ANSWER: **Addition**:

Scalar multiplication:

<><><><><><><><><><><><><><><><><><><><><><><><><><><><><><>

h) The subset of vectors $t\cdot\begin{pmatrix}1\\0\\-5\end{pmatrix} + s\cdot\begin{pmatrix}0\\1\\3\end{pmatrix}$ where s and t are any scalars.

ANSWER: **Addition**:

Scalar multiplication:

<><><><><><><><><><><><><><><><><><><><><><><><><><><><><><>

i) All 2-vectors of the form $(x, x^2 + 1)$. These points are called the graph of the function $f(x) = x^2 + 1$.

ANSWER: **Addition**:

Scalar multiplication:

<><><><><><><><><><><><><><><><><><><><><><><><><><><><><><>

There are three ***special*** subsets in $\mathbf{R}^n$:

i) The subset that has no elements - called the **empty set**.

ii) The subset that consists of the zero element: (0, 0, 0, ..., 0).

iii) The subset that consists of all elements of $\mathbf{R}^n$.

Be sure not to confuse the empty set with the zero subset. They have a different number of elements. The empty set has no elements, the zero subset has one element.

Is the **empty set** closed under addition? under scalar multiplication?

ANSWER: **Addition**:

Scalar multiplication:

<><><><><><><><><><><><><><><><><><><><><><><><><><><><><><><><><><><><><>

Is the zero subset closed under addition? under scalar multiplication?

ANSWER: **Addition**:

Scalar multiplication:

<><><><><><><><><><><><><><><><><><><><><><><><><><><><><><><><><><><><><>

Is the set $\mathbf{R}^n$ closed under addition? under scalar multiplication?

ANSWER: **Addition**:

Scalar multiplication:

<><><><><><><><><><><><><><><><><><><><><><><><><><><><><><><><><><><><><>

Since $\mathbf{R}^n$ is closed under both scalar multiplication and addition every subset is contained in a larger subset that is closed under these operations. For a given set, $\mathcal{S}$, the smallest set that contains $\mathcal{S}$ and is closed under addition is called the **closure of $\mathcal{S}$ under addition** and we denote this here as Closure($\mathcal{S}$, +). Closure($\mathcal{S}$, ·) and Closure($\mathcal{S}$, +, ·) are defined similarly.

Carefully write out the meaning of Closure($\mathcal{S}$, ·) and Closure($\mathcal{S}$, +, ·).

ANSWER:

<><><><><><><><><><><><><><><><><><><><><><><><><><><><><><><><><><><><><>

Exercise

For each of the following sets determine Closure($\mathcal{S}$, +, ·); i.e., the smallest set that contains the given subset and is closed under both addition and scalar multiplication. The first example is completed as a model of the type of discussion that is required.

a) $\mathcal{S}$ is the subset of $\mathbf{R}^3$ consisting of the single vector: $\begin{pmatrix} 1 \\ -1 \\ 1 \end{pmatrix}$

ANSWER: To be closed under addition the set must contain:

$$\begin{pmatrix} 1 \\ -1 \\ 1 \end{pmatrix} \quad \text{and} \quad \begin{pmatrix} 1 \\ -1 \\ 1 \end{pmatrix} + \begin{pmatrix} 1 \\ -1 \\ 1 \end{pmatrix} = \begin{pmatrix} 2 \\ -2 \\ 2 \end{pmatrix} \quad \text{and} \quad \begin{pmatrix} 2 \\ -2 \\ 2 \end{pmatrix} + \begin{pmatrix} 1 \\ -1 \\ 1 \end{pmatrix} = \begin{pmatrix} 3 \\ -3 \\ 3 \end{pmatrix}$$

and continuing in this way all vectors: $\begin{pmatrix} n \\ -n \\ n \end{pmatrix}$ where n is a positive integer.

To be closed under scalar multiplication the set must contain all vectors of the form:

$$t \cdot \begin{pmatrix} 1 \\ -1 \\ 1 \end{pmatrix} \quad \text{where t is a scalar.}$$

Thus the closure must contain all vectors of the form $n \cdot \begin{pmatrix} 1 \\ -1 \\ 1 \end{pmatrix}$

all vectors of the form $t \cdot \begin{pmatrix} 1 \\ -1 \\ 1 \end{pmatrix}$ and all sums from these

two sets. The first set is contained in the second and it is clear that the second is closed under both scalar multiplication and addition. Thus the closure is all scalar multiples of (1,-1,1). This is a line in 3-space.

<><><><><><><><><><><><><><><><><><><><><><><><><><><><><><><><><><>

b) $\mathcal{S}$ is the subset of $\mathbf{R}^3$ consisting of the single vector: $\begin{pmatrix} 0 \\ 0 \\ 0 \end{pmatrix}$

ANSWER:

<><><><><><><><><><><><><><><><><><><><><><><><><><><><><><><><><><>

c) $\mathcal{S}$ is the subset of $\mathbf{R}^3$ consisting of the vectors: $\begin{pmatrix}1\\0\\0\end{pmatrix}$ and $\begin{pmatrix}0\\1\\0\end{pmatrix}$

ANSWER:

◇◇◇◇◇◇◇◇◇◇◇◇◇◇◇◇◇◇◇◇◇◇◇◇◇◇◇◇◇◇◇

d) $\mathcal{S}$ is the subset of $\mathbf{R}^3$ consisting of the vectors $\begin{pmatrix}1\\0\\0\end{pmatrix}$ $\begin{pmatrix}0\\1\\0\end{pmatrix}$ and $\begin{pmatrix}0\\0\\1\end{pmatrix}$

***ANSWER*:**

◇◇◇◇◇◇◇◇◇◇◇◇◇◇◇◇◇◇◇◇◇◇◇◇◇◇◇◇◇◇◇

e) $\mathcal{S}$ is the subset of $\mathbf{R}^3$ consisting of the vectors $\begin{pmatrix}1\\1\\1\end{pmatrix}$ $\begin{pmatrix}0\\1\\1\end{pmatrix}$ and $\begin{pmatrix}1\\0\\0\end{pmatrix}$

***ANSWER*:**

◇◇◇◇◇◇◇◇◇◇◇◇◇◇◇◇◇◇◇◇◇◇◇◇◇◇◇◇◇◇◇

f) $\mathcal{S}$ is the subset of $\mathbf{R}^3$ consisting of the vectors $\begin{pmatrix}1\\1\\1\end{pmatrix}$ $\begin{pmatrix}0\\1\\1\end{pmatrix}$ and $\begin{pmatrix}1\\0\\1\end{pmatrix}$

***ANSWER*:**

◇◇◇◇◇◇◇◇◇◇◇◇◇◇◇◇◇◇◇◇◇◇◇◇◇◇◇◇◇◇◇

Let $\mathcal{S}$ be a set consisting of the k vectors $\mathbf{v}_1, \mathbf{v}_2, \ldots, \mathbf{v}_k$ in $\mathbf{R}^n$. Span($\mathcal{S}$) = Span($\mathbf{v}_1, \mathbf{v}_2, \ldots, \mathbf{v}_k$) is the set of all linear combinations of the vectors $\mathbf{v}_1, \mathbf{v}_2, \ldots, \mathbf{v}_k$. Explain why Closure($\mathcal{S}$, +, ·) equals Span($\mathbf{v}_1, \mathbf{v}_2, \ldots, \mathbf{v}_k$) where $\mathcal{S}$ is the set of vectors $\mathbf{v}_1, \mathbf{v}_2, \ldots, \mathbf{v}_k$.

ANSWER:

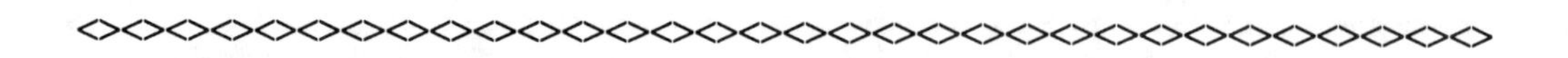

> **A subset of R^n is called a SUBSPACE if**
> **a) it is not empty**
> **b) it is closed under scalar multiplication**
> **and c) it is closed under addition.**

Let U be a subspace of $\mathbf{R}^n$. Explain why the n-vector $\mathbf{0}$ = (0,0,...,0) is in U. Which of the properties a), b), and c) above do you need to ensure that $\mathbf{0}$ is in U?

ANSWER:

Indicate which of the subsets we examined above are subspaces and which are not by putting a YES or NO on the line before the example letter.

___a) All vectors $\mathbf{x}$ in $\mathbf{R}^5$ such that $|\mathbf{x}| \leq 1$.

___b) All 4-tuples, (x_1, x_2, x_3, x_4), such that $x_1 + x_2 - x_3 - x_4 = 1$.

___c) All vectors $\begin{pmatrix} x \\ y \\ z \end{pmatrix}$ such that $\begin{pmatrix} 1 & 5 & 3 \\ -5 & 1 & 0 \end{pmatrix} \cdot \begin{pmatrix} x \\ y \\ z \end{pmatrix} = \begin{pmatrix} 0 \\ 0 \end{pmatrix}$

___d) The subset of all vectors in 5-space such that the product of the components is zero.

___e) The subset of 3-vectors (x, y, z) such that

$$2\cdot x + y - z = 0$$
$$-x + 5\cdot y + 12\cdot z = 0$$
$$11\cdot x - 3\cdot y + 5\cdot z = 0$$

___f) The subset of pairs (x, y) such that $3 \cdot x - 3 \cdot y = 1$ and $x - y = 1$.

___g) The subset of vectors $t \cdot \begin{pmatrix} 1 \\ 0 \\ -5 \end{pmatrix}$ where t is any scalar.

___h) The subset of vectors $t \cdot \begin{pmatrix} 1 \\ 0 \\ -5 \end{pmatrix} + s \cdot \begin{pmatrix} 0 \\ 1 \\ 3 \end{pmatrix}$ where s and t are any scalars.

___i) All 2-vectors of the form $(x, x^2 + 1)$. These points are called the graph of the function $f(x) = x^2 + 1$.

<><><><><><><><><><><><><><><><><><><><><><><><><><><><><><>

Suppose that U is a subspace of $\mathbf{R}^n$. Then U = closure(U, + , ·). That is simply another way of stating the definition of a subspace. Informally, we could say that a subset V of U *generates a subspace U* if U = closure(V, +, ·); i.e., every vector in U is a linear combination of vectors in the subset V.

Example

Let U be all vectors $\begin{pmatrix} t \\ t \\ t \end{pmatrix}$ where t is a scalar.

(U is the set of all vectors in $\mathbf{R}^3$ that have the property that all of their components are equal.) This is a subspace since if we add two vectors of this form we get another vector, all of whose components are equal. Similarly scalar multiplication preserves the property that all components are equal. This subspace is a line in $\mathbf{R}^3$.

Which of the following subsets generate (in the above sense) U?

a) $\begin{pmatrix} 0 \\ 0 \\ 0 \end{pmatrix}$ and $\begin{pmatrix} 1 \\ 1 \\ 1 \end{pmatrix}$

b) $\begin{pmatrix} 1 \\ 0 \\ 0 \end{pmatrix}$ $\begin{pmatrix} 0 \\ 1 \\ 0 \end{pmatrix}$ and $\begin{pmatrix} 0 \\ 0 \\ 1 \end{pmatrix}$

c) $\begin{pmatrix} 1 \\ 1 \\ 1 \end{pmatrix}$ $\begin{pmatrix} -1 \\ -1 \\ -1 \end{pmatrix}$ $\begin{pmatrix} 1 \\ 0 \\ -1 \end{pmatrix}$

d) $\begin{pmatrix} \pi \\ \pi \\ \pi \end{pmatrix}$

e) $\begin{pmatrix} 17 \\ 17 \\ 17 \end{pmatrix}$

f) $\begin{pmatrix} 0 \\ 0 \\ 0 \end{pmatrix}$

ANSWER: (List the letter of the subsets that generate U.)

<><><><><><><><><><><><><><><><><><><><><><><><><><><><><><>

What you should have concluded from the above exercise is that there are many subsets that *generate* a given subspace and that there is no unique smallest generating subset. However we can talk about 'minimal generating' subsets of a subspace, U, in the sense that a subset, M, is a **minimal generating subset** if M generates U and no subset of M generates U. (We also use the terminology **minimal spanning set**.)

If $\mathbf{v}_1, \mathbf{v}_2, ..., \mathbf{v}_k$ is a minimal spanning set for the subspace U is it true that $\text{Span}(\mathbf{v}_1, \mathbf{v}_2, ..., \mathbf{v}_k) = U$? Explain.

ANSWER:

Make a list of the subsets in the previous exercise that are minimal generating subsets?

ANSWER:

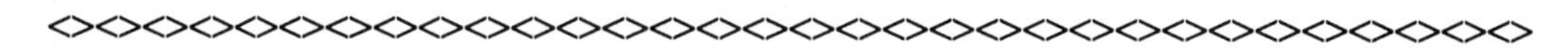

Exercises

1. Find a minimal generating subset for the subspace of solutions to

$$x + 5 \cdot y + 3 \cdot z = 0$$

$$7 \cdot x + 34 \cdot y + 10 \cdot z = 0$$

ANSWER:

2. Let $\quad \mathbf{A} := \begin{pmatrix} 1 & -5 & 0 & 0 & 7 & 0 \\ 0 & 0 & 1 & 0 & 2 & 0 \\ 0 & 0 & 0 & 1 & -5 & 0 \\ 0 & 0 & 0 & 0 & 0 & 1 \end{pmatrix}$

Find a minimal spanning set for the set of all solutions to $\mathbf{A} \cdot \mathbf{x} = \mathbf{0}$.

ANSWER:

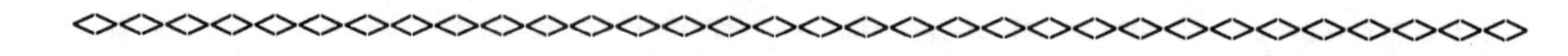

3. (Individual) Tell what each of the following terms means in your own words.

a) $\mathbf{R}^7$

b) subset of $\mathbf{R}^7$

c) closed under vector addition

d) Closure($\mathcal{S}$, +, ·)

e) Span($\mathbf{v}_1$, $\mathbf{v}_2$, ..., $\mathbf{v}_n$)

f) minimal spanning set

g) subspace

h) empty set

i) zero subset

Section 2-6 The Set of Solutions

Name:________________

Score:______________

Overview

The goal of this section is to describe the structure of the solution set of a system of linear equations. Systems of the form $\mathbf{A}\cdot\mathbf{x} = \mathbf{0}$ have solution sets with special properties that can be described easily from the RREF of the coefficient matrix.

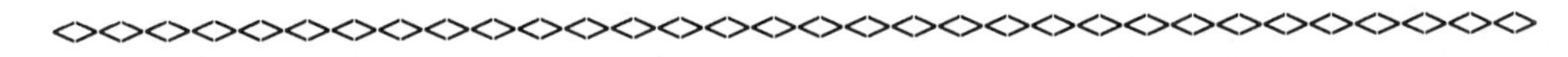

A system of linear equations $\mathbf{Ax} = \mathbf{b}$ is called **HOMOGENEOUS** if $\mathbf{b} = \mathbf{0}$; otherwise it is called **NONHOMOGENEOUS**.

Homogeneous systems

A homogeneous linear system of equations always has at least one solution. What is that solution?

ANSWER:

Explain how one can determine from the RREF of the augmented matrix $[\mathbf{A} \mid \mathbf{0}]$ if the homogeneous system of equations $\mathbf{A}\cdot\mathbf{x} = \mathbf{0}$ has more than one solution.

ANSWER:

If the homogeneous system of equations $\mathbf{A}\cdot\mathbf{x} = \mathbf{0}$ has more than one solution, it has an infinite number of solutions. Explain why this is true by referring to the RREF of the augmented matrix $[\mathbf{A} \mid \mathbf{0}]$.

ANSWER:

Let S be the subset of $\mathbf{R}^n$ that consists of all solutions to the homogeneous system of linear equations $\mathbf{A}\cdot\mathbf{x} = \mathbf{0}$. Write the matrix equation that 'says' that **s1** is a member of the set S (i.e., **s1** is a solution to $\mathbf{A}\cdot\mathbf{x} = \mathbf{0}$.)

ANSWER:

Let a be a scalar and let **s1** be as above. For what values of a is $a \cdot \mathbf{s1}$ a solution to $\mathbf{A} \cdot \mathbf{x} = \mathbf{0}$?

ANSWER:

Is the set S of all solutions to $\mathbf{A} \cdot \mathbf{x} = \mathbf{0}$ closed under scalar multiplication? Explain.

ANSWER:

Let **s2** be a another element of S. Show why **s1** + **s2** belongs to S.

ANSWER:

Is S closed under addition? Explain.

ANSWER:

Explain why the subset of $\mathbf{R}^n$ that consists of all solutions to the homogeneous system of equations, $\mathbf{A} \cdot \mathbf{x} = \mathbf{0}$ is a subspace.

ANSWER:

<><><><><><><><><><><><><><><><><><><><><><><><><><><><><><><><><>

Find a minimal generating set for the subspace of all solutions to

$$\begin{pmatrix} 1 & 5 & 1 \\ 3 & 0 & 2 \end{pmatrix} \cdot \begin{pmatrix} x \\ y \\ z \end{pmatrix} = \begin{pmatrix} 0 \\ 0 \end{pmatrix}$$

ANSWER:

<><><><><><><><><><><><><><><><><><><><><><><><><><><><><><><><><>

Find a minimal generating set for the subspace of all solutions to

$$\begin{pmatrix} 1 & -1 \\ 5 & 2 \end{pmatrix} \cdot \begin{pmatrix} x \\ y \end{pmatrix} = \begin{pmatrix} 0 \\ 0 \end{pmatrix}$$

ANSWER:

<><><><><><><><><><><><><><><><><><><><><><><><><><><><><><><><><>

Find a minimal generating set for the subspace of all solutions to

$$3 \cdot x - 5 \cdot y = 0$$

ANSWER:

<><><><><><><><><><><><><><><><><><><><><><><><><><><><><><>

If the set of vectors $\mathbf{v}_1, \mathbf{v}_2, \ldots, \mathbf{v}_k$ is a minimal generating set for the subspace of solutions of $\mathbf{A} \cdot \mathbf{x} = \mathbf{0}$, explain why $\text{Span}(\mathbf{v}_1, \mathbf{v}_2, \ldots, \mathbf{v}_k)$ is the set of all solutions of $\mathbf{A} \cdot \mathbf{x} = \mathbf{0}$.

ANSWER:

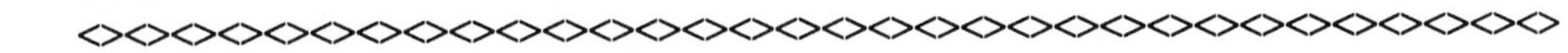

Nonhomogeneous systems

Let N be the subset of all solutions to the nonhomogeneous linear system of equations $\mathbf{A} \cdot \mathbf{x} = \mathbf{b}$ where $\mathbf{b}$ is not the zero vector.

Write the matrix equation that *says* that **s1** is a member of the set N (i.e., **s1** is a solution to $\mathbf{A} \cdot \mathbf{x} = \mathbf{b}$.)

ANSWER:

Let **s1** be a solution to $\mathbf{Ax} = \mathbf{b}$ and let a be a scalar. For what values of a is $a \cdot \mathbf{s1}$ a solution to $\mathbf{A} \cdot \mathbf{x} = \mathbf{b}$?

ANSWER:

Is N closed under scalar multiplication? Explain.

ANSWER:

Let **s2** be another solution to $\mathbf{Ax} = \mathbf{b}$. Is $\mathbf{s1} + \mathbf{s2}$ a solution of $\mathbf{Ax} = \mathbf{b}$?

ANSWER:

Is N closed under addition? Explain.

ANSWER:

Is the solution set N of a nonhomogeneous linear system a subspace? Explain.
ANSWER:

For each system of equations $\mathbf{A \cdot x = b}$, the system $\mathbf{A \cdot x = 0}$ is called the **corresponding homogeneous system**. Assume that vector $\mathbf{c}$ is a solution to the system $\mathbf{A \cdot x = b}$ (i.e., $\mathbf{A \cdot c = b}$) and let $\mathbf{u}$ be any solution to the corresponding homogeneous system.

Show that $\mathbf{c + u}$ is a solution to the nonhomogeneous linear system. (Hint: What does it mean *to be a solution*?)
ANSWER:

Let $\mathbf{c}$ be a solution to $\mathbf{A \cdot x = b}$. We will call $\mathbf{c}$ a **particular solution**.

Let $\mathbf{v}$ be any other solution to $\mathbf{A \cdot x = b}$. Show that $\mathbf{v - c}$ is a solution to the associated homogeneous system, $\mathbf{A \cdot x = 0}$. (This says that any solution of $\mathbf{A \cdot x = b}$ minus a particular solution of $\mathbf{A \cdot x = b}$ is a solution of $\mathbf{A \cdot x = 0}$.)
ANSWER:

We have seen that if $\mathbf{c}$ is a fixed particular solution to $\mathbf{A \cdot x = b}$ then

(1) Every solution to $\mathbf{A \cdot x = b}$ is equal to $\mathbf{c + u}$ where $\mathbf{u}$ is a solution to the associated homogeneous system.

(2) If $\mathbf{u}$ is a solution to the associated homogeneous system then $\mathbf{c + u}$ is a solution to $\mathbf{A \cdot x = b}$.

We conclude that the set of all solutions to the system $\mathbf{A \cdot x = b}$ is the set of all sums $\mathbf{c + u}$ where $\mathbf{c}$ is a particular solution and $\mathbf{u}$ can be any solution to the associated homogeneous system.

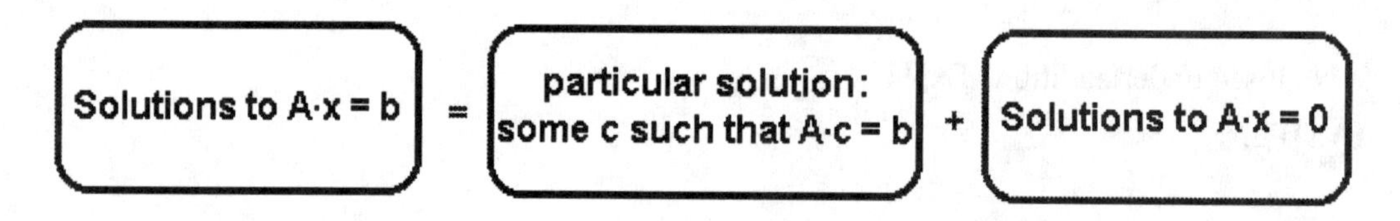

Explain the difference between statements (1) and (2) above.

ANSWER:

If S is a set of vectors and **c** is a particular vector then we write **c** + S to mean the set of all sums **c** + **s** where **s** is a member of S.

It might be possible for the associated homogeneous system to have solutions and the system $\mathbf{A}\cdot\mathbf{x} = \mathbf{b}$ to have no solutions. How could you tell that this is the case from the RREF of [**A** | **b**]?

ANSWER:

<><><><><><><><><><><><><><><><><><><><><><><><><><><><><><><>

For the linear system $\mathbf{A}\cdot\mathbf{x} = \mathbf{b}$, where $\mathbf{A} := \begin{pmatrix} 2 & 1 & 3 & 2 \\ -1 & 0 & 2 & -1 \\ 3 & 2 & 8 & 3 \end{pmatrix}$ and $\mathbf{b} := \begin{pmatrix} 1 \\ 1 \\ 3 \end{pmatrix}$

the RREF of [**A** | **b**] is $\begin{pmatrix} 1 & 0 & -2 & 1 & -1 \\ 0 & 1 & 7 & 0 & 3 \\ 0 & 0 & 0 & 0 & 0 \end{pmatrix}$

How many free variables are there in the general solution to $\mathbf{A}\cdot\mathbf{x} = \mathbf{b}$?

ANSWER:

Find a particular solution to $\mathbf{A}\cdot\mathbf{x} = \mathbf{b}$.

ANSWER:

Find a generating set for the set of all solutions to the associated homogeneous system.

ANSWER:

Describe the set of all solutions to $\mathbf{A}\cdot\mathbf{x} = \mathbf{b}$.

ANSWER:

<><><><><><><><><><><><><><><><><><><><><><><><><><><><><><><>

Complete the following statement:

The solution set of the linear system $\mathbf{A}\cdot\mathbf{x} = \mathbf{b}$ *consists of*

__________________ *solution of* $\mathbf{A}\cdot\mathbf{x} = \mathbf{b}$ *plus the*

__________________ *of the corresponding homogeneous linear system* $\mathbf{A}\cdot\mathbf{x} = \mathbf{0}$.

<><><><><><><><><><><><><><><><><><><><><><><><><><><><><><><>

Let the coefficient matrix $\mathbf{A}$ be an m by n matrix. Assume that the system $\mathbf{A}\cdot\mathbf{x} = \mathbf{b}$ is consistent. Is the solution set S a subset of $\mathbf{R}^m$ or $\mathbf{R}^n$? Explain.

ANSWER:

Is the solution set S of a consistent nonhomogeneous linear system $\mathbf{A}\cdot\mathbf{x} = \mathbf{b}$ a subspace? Explain.

ANSWER:

Is the solution set T of the corresponding homogeneous linear system $\mathbf{A}\cdot\mathbf{x} = \mathbf{0}$ a subspace? Explain.

ANSWER:

<><><><><><><><><><><><><><><><><><><><><><><><><><><><><><><>

In Section 2-3 we saw that the general solution of a linear system could be written as the sum of a particular solution **c** plus all linear combinations of vectors $\mathbf{v}_1, \mathbf{v}_2, \ldots, \mathbf{v}_j$.

$$\{ c + t_1 \cdot v_1 + t_2 \cdot v_2 + \ldots + t_j \cdot v_j \mid -\infty < t_j < \infty \}$$

We stated above that the set of all solutions of a nonhomogeneous system of linear equations could be expressed as all sums **c + u** where **c** is a particular solution and **u** is a solution to the associated homogeneous system.

How are these two descriptions related?

ANSWER:

<><><><><><><><><><><><><><><><><><><><><><><><><><><><><><><><><><><><><><><><>

Examples

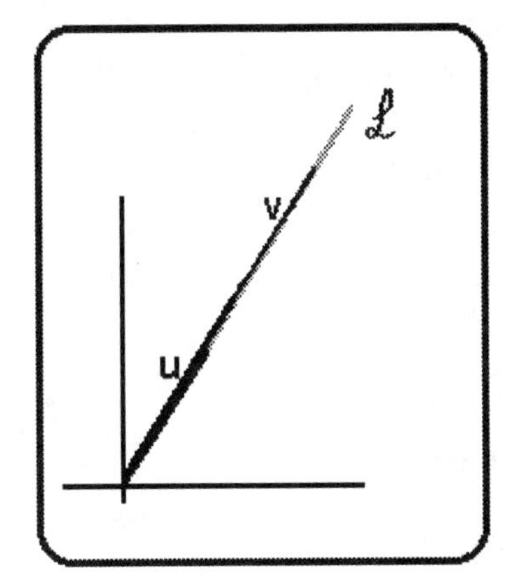

If we add two vectors on the line $\mathcal{L}$, do we stay on the line?

ANSWER:

Is this line a subspace of $\mathbf{R}^2$?

ANSWER:

If we add two vectors, u and v, on the line $\mathcal{L}$, do we stay on the line?

ANSWER:

Is this line a subspace of $\mathbf{R}^2$?

ANSWER:

The point (1,2) is on this line.
What is the equation of the line?

ANSWER:

The points (3,0) and (0,-2) are on this line.
What is the equation of the line?

ANSWER:

Is the equation of this line a homogeneous equation?

ANSWER:

Is the equation of this line a homogeneous equation?

ANSWER:

<><><><><><><><><><><><><><><><><><><><><><><><><><><><><><><><><><><><><>

We write **0** to stand for the zero vector. Of course there is a different zero vector in each $\mathbf{R}^n$. Is **0** a member of every subspace in $\mathbf{R}^n$? Explain.

ANSWER:

<><><><><><><><><><><><><><><><><><><><><><><><><><><><><><><><><><><><><>

In Section 2-3 we saw that the general solution of a linear system of equations could be expressed as a particular solution **c** plus all linear combination of the vectors $\mathbf{v}_1, \mathbf{v}_2, \ldots, \mathbf{v}_j$. More precisely, the set of all vectors of the form:

$$\{ \mathbf{c} + t_1 \cdot \mathbf{v}_1 + t_2 \cdot \mathbf{v}_2 + \ldots + t_j \cdot \mathbf{v}_j \mid -\infty < t_j < \infty \}$$

For the linear system: $\mathbf{A} \cdot \mathbf{x} = \mathbf{0}$, what is the particular solution **c**?

ANSWER:

<><><><><><><><><><><><><><><><><><><><><><><><><><><><><><><><><><><><><>

Let $\mathbf{A} := \begin{pmatrix} 2 & 1 & 3 & 2 \\ -1 & 0 & 2 & -1 \\ 3 & 2 & 8 & 3 \end{pmatrix}$ the RREF of $[\mathbf{A} \mid \mathbf{0}]$ is $\begin{pmatrix} 1 & 0 & -2 & 1 & 0 \\ 0 & 1 & 7 & 0 & 0 \\ 0 & 0 & 0 & 0 & 0 \end{pmatrix}$

Describe the general solution.

ANSWER:

<><><><><><><><><><><><><><><><><><><><><><><><><><><><><><><><><><><><><>

Label the variables as shown below. Choose the free variables to be those variables that do not have a leading 1 in the corresponding column. Here y and z are chosen to be the free variables and w and x are expressed in terms of y and z.

$$\begin{array}{cccc|c} w & x & y & z & \\ \end{array}$$
$$\left(\begin{array}{cccc|c} 1 & 0 & -2 & 1 & 0 \\ 0 & 1 & 7 & 0 & 0 \\ 0 & 0 & 0 & 0 & 0 \end{array}\right)$$

Since there are two free variables there are two **v**'s in the general solution. Determine the **v**'s and list them below.

ANSWER:

<><><><><><><><><><><><><><><><><><><><><><><><><><><><><><><><><><><><>

1. The set of all solutions to $A \cdot x = 0$ is a subspace of R^n.

2. For a given subspace, V, there are vectors $v_1, v_2, ..., v_n$ such that $V = \text{Span}(v_1, v_2, ..., v_n)$.

3. Therefore, the set of all solutions to $A \cdot x = 0$ can be described as the span of a set of vectors.

Find a spanning set for the solution space of the linear system $\mathbf{A} \cdot \mathbf{x} = \mathbf{0}$, where the RREF of $[\mathbf{A} \mid \mathbf{0}]$ is

$$\begin{pmatrix} 1 & -2 & 0 & 3 & 0 & 0 & 0 \\ 0 & 0 & 1 & -4 & 0 & 0 & 0 \\ 0 & 0 & 0 & 0 & 1 & -5 & 0 \\ 0 & 0 & 0 & 0 & 0 & 0 & 0 \end{pmatrix}$$

ANSWER:

<><><><><><><><><><><><><><><><><><><><><><><><><><><><><><><><><><><><>

In the preceding work with homogeneous system $\mathbf{A}\cdot\mathbf{x} = \mathbf{0}$ we used the augmented matrix $[\mathbf{A} \mid \mathbf{0}]$. Since the augmented column is all zeros, row operations will not change it. Hence we could have performed row operations on just the coefficient matrix **A** alone and interpreted the output in a similar manner.

Let $\mathbf{A}\cdot\mathbf{x} = \mathbf{0}$ be a homogeneous system of 2 equations in 2 unknowns. Describe the two possible types of solution spaces. Assume **A** is not the matrix with all zeros.

Algebraic Description as a set of vectors	**Geometric Description**
Set # 1 ____________________	____________________
Set # 2 ____________________	____________________

Describe a spanning set for each solution space.

ANSWER:

When the solution space is exactly one vector, we call the solution space the **trivial subspace** (of $\mathbf{R}^2$). What is the trivial subspace of $\mathbf{R}^3$? of $\mathbf{R}^{29}$?

ANSWER:

<><><><><><><><><><><><><><><><><><><><><><><><><><><><><><><><><><>

For each of the following matrices **A** find a spanning set for the solution space of $\mathbf{A}\cdot\mathbf{x} = \mathbf{0}$.

$$\mathbf{A} := \begin{pmatrix} 1 & -3 \\ -2 & 6 \end{pmatrix} \qquad\qquad \mathbf{A} := \begin{pmatrix} -4 & .5 \\ -.8 & .1 \end{pmatrix}$$

ANSWER: ***ANSWER:***

<><><><><><><><><><><><><><><><><><><><><><><><><><><><><><><><><><>

In each case above the entire solution space of infinitely many vectors can be described as multiples of a single vector. That is about as efficient as we can get to describe a set with infinitely many vectors. In Section 2-7 we explore the general concept of *efficiency* for spanning sets in detail.

<><><><><><><><><><><><><><><><><><><><><><><><><><><><><><>

Let $\mathbf{A}\cdot\mathbf{x} = \mathbf{0}$ represent a homogeneous system of 2 equations in 3 unknowns. Further assume that column 1 of the coefficient matrix $\mathbf{A}$ is not all zeros. List the possible forms of the RREF and for each form tell the smallest number of vectors in $\mathbf{R}^3$ that can be in a spanning set of the solution space. Then describe each of the solution spaces geometrically.

ANSWER:

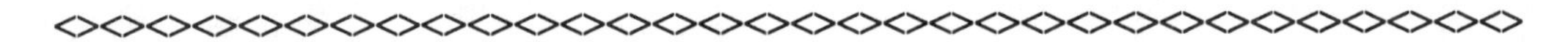

Exercise: (Individual) Write a summary of the MATHEMATICS that you learned in this section.

Section 2-7 Bases

Name:_______________

Score:____________

Overview

In Section 2-6 the solution space of a homogeneous system of equations, $\mathbf{A} \cdot \mathbf{x} = \mathbf{0}$, was described as the span of a set of vectors computed from the RREF of the augmented matrix [**A** | **0**]. In this section we investigate spanning sets for subspaces and ask:

> ***When is the set of spanning vectors minimal; that is to say, as small as possible for a given subspace?***

We develop a criterion to determine when the spanning set is minimal.

<><><><><><><><><><><><><><><><><><><><><><><><><><><><><><>

In Section 2-5 we defined **the span of a set of vectors** as the set of all linear combinations of the vectors. In Section 2-6 you saw that the solution space of the homogeneous system $\mathbf{A} \cdot \mathbf{x} = \mathbf{0}$ could be given as the span of a set of vectors computed from the RREF of [**A** | **0**]. In this section we examine other spanning sets for the solution space.

Let $\mathbf{A} := \begin{pmatrix} \frac{1}{2} & \frac{3}{2} & 2 & -3 & -6 \\ 3 & 9 & -10 & 4 & 52 \\ 1 & 3 & -4 & 2 & 20 \end{pmatrix}$

For the linear system $\mathbf{A} \cdot \mathbf{x} = \mathbf{0}$ the RREF of [**A** | **0**] is $\left(\begin{array}{ccccc|c} 1 & 3 & 0 & -2 & 4 & 0 \\ 0 & 0 & 1 & -1 & -4 & 0 \\ 0 & 0 & 0 & 0 & 0 & 0 \end{array}\right)$

How many free variables are there in the general solution to this system?

ANSWER:

Find a spanning set for the solution space of $\mathbf{A} \cdot \mathbf{x} = \mathbf{0}$. Name the vectors **v1**, **v2**, ... etc.

ANSWER:

Describe the solution space in terms of the **v**'s.

ANSWER:

Verbal description:

Algebraic description:

Explain why, if each **si** is a member of Span(**t1**, ..., **tm**) and each **tj** is a member of Span(**s1**, ..., **sk**), then it follows that Span(**s1**, ..., **sk**) equals Span(**t1**, ..., **tm**).

ANSWER:

Apply the previous result to each of the following:

Does Span(**v1**, **v2**, **v3**) = Span(**v1**, **v2**, **v3**, **v1** + **v3**)? Explain.

ANSWER:

Does Span(**v2, v3, v1** + **v3**) = Span(**v1**, **v2**, **v3**)? Explain.

ANSWER:

Does Span(**v1, v3, v1** + **v3**) = Span(**v1**, **v2**, **v3**)? Explain.

ANSWER:

Does Span(**v1, v2, v1** + **v3**) = Span(**v1**, **v2**, **v3**)? Explain.

ANSWER:

Does Span(**v1**, **v2**, **v3**) = Span(**v1**, **v2**, **v3**, 4·**v2** - 3·**v1,** 17·**v3**)? Explain.

ANSWER:

Does Span(**v1**, **v2**, **v3**) = Span(**v1**, **v2**)? Explain.

ANSWER:

Under what circumstances will the solution space for $\mathbf{A} \cdot \mathbf{x} = \mathbf{0}$ have a unique spanning set? Explain.

ANSWER:

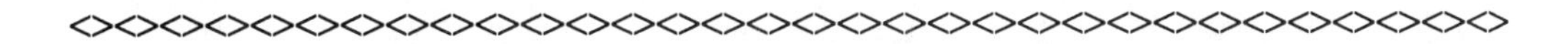

If two sets of vectors span the same subspace, then the set with the smaller number of vectors is the **more efficient spanning set**. A spanning set with the property that no subset also spans the same subspace is called **minimal**.

For each spanning set **u1**, **u2**, ..., **uk** of a subspace **V** we need a way to determine if it is minimal. That is, is there a subset of **u1**, **u2**, ..., **uk** (other than **u1**, **u2**, ..., **uk** itself) that spans the subspace? *In terms of information, we want to eliminate redundancy by finding the smallest number of vectors that contain all the information.*

<><><><><><><><><><><><><><><><><><><><><><><><><><><><><><><>

Let $\mathbf{A} := \begin{pmatrix} 0 & 0 \\ 0 & 0 \end{pmatrix}$

Describe in words the solution space of the linear system $\mathbf{A} \cdot \mathbf{x} = \mathbf{0}$.

ANSWER:

Each of the following sets spans the solution space of $\mathbf{A} \cdot \mathbf{x} = \mathbf{0}$. Determine which, if any, are minimal and explain why.

$$\mathbf{v1} := \begin{pmatrix} 1 \\ 0 \end{pmatrix} \qquad \mathbf{v2} := \begin{pmatrix} -2 \\ 0 \end{pmatrix} \qquad \mathbf{v3} := \begin{pmatrix} 0 \\ 1 \end{pmatrix}$$

ANSWER:

$$\mathbf{v1} := \begin{pmatrix} 0 \\ 1 \end{pmatrix} \qquad \mathbf{v2} := \begin{pmatrix} -1 \\ 0 \end{pmatrix} \qquad \mathbf{v3} := \begin{pmatrix} -4 \\ 7 \end{pmatrix}$$

ANSWER:

$$\mathbf{v1} := \begin{pmatrix} 1 \\ 1 \end{pmatrix} \qquad \mathbf{v2} := \begin{pmatrix} 3 \\ 3 \end{pmatrix} \qquad \mathbf{v3} := \begin{pmatrix} 0 \\ -1 \end{pmatrix} \qquad \mathbf{v4} := \begin{pmatrix} 0 \\ 1 \\ \overline{2} \end{pmatrix}$$

ANSWER:

$$\mathbf{v1} := \begin{pmatrix} 0 \\ 0 \end{pmatrix} \qquad \mathbf{v2} := \begin{pmatrix} 1 \\ 0 \end{pmatrix} \qquad \mathbf{v3} := \begin{pmatrix} 0 \\ -4 \end{pmatrix}$$

ANSWER:

<><><><><><><><><><><><><><><><><><><><><><><><><><><><><><><>

The set of vectors $\begin{pmatrix} 1 \\ 0 \end{pmatrix}$ $\begin{pmatrix} 0 \\ 1 \end{pmatrix}$ is called the ***natural*** spanning set for $\mathbf{R}^2$

since any vector $\begin{pmatrix} a \\ b \end{pmatrix}$ is equal to $\mathbf{a} \cdot \begin{pmatrix} 1 \\ 0 \end{pmatrix} + \mathbf{b} \cdot \begin{pmatrix} 0 \\ 1 \end{pmatrix}$

and no smaller subset of vectors of $\begin{pmatrix} 1 \\ 0 \end{pmatrix}$ $\begin{pmatrix} 0 \\ 1 \end{pmatrix}$ spans $\mathbf{R}^2$.

<><><><><><><><><><><><><><><><><><><><><><><><><><><><><><><>

What is the natural spanning set for $\mathbf{R}^3$? for $\mathbf{R}^5$?

ANSWER:

Describe the natural spanning set for $\mathbf{R}^k$ in terms of an identity matrix.

ANSWER:

<><><><><><><><><><><><><><><><><><><><><><><><><><><><><><><>

Explain why Span(**u1**, **u2**, ... ,**uk**) is closed under addition and under scalar multiplication and hence is a subspace.
ANSWER:

Suppose that one of the vectors, say **u3**, can be written as a linear combination of the other vectors. Is **u1**, **u2**, ... ,**uk** minimal? Explain.

ANSWER:

If **u3** can be written as a linear combination of the other vectors, then there exist coefficients c1, c2, c4, ..., ck so that **u3** = c1·**u1** + c2·**u2** + c4·**u4** + ... + ck·**uk**.

Then any vector that can be written as a linear combination of the vectors **u1**, **u2**, **u3**, **u4**, ..., **uk** can also be written as a linear combination of the vectors **u1**, **u2**, **u4**, ..., **uk**. Explain why this is true.
ANSWER:

If a set has only the zero vector in it, or if some vector in the set is a linear combination of the other vectors in the set, we say that the set is LINEARLY DEPENDENT. A set that is not linearly dependent is called LINEARLY INDEPENDENT.

If a spanning set is linearly dependent, is it minimal? Explain.

ANSWER:

Testing each vector to see if it is a linear combination of the other vectors in a set is time consuming. You can shorten the procedure to determine if a set is linearly dependent as follows:

If there are coefficients c1, c2, ..., ck , **not all zero**, so that
c1·**u1** + c2·**u2** + ... + ck·**uk** = **0** then the set S is linearly dependent. Explain.

ANSWER:

The preceding says that if there is a nontrivial linear combination of **u1**, **u2**, ..., **uk** that gives the zero vector, then **u1**, **u2**, ..., **uk** are linearly dependent. (By nontrivial we mean that some coefficient is not zero.)

Let **U** be the matrix whose jth column is **uj**; i.e., $\mathbf{U}^{<j>} = \mathbf{uj}$ for j = 1, ..., k and let **c** be the column vector such that $\mathbf{c}_j = cj$. Then $c1 \cdot \mathbf{u1} + c2 \cdot \mathbf{u2} + ... + ck \cdot \mathbf{uk} = \mathbf{0}$ is the homogeneous system, $\mathbf{U} \cdot \mathbf{c} = \mathbf{0}$.

> The set, **u1**, **u2**, ..., **uk**, is linearly dependent if there is a nontrivial solution to the homogeneous system $\mathbf{U} \cdot \mathbf{c} = \mathbf{0}$. Otherwise, the set **u1**, **u2**, ..., **uk** is linearly independent.

Recall that $\mathbf{U} \cdot \mathbf{c} = \mathbf{0}$ will have a nontrivial solution if and only if there is at least one free variable. In terms of the RREF of [**U** | **0**], how do we recognize a linearly independent set?
ANSWER:

In terms of the RREF of [**U** | **0**], how do we recognize a linearly dependent set?
ANSWER:

Use the RREF to determine which of the following sets of vectors are linearly dependent.

$$\mathbf{u1} := \begin{pmatrix} 1 \\ 2 \end{pmatrix} \qquad \mathbf{u2} := \begin{pmatrix} -1 \\ -2 \end{pmatrix} \qquad \mathbf{u3} := \begin{pmatrix} 2 \\ 2 \end{pmatrix}$$

ANSWER:

$$\mathbf{u1} := \begin{pmatrix} 1 \\ -2 \\ 0 \end{pmatrix} \qquad \mathbf{u2} := \begin{pmatrix} 0 \\ 3 \\ -1 \end{pmatrix} \qquad \mathbf{u3} := \begin{pmatrix} 2 \\ -1 \\ -1 \end{pmatrix}$$

ANSWER:

$$u1 := \begin{pmatrix} 2 \\ 0 \\ 0 \\ 1 \end{pmatrix} \qquad u2 := \begin{pmatrix} 2 \\ 1 \\ 0 \\ -1 \end{pmatrix} \qquad u3 := \begin{pmatrix} 5 \\ 1 \\ 0 \\ 2 \end{pmatrix}$$

ANSWER:

<><><><><><><><><><><><><><><><><><><><><><><><><><><><><><><>

What does it mean for a set of two vectors to be linearly dependent?

ANSWER:

What does it mean for a set of two vectors to be linearly independent?

ANSWER:

Any set containing the zero vector is linearly dependent. Explain why.

ANSWER:

A linearly independent spanning set is called a BASIS.

A basis contains the smallest number of vectors that can span a subspace, and is the most ***efficient*** spanning set.

The number of vectors in a basis of a subspace is called the DIMENSION of the subspace.

The dimension of the subspace S is denoted by DIM(S).

We must treat the subspace consisting of the zero vector a little differently since that subspace does not contain a linearly independent set. Consequently there is no basis for that subspace and its dimension equals 0.

<><><><><><><><><><><><><><><><><><><><><><><><><><><><><><><><><><><><>

Give a basis for $\mathbf{R}^2$.

ANSWER: DIM($\mathbf{R}^2$) = ______

Give a basis for $\mathbf{R}^5$.

ANSWER: DIM($\mathbf{R}^5$) = ______

Give a basis for $\mathbf{R}^n$.

ANSWER: DIM($\mathbf{R}^n$) = ______

Is there a basis for the subspace of $\mathbf{R}^n$ that consists of the zero vector (we call this the zero subspace and denote it by $\mathbf{0}$)?

ANSWER: DIM($\mathbf{0}$) = ______

<><><><><><><><><><><><><><><><><><><><><><><><><><><><><><><><><><><><>

Can a subspace have more than one basis? Explain.

ANSWER:

<><><><><><><><><><><><><><><><><><><><><><><><><><><><><><><><><><><><>

Implicit in the notion of the dimension of a subspace is the fact that all bases for the subspace contain the same number of vectors. This is true although we will not prove it here. ***If the dimension of a subspace is n then there will be exactly n vectors in any basis for the subspace.***

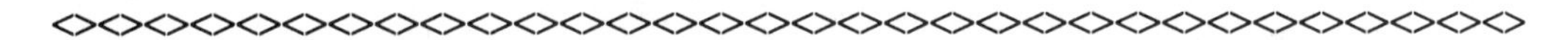

Basis for the solution space of Ax = 0

You saw above that the solution space of a homogeneous system of linear equations is a subspace and therefore can be described as the **SPAN** of certain vectors that can be computed from the RREF of the matrix corresponding to the system. We now show that these vectors are linearly independent and are therefore a basis for the solution space.

Consider the following example.

Let the RREF of **A** corresponding to a system $\mathbf{A}\cdot\mathbf{x} = \mathbf{0}$ be:

$$\begin{matrix} x_1 & x_2 & x_3 & x_4 & x_5 & x_6 \end{matrix}$$
$$\begin{bmatrix} 1 & 2 & 0 & 3 & -2 & 0 \\ 0 & 0 & 1 & 5 & 1 & 0 \\ 0 & 0 & 0 & 0 & 0 & 1 \end{bmatrix}$$

The corresponding system of equations is:

$$x_1 = -2\cdot x_2 - 3\cdot x_4 + 2\cdot x_5$$
$$x_2 = x_2$$
$$x_3 = -5\cdot x_4 - x_5$$
$$x_4 = x_4$$
$$x_5 = x_5$$
$$x_6 = 0$$

The free variables are x_2, x_4 and x_5 and the general solution is

$$x_2\cdot\begin{bmatrix} -2 \\ 1 \\ 0 \\ 0 \\ 0 \\ 0 \end{bmatrix} + x_4\cdot\begin{bmatrix} -3 \\ 0 \\ -5 \\ 1 \\ 0 \\ 0 \end{bmatrix} + x_5\cdot\begin{bmatrix} 2 \\ 0 \\ -1 \\ 0 \\ 1 \\ 0 \end{bmatrix} = \begin{bmatrix} -2\cdot x_2 - 3\cdot x_4 + 2\cdot x_5 \\ x_2 \\ -5\cdot x_4 - x_5 \\ x_4 \\ x_5 \\ 0 \end{bmatrix}$$

Are the vectors $\begin{bmatrix} -2 \\ 1 \\ 0 \\ 0 \\ 0 \\ 0 \end{bmatrix}$ $\begin{bmatrix} -3 \\ 0 \\ -5 \\ 1 \\ 0 \\ 0 \end{bmatrix}$ and $\begin{bmatrix} 2 \\ 0 \\ -1 \\ 0 \\ 1 \\ 0 \end{bmatrix}$ linearly dependent?

If so there must be nonzero a, b, and c such that

$$a \cdot \begin{bmatrix} -2 \\ 1 \\ 0 \\ 0 \\ 0 \\ 0 \end{bmatrix} + b \cdot \begin{bmatrix} -3 \\ 0 \\ -5 \\ 1 \\ 0 \\ 0 \end{bmatrix} + c \cdot \begin{bmatrix} 2 \\ 0 \\ -1 \\ 0 \\ 1 \\ 0 \end{bmatrix} = \begin{bmatrix} 0 \\ 0 \\ 0 \\ 0 \\ 0 \\ 0 \end{bmatrix} \quad \text{or} \quad \begin{bmatrix} -2 \cdot a - 3 \cdot b + 2 \cdot c \\ a \\ -5 \cdot b - c \\ b \\ c \\ 0 \end{bmatrix} = \begin{bmatrix} 0 \\ 0 \\ 0 \\ 0 \\ 0 \\ 0 \end{bmatrix}$$

Explain why a = b = c = 0 and, hence, the vectors are linearly independent.
ANSWER:

Explain why the spanning set that comes from the RREF is always a basis.
ANSWER:

What is the dimension of the solution space of the linear system given above?
ANSWER:

Summary: To find a basis for the solution space of A·x = 0, compute the RREF of [A | 0] and determine the general solution. Express the general solution as a linear combination of the vectors vj that 'come' from the RREF. The vj form a basis for the solution space.

Explain how to compute the dimension of the solution space of $\mathbf{A} \cdot \mathbf{x} = \mathbf{0}$.
ANSWER:

For any matrix, the subspace spanned by the rows of the matrix is called the **ROW SPACE** of the matrix. If the matrix has m rows and n columns, the row space is a subspace of $\mathbf{R}^n$.

Row-reduction is a process that eliminates linearly dependent rows.

As a result, the nonzero rows in the REF or RREF of the matrix are a basis for the row space of the matrix. The number of such rows is the dimension of the row space.

The **dimension of the row space** of the matrix is called the **RANK** of the matrix.

The nonzero rows remaining in the REF or RREF of the matrix span the row space since the original rows can be reconstructed from the remaining rows. Why are they linearly independent?
ANSWER:

How is the dimension of the row space of a matrix **A** (i.e., the rank of the matrix) related to the number of leading 1's in the RREF of **A**?
ANSWER:

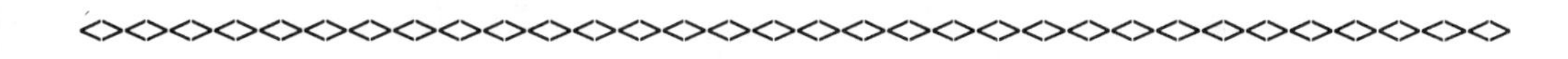

Let **A** be an m by n matrix (m rows and n columns) and consider the homogeneous system of linear equations $\mathbf{A}\cdot\mathbf{x} = \mathbf{0}$. Let Row(**A**) be the row space of **A** and Solution(**A**) be the solution space of the homogeneous system then

$$\text{DIM (Row (A)) + DIM (Solution(A)) = n}$$

or

$$\text{Rank(A) + DIM(Solution(A)) = n}$$

Explain why this is true. (Hint: Think about the dimensions in terms of the number of variables they determine.)

ANSWER:

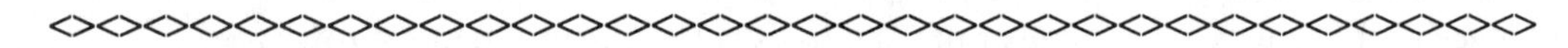

Exercises

For each of the following matrices compute (1) a basis for the row space and (2) a basis for the solution space.

a) $\begin{pmatrix} 1 & -4 & 0 & 1 & 7 & 4 \\ 2 & 5 & 3 & 1 & 1 & -7 \\ 3 & 2 & 5 & 1 & 2 & 0 \\ 0 & -1 & -2 & 1 & 6 & -3 \end{pmatrix}$

ANSWER:

b) $\begin{pmatrix} 1 & 8 & -9 & 0 \end{pmatrix}$

ANSWER:

c) $\begin{pmatrix} 1 & 0 & 3 & 2 \\ 2 & 0 & 7 & 5 \end{pmatrix}$

ANSWER:

For any matrix, **A**, the subspace spanned by the columns of **A** is called the **COLUMN SPACE** of **A** and is denoted by Col(**A**). If the matrix has m rows and n columns, the column space is a subspace of $\mathbf{R}^m$. Explain in terms of **row operations** how to compute a basis for the column space and therefore how to compute the dimension of the column space.

ANSWER:

Let $\quad \mathbf{A} := \begin{pmatrix} 0 & 1 & -2 & 3 & 2 \\ -9 & -6 & 13 & 8 & 6 \\ 2 & 1 & 5 & 0 & 3 \\ 3 & 2 & 0 & -1 & 1 \end{pmatrix}$

Compute a basis for the column space of **A**.

ANSWER:

Compute a basis for the row space of **A**.

ANSWER:

Compare the dimension of the column space with the dimension of the row space.

ANSWER:

Exercise: (Individual) Each of the following terms will be used throughout the remainder of the course. Explain in your own words what each one means.

a) subspace of $\mathbf{R}^n$

b) spanning set

c) linear independence

d) solution set of a system of linear equations

e) basis of a subspace

f) dimension of a subspace.

Write a summary of the MATHEMATICS that you learned in this section.

Section 2-8 Inverses

Name:________________

Score:_____________

Overview

In Section 2-2 you saw that each row operation is reversible by applying another row operation of the same type. Since a row operation is equivalent to multiplying (on the left) by a corresponding elementary matrix, every elementary matrix **M** is reversible in the sense that there exists another elementary matrix **N** of the same kind so that $\mathbf{M \cdot N = N \cdot M = I}$. Matrices **M** and **N** with this property are called ***inverses*** of one another. In this section we study square matrices and seek ways to determine if a matrix has an inverse.

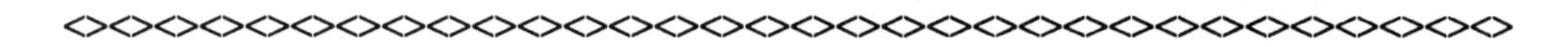

Let **A** be an n by n matrix. A matrix **B** is called an ***inverse*** to **A** if

$$\mathbf{B \cdot A = I} \qquad \text{and} \qquad \mathbf{A \cdot B = I}$$

where **I** is the (n by n) identity matrix. If there is such a matrix, we write: $\mathbf{B = A^{-1}}$ and say that **A** is **INVERTIBLE.**

A matrix, **A**, is called **square** if the number of rows of **A** equals the number of columns of **A**. To be invertible a matrix must be square; but not all square matrices are invertible.

Let **A** be a square matrix and assume the elements of the jth row are all zero. If **A** has an inverse **B** then the jth row of **A** times the jth column of **B** must be $\mathbf{I}_{j,j} = 1$. Explain why this is impossible.

ANSWER:

Let $\mathbf{A \cdot x = b}$ be a system of n linear equations in n unknowns. If **A** is invertible, $\mathbf{A^{-1} \cdot b}$ is a solution to the system of equations. Explain why.

ANSWER:

Let **c** be a solution to $\mathbf{A \cdot x = b}$, Must $\mathbf{c = A^{-1} \cdot b}$? (Hint: $\mathbf{A^{-1} \cdot (A \cdot c) = (A^{-1} \cdot A) \cdot c}$)

ANSWER:

If the coefficient matrix **A** is invertible, what is the solution space of $\mathbf{A}\cdot\mathbf{x} = \mathbf{0}$? What is the dimension of this solution space?

ANSWER:

If the matrix **A** has an inverse, what is the dimension of the row space of **A**?

ANSWER:

Let **A** be an invertible n by n matrix. What is the RREF of **A**?

ANSWER:

Summary (complete the following)

If **A** is invertible, then **RREF(A)** = ______.

If **A** is invertible, then $\mathbf{A}\cdot\mathbf{x} = \mathbf{b}$ has a _________ solution.

If **A** is invertible, then $\mathbf{A}\cdot\mathbf{x} = \mathbf{0}$ has only the ________ solution.

The preceding statements provide ways to recognize matrices with inverses in terms of the RREF and solution sets. The converse of each of the Summary statements above is also true. (If ***p*** and ***q*** are statements, the converse of ***if p then q*** is ***if q then p***.) Write the converse of each of the preceding statements below. (Warning: The converse of a proposition is not always true.)

ANSWER:

Show that the inverse, if it exists, is unique. This means that if **B** and **C** are both inverses of **A** then **B** must equal **C**. Hint: Expand the left and right sides of

$$\mathbf{B}\cdot(\mathbf{A}\cdot\mathbf{C}) = (\mathbf{B}\cdot\mathbf{A})\cdot\mathbf{C}$$

ANSWER:

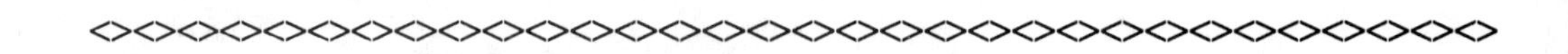

An invertible matrix is also called **nonsingular**. A square matrix that does **not** have an inverse is called **non-invertible** or **singular**.

Explain why each of the following matrices is singular. (Hint: Use the statements completed above.)

$$\begin{pmatrix} 0 & 0 & 0 \\ 0 & 0 & 0 \\ 0 & 0 & 0 \end{pmatrix} \qquad \begin{pmatrix} 1 & 2 \\ 0 & 0 \end{pmatrix} \qquad \begin{pmatrix} 1 & 0 & 3 & -4 \\ 2 & 0 & 1 & 0 \\ 1 & 0 & 2 & -2 \\ 1 & 0 & -1 & 5 \end{pmatrix}$$

ANSWER: ***ANSWER:*** ***ANSWER:***

<><><><><><><><><><><><><><><><><><><><><><><><><><><><><>

As we saw above, it is sometimes easy to determine if a matrix **A** has an inverse. That still leaves the question of computing the inverse. One way to proceed is by *brute force* as illustrated below:

Let $\mathbf{A} := \begin{pmatrix} 1 & 2 \\ 3 & 5 \end{pmatrix}$ Assume that $\mathbf{A}^{-1} = \begin{pmatrix} w & y \\ x & z \end{pmatrix}$

Then $$\begin{pmatrix} 1 & 2 \\ 3 & 5 \end{pmatrix} \cdot \begin{pmatrix} w & y \\ x & z \end{pmatrix} = \begin{pmatrix} 1 & 0 \\ 0 & 1 \end{pmatrix}$$

Evaluate the product symbolically. From the equality derive four equations in four unknowns (one for each matrix entry). Solve this using rref.

ANSWER:

Verify that your answer is $\mathbf{A}^{-1}$ by multiplying your answer times **A**.

ANSWER:

This method of finding the inverse will work in general although as you can tell the number of variables increases as n^2 grows and it rapidly becomes much too cumbersome. We need to develop a more efficient computational procedure for finding the inverse of a matrix. The word **algorithm** is often used in place of *computational procedure*.

To determine if **A** is nonsingular compute its RREF.

A is nonsingular if and only if RREF**(A)** = **I**.

To compute RREF(**A**) from **A** we can apply row operations or equivalently multiply on the left by elementary matrices. This process results in a matrix equation of the form $\mathbf{E}_k \cdot \mathbf{E}_{k-1} \cdot \ldots \cdot \mathbf{E}_2 \cdot \mathbf{E}_1 \cdot \mathbf{A} = \mathbf{I}$ where each $\mathbf{E}_j$ is an elementary matrix.

Let $\mathbf{Q} = \mathbf{E}_k \cdot \mathbf{E}_{k-1} \cdot \ldots \cdot \mathbf{E}_2 \cdot \mathbf{E}_1$. What is $\mathbf{Q} \cdot \mathbf{A}$? How is **Q** related to **A**?

ANSWER:

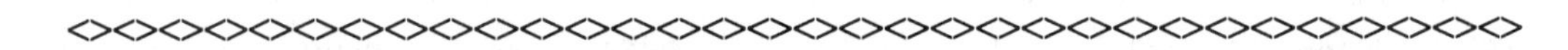

Computation of A^{-1}

As you saw above, $\mathbf{A}^{-1}$ is the product of the elementary matrices needed to reduce **A** to its RREF. Since $\mathbf{A}^{-1} = \mathbf{A}^{-1} \cdot \mathbf{I} = \mathbf{E}_k \cdot \mathbf{E}_{k-1} \cdot \ldots \cdot \mathbf{E}_2 \cdot \mathbf{E}_1 \cdot \mathbf{I}$ the inverse can be computed by applying the row operations to the identity matrix. From an accounting point of view this is done most efficiently by forming a new matrix [**A** | **I**] (This means forming the n by 2n matrix whose first n columns are **A** and whose last n columns are **I**.) and operating on the new matrix by the operations needed to reduce **A** to reduced row echelon form.

Complete the following matrix equation below.

ANSWER:

$$\mathbf{E}_k \cdot \mathbf{E}_{k-1} \cdot \ldots \cdot \mathbf{E}_2 \cdot \mathbf{E}_1 \cdot [\mathbf{A} \mid \mathbf{I}] = [\quad \mid \quad]$$

Another way to look at this computation is as follows: You know that to compute the solution to $\mathbf{A} \cdot \mathbf{x} = \mathbf{b1}$ you form the augmented matrix [**A** | **b1**] and compute its RREF. Similarly to compute the solution to $\mathbf{A} \cdot \mathbf{x} = \mathbf{b2}$ you form the augmented matrix [**A** | **b2**] and compute its RREF. You don't have to solve each of these equations separately. We can find both solutions at once by forming the augmented matrix [**A** | **b1 b2**]. Obviously this procedure can be extended to any number of vectors.

Suppose you solved the n equations: $\mathbf{A}\cdot\mathbf{x} = \mathbf{ej}$ for $j = 1, ..., n$ where **ej** is the n-vector all of whose components except the jth are zero and the jth is one. Suppose the solution to the jth equation is the vector **wj** (i.e., $\mathbf{A}\cdot\mathbf{wj} = \mathbf{ej}$ for $j = 1, ..., n$). Express $\mathbf{A}^{-1}$ in terms of the vectors **wj**.
ANSWER:

What augmented matrix would you set up to calculate the **wj**?
ANSWER:

Summary:
To determine if a matrix **A** is nonsingular and compute its inverse find the RREF of **[A | I]**. If the result is **[I | Q]** then **A** is nonsingular and $\mathbf{Q} = \mathbf{A}^{-1}$.

Try this computation for the 2 by 2 matrix

$$\mathbf{A} := \begin{pmatrix} 1 & 2 \\ 3 & 5 \end{pmatrix}$$ whose inverse we computed above.

Compute the RREF of $\begin{pmatrix} 1 & 2 & 1 & 0 \\ 3 & 5 & 0 & 1 \end{pmatrix}$

What is $\mathbf{A}^{-1}$? Is this the same as the computation above?
ANSWER:

Repeat this computation for the symbolic matrix: $\mathbf{A} := \begin{pmatrix} a & b \\ c & d \end{pmatrix}$

Form the augmented matrix [**A** | **I**] $\begin{pmatrix} a & b & 1 & 0 \\ c & d & 0 & 1 \end{pmatrix}$

and determine its RREF by using elementary matrices and symbolic operations. Keep track of any assumptions made during the computation.
ANSWER:

Check your answer symbolically by multiplying the answer by **A** on the left and then on the right.

ANSWER:

Compute $\begin{pmatrix} a & b \\ c & d \end{pmatrix}^{-1}$ by boxing the expression on the left and pressing Shift F9.

ANSWER:

How would the computation change if a = 0? if c = 0? if both a = 0 and c = 0?

ANSWER:

What would happen if ad = bc?

ANSWER:

What is the dimension of the row space of **A** when ad = bc? What does this imply about the matrix **A**?

ANSWER:

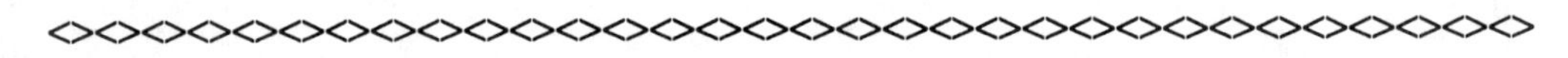

Summary

For the 2 by 2 matrix $\mathbf{A} := \begin{pmatrix} a & b \\ c & d \end{pmatrix}$

if ad - bc is not zero then we can compute the inverse by

a) interchanging elements on the main diagonal,
b) multiplying non-diagonal elements by -1,
c) dividing all elements by ad - bc.

We illustrate this for the matrix **A**.

$$A := \begin{pmatrix} 3 & -5 \\ 6 & 0 \end{pmatrix} \qquad B := \frac{1}{30}\cdot\begin{pmatrix} 0 & 5 \\ -6 & 3 \end{pmatrix} \qquad A\cdot B = \begin{pmatrix} 1 & 0 \\ 0 & 1 \end{pmatrix}$$

Use the above procedure to find the inverse for each of the following matrices, if possible.

$$\begin{pmatrix} 1 & 5 \\ 3 & 7 \end{pmatrix}\cdot\begin{pmatrix} \blacksquare & \blacksquare \\ \blacksquare & \blacksquare \end{pmatrix} = \qquad \begin{pmatrix} 2 & -1 \\ 4 & -2 \end{pmatrix}\cdot\begin{pmatrix} \blacksquare & \blacksquare \\ \blacksquare & \blacksquare \end{pmatrix} =$$

$$\begin{pmatrix} 0 & -3 \\ 1 & 5 \end{pmatrix}\cdot\begin{pmatrix} \blacksquare & \blacksquare \\ \blacksquare & \blacksquare \end{pmatrix} = \qquad \begin{pmatrix} 4 & 9 \\ 7 & 2 \end{pmatrix}\cdot\begin{pmatrix} \blacksquare & \blacksquare \\ \blacksquare & \blacksquare \end{pmatrix} =$$

There are similar formulas for matrices of other dimensions but they are much too complicated to use effectively. To see this, compute the following symbolically.

$$\begin{pmatrix} a & b & c & d \\ e & f & g & h \\ i & j & k & l \\ m & n & o & p \end{pmatrix}$$

Box and then choose **Invert Matrix** from the **Symbolic** menu.

Scroll horizontally to look at the result. Then box it and cut it.

Example: Best Fit Line

Suppose that an experiment is run that computes some dependent variable, y, in terms of an independent variable x. Suppose further that the experiment produces ten pairs of data. These are given below as the vectors **x** and **y**.

$$x := (-3\ \ 2\ \ 4\ \ 5\ \ 6\ \ 9\ \ 10\ \ 10.5\ \ 12\ \ 14\ )^T \qquad j := 1..10$$

$$y := (12\ \ 9\ \ 3\ \ 2\ \ 0\ \ 1\ \ 11\ \ 15\ \ 17\ \ 25\ )^T$$

Press F9 to see the plot of these points.

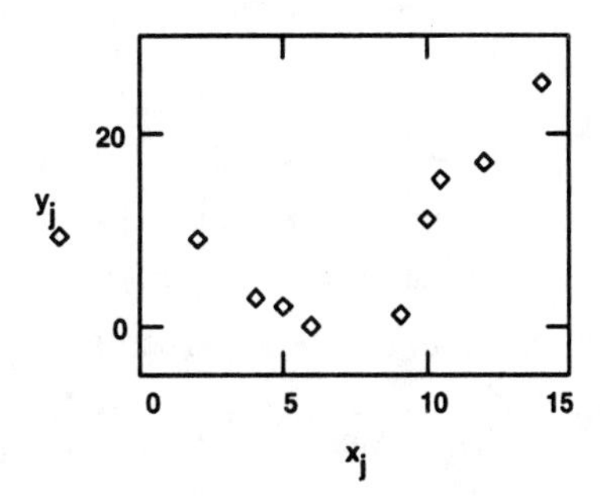

What is the line that provides the ***best fit*** for these data? Experiment by changing the values of m and b and guess which values you think provide the best fit.

$t := 0..15$

$m := 2 \qquad b := -4$

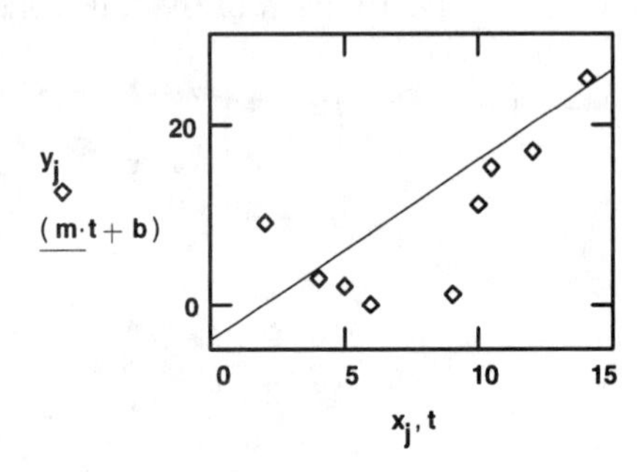

What does ***best fit*** mean? You shall discover later in this text that there are several possible meanings of the phrase. What is meant here is that the sum of the squares of the distance from the point ***predicted*** by the best fit line and the data point itself is a minimum. In the picture to the right this means that the sum of the squares of the lengths of the vertical lines is a minimum.

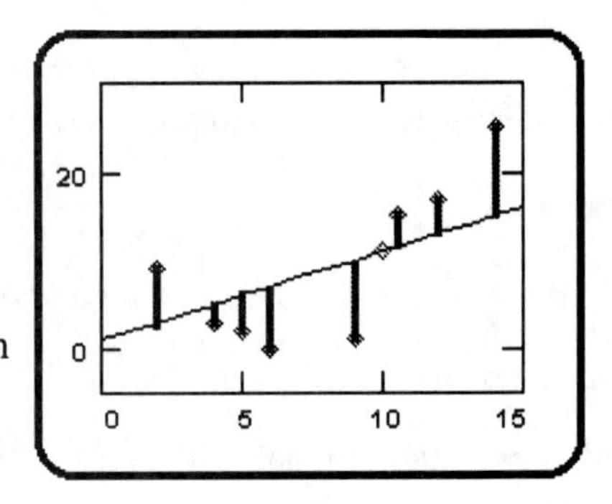

The length of the ith line is $|y_i - (mx_i + b)|$. The sum of the squares can be expressed in terms of the dot product as

$$E(m, b) := (y - (m \cdot x + b)) \cdot (y - (m \cdot x + b))$$

Calculus tells us that to minimize this quantity as a function of m and b we set

$$\frac{\partial E}{\partial m} = 0 \quad \text{and} \quad \frac{\partial E}{\partial b} = 0$$

(If the partial derivative symbols are not familiar to you don't worry. The partial of E with respect to m simply means take the derivative with respect to m and treat b as a constant. The partial derivative with respect to b is similarly defined.)

The fact that Maple will not substitute makes it complicated to take these partial derivatives in the general case. As a result we solve this problem for the particular values of the vectors **x** and **y**. We leave it to you to solve the general case.

The quantity to be minimized is:

$$\left[\begin{bmatrix}12\\9\\3\\2\\0\\1\\11\\15\\17\\25\end{bmatrix} - \left[m\cdot\begin{bmatrix}-3\\2\\4\\5\\6\\9\\10\\10.5\\12\\14\end{bmatrix} + b\right]\right]\cdot\left[\begin{bmatrix}12\\9\\3\\2\\0\\1\\11\\15\\17\\25\end{bmatrix} - \left[m\cdot\begin{bmatrix}-3\\2\\4\\5\\6\\9\\10\\10.5\\12\\14\end{bmatrix} + b\right]\right]$$

Use **Simplify** from the **Symbolic** menu to evaluate this.

In the the simplified expression place the cursor next to m and choose **Differentiate on Variable** from the **Symbolic** Menu. Then take the derivative of the original expression with respect to b. If you did this correctly you should have two expressions, each of which is linear with respect to both m and b. Set these two expressions equal to zero. You now have two equations in two unknowns.

Solve for m and b **using the inverse of the coefficient matrix**. (Do not use **rref**.)

ANSWER:

Substitute the values of m and b that you computed in the placeholders below to see the best fit line.

$m := \blacksquare \qquad b := \blacksquare \qquad t := 0..15$

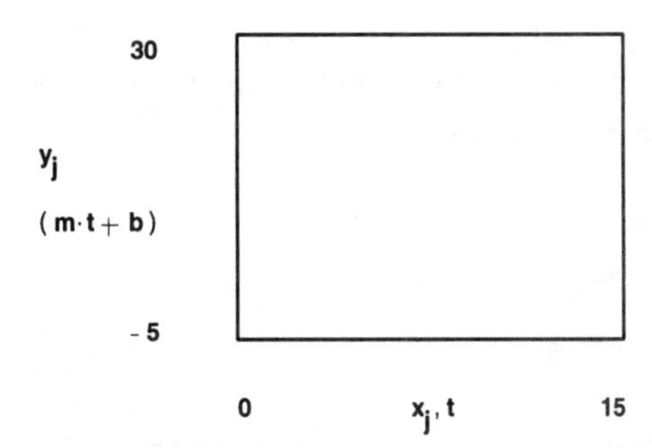

To measure the *goodness of fit* of the line you computed, calculate the correlation coefficient of **x** and **y**.

ANSWER:

<><><><><><><><><><><><><><><><><><><><><><><><><><><><><><>

Find the best fit line for

$\mathbf{x} := (-5\ \ -3\ \ 1\ \ 4\ \ 2\ \ 4\ \ 0)^T$ $\qquad$ $\mathbf{y} := (2\ \ 3\ \ 9\ \ 12\ \ 5\ \ 9\ \ 1)^T$

ANSWER:

<><><><><><><><><><><><><><><><><><><><><><><><><><><><><><>

Determine if the matrix $\mathbf{A} := \begin{pmatrix} 2 & 3 & 1 \\ 1 & 1 & 1 \\ 1 & 2 & 1 \end{pmatrix}$ is invertible. If it is, find the inverse.

Begin by setting $\mathbf{M} := \mathbf{augment}(\mathbf{A}, \mathbf{identity}(3))$

Display $\mathbf{M} = \begin{pmatrix} 2 & 3 & 1 & 1 & 0 & 0 \\ 1 & 1 & 1 & 0 & 1 & 0 \\ 1 & 2 & 1 & 0 & 0 & 1 \end{pmatrix}$

Reminder: We have provided an **rref** function for Mathcad as well as the row operations, **rswap, rmult**, and **rcomb**. Check with your instructor to make sure that these are installed on your system. You may use them in the same way you would use any other Mathcad function.

Reduce this to RREF either by using elementary matrices, the row operation functions, **rswap**, **rmult**, and **rcomb** or the **rref** function.

ANSWER:

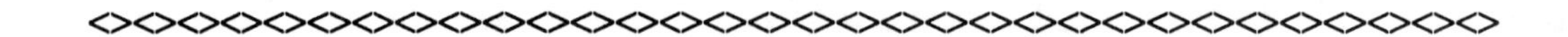

Determine if the matrix $\mathbf{A} := \begin{pmatrix} 1 & 2 & -1 \\ 2 & 0 & 2 \\ 1 & -2 & 3 \end{pmatrix}$ is invertible. If it is compute $\mathbf{A}^{-1}$.

ANSWER:

For the matrix **A** in the previous exercise, compute its inverse symbolically.

$$\begin{pmatrix} 1 & 2 & -1 \\ 2 & 0 & 2 \\ 1 & -2 & 3 \end{pmatrix}$$

<== Box and **Invert Matrix** from the **Symbolic** Menu.

ANSWER:

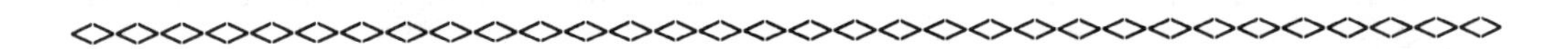

WARNING: If we used row operations in the preceding examples the computations are done numerically using what is called ***floating point arithmetic***. Floating point arithmetic uses a fixed number of digits in its operations, hence errors due to loss of accuracy can and will occur. Hence if matrix entries are displayed as decimals what we see is probably only a portion of the actual entry that is stored in memory.

This topic is discussed further in Section 2-9. Save your work and read Section 2-9 now.

The problem of numerical accuracy also occurs in Mathcad itself as illustrated by the following example.

$$\mathbf{C} := \begin{pmatrix} \frac{1}{3} & \frac{1}{5} \\ \frac{5}{7} & \frac{3}{7} \end{pmatrix} \qquad \mathbf{C}^{-1} = \begin{pmatrix} -1.158 \cdot 10^{16} & 5.404 \cdot 10^{15} \\ 1.93 \cdot 10^{16} & -9.007 \cdot 10^{15} \end{pmatrix}$$

$$\mathbf{C} \cdot \mathbf{C}^{-1} = \begin{pmatrix} 0.5 & 0 \\ 0 & 1 \end{pmatrix} \qquad \mathbf{C}^{-1} \cdot \mathbf{C} = \begin{pmatrix} 0.5 & 0 \\ 1 & 1 \end{pmatrix}$$

One can ask why Mathcad computes an incorrect answer. As we mentioned in Section 2-9, this is related to roundoff error. We saw earlier that the matrix

$\begin{pmatrix} a & c \\ b & d \end{pmatrix}$ will not have an inverse if ad = bc since in that case the rows are proportional and the dimension of the row space is one.

For the matrix **C**, let's compute ad - bc.

$x := \frac{1}{3}\cdot\frac{3}{7} - \frac{1}{5}\cdot\frac{5}{7}$ $\qquad x = 0$

From Numerical Format in the Math menu set the accuracy of x above to 15 places. You will see that it remains equal to zero. However if you compute 1/x below you will see that x is not really 0 and this results in the bizarre behavior evidenced above. This should serve as a good warning to always have an understanding of any computational package that you use.

Evaluate $\frac{1}{x}$

To restore your faith in computers, compute $\mathbf{C}^{-1}$ using the invert matrix function in Maple.

ANSWER:

<>\<>\<>\<>\<>\<>\<>\<>\<>\<>\<>\<>\<>\<>\<>\<>\<>\<>\<>\<>\<>\<>\<>\<>\<>\<>\<>\<>\<>\<>\<>\<>\<>\<>\<>\<>\<>

Inverses and Linear Systems: Matrix Equations

Let $\mathbf{A} := \begin{pmatrix} 1 & 3 & 2 & 6 \\ -5 & 1 & 5 & 0 \\ 0 & 9 & 1 & 3 \\ 2 & 11 & 0 & 5 \end{pmatrix}$ $\qquad \mathbf{b} := \begin{pmatrix} 1 \\ 0 \\ 0 \\ 2 \end{pmatrix}$

Evaluate symbolically $\begin{pmatrix} 1 & 3 & 2 & 6 \\ -5 & 1 & 5 & 0 \\ 0 & 9 & 1 & 3 \\ 2 & 11 & 0 & 5 \end{pmatrix}^{-1} \cdot \begin{pmatrix} 1 \\ 0 \\ 0 \\ 2 \end{pmatrix}$

Why is this vector the solution to the equation $\mathbf{A}\cdot\mathbf{x} = \mathbf{b}$? Are there any other solutions?

ANSWER:

Solve the linear system **A·x = b** where

$$\mathbf{A} := \begin{pmatrix} 1 & 3 & 2 & 1 \\ 9 & 4 & 4 & 4 \\ 10 & -1 & 1 & 3 \\ 1 & 0 & 1 & 0 \end{pmatrix} \qquad \text{and} \qquad \mathbf{b} := \begin{bmatrix} \frac{3}{2} \\ \frac{25}{2} \\ 13 \\ -\frac{1}{2} \end{bmatrix}$$

ANSWER:

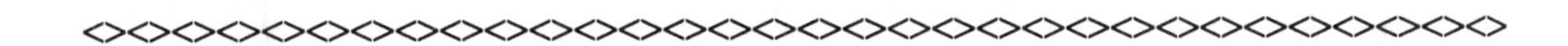

Recall the Leontief Model:

	Supplies		Energy		Constr		Transp		consumer demand
Energy	x	=	0.4 x	+	0.2 y	+	0.1 z	+	100
Construction	y	=	0.2 x	+	0.4 y	+	0.1 z	+	50
Transportation	z	=	0.15 x	+	0.2 y	+	0.2 z	+	100

These equations can be written in the form **V = A·V + c**.

Since **I·V = V** where **I** is the identity matrix this system can be rewritten as

$$\mathbf{I} \cdot \mathbf{V} = \mathbf{A} \cdot \mathbf{V} + \mathbf{c} \qquad \text{or} \qquad (\mathbf{I} - \mathbf{A}) \cdot \mathbf{V} = \mathbf{c}$$

Thus the solution is $\mathbf{V} = (\mathbf{I} - \mathbf{A})^{-1} \cdot \mathbf{c}$

Perform this calculation and verify that the solution is the same as the solution you computed in Section 2-4.
ANSWER:

The Leontief Input Constraint ensures that the matrix (**I** - **A**) is nonsingular.

<><><><><><><><><><><><><><><><><><><><><><><><><><><><><><><><>

We conclude this section with the statement of several important theoretical relationships.

> Let **A** be an n by n matrix.
>
> The following are **equivalent**:
>
> - **A** is **nonsingular**.
> - $\mathbf{A}\cdot\mathbf{x} = \mathbf{b}$ has a unique solution, $\mathbf{x} = \mathbf{A}^{-1}\cdot\mathbf{b}$.
> - $\mathbf{A}\cdot\mathbf{x} = \mathbf{0}$ has a unique solution, $\mathbf{x} = \mathbf{0}$.
>
> If the system $\mathbf{A}\cdot\mathbf{x} = \mathbf{b}$ or $\mathbf{A}\cdot\mathbf{x} = \mathbf{0}$ has infinitely many solutions then **A** is **singular**.
>
> If the system $\mathbf{A}\cdot\mathbf{x} = \mathbf{b}$ has no solution, then **A** is **singular**.

Exercises on Matrix Inverses

1. Find the inverse of the following matrix by computing the RREF of [**A** | **I**].

$$\mathbf{A} := \begin{pmatrix} 1 & -2 & 2 \\ -1 & 0 & 1 \\ 3 & 4 & -1 \end{pmatrix}$$

Display the inverse of A here. ==>

<><><><><><><><><><><><><><><><><><><><><><><><><><><><><><><><><><><><><><><><>

2. Solve each of the following linear systems $\mathbf{A}\cdot\mathbf{x} = \mathbf{b}$ by computing $\mathbf{x} = \mathbf{A}^{-1}\cdot\mathbf{b}$ numerically and symbolically (Use the built-in Mathcad and Maple functions. Do not compute the RREF.) Check your solution **x** by computing $\mathbf{A}\cdot\mathbf{x} - \mathbf{b}$.

(a) $$\mathbf{A} := \begin{pmatrix} 3 & -1 & 0 & 0 \\ -1 & 3 & -1 & 0 \\ 0 & -1 & 3 & -1 \\ 0 & 0 & -1 & 3 \end{pmatrix} \qquad \mathbf{b} := \begin{pmatrix} 0 \\ 7 \\ 2 \\ -7 \end{pmatrix}$$

ANSWER:

<><><><><><><><><><><><><><><><><><><><><><><><><><><><><><><><><><><><><><><><>

(b)

$$A := \begin{bmatrix} 1 & \frac{1}{2} & \frac{1}{3} & \frac{1}{4} & \frac{1}{5} \\ \frac{1}{2} & \frac{1}{3} & \frac{1}{4} & \frac{1}{5} & \frac{1}{6} \\ \frac{1}{3} & \frac{1}{4} & \frac{1}{5} & \frac{1}{6} & \frac{1}{7} \\ \frac{1}{4} & \frac{1}{5} & \frac{1}{6} & \frac{1}{7} & \frac{1}{8} \\ \frac{1}{5} & \frac{1}{6} & \frac{1}{7} & \frac{1}{8} & \frac{1}{9} \end{bmatrix} \qquad b := \begin{bmatrix} 1 \\ 0 \\ 0 \\ 0 \\ 0 \end{bmatrix}$$

Check your solution **x** by computing **A**·**x** - **b**.

ANSWER:

<><><><><><><><><><><><><><><><><><><><><><><><><><><><><><><><><>

3. Let **A** be the matrix in Exercise 2a and $b := \begin{pmatrix} 1 \\ 0 \\ 0 \\ 0 \end{pmatrix}$

Solve the matrix equation **A**·**z** = **z** + **b** for **z**. Show your steps and explain what you did.

<><><><><><><><><><><><><><><><><><><><><><><><><><><><><><><><><>

4. An inheritance of \$12,000 is divided into three parts, with the second part twice as large as the first part. The three parts are invested and earn interest at the rates of 9%, 10%, and 6% respectively. If the interest returned at the end of the first year is \$1,105 what were the three amounts invested?

Display the coefficient matrix, right-hand side, and solution to the system you create to serve as a model for this problem.

ANSWER:

<><><><><><><><><><><><><><><><><><><><><><><><><><><><><><><><><><>

5. (Individual) Write a summary of the MATHEMATICS that you learned in this section.

Section 2-9 Roundoff Error

As you compute in Mathcad you should be aware that if you row-reduce a matrix either using row operations or elementary matrices and you do this by using the precision displayed on the screen then there may be a significant loss of accuracy due to ***rounding off***. We illustrate this and suggest several ways to overcome this problem.

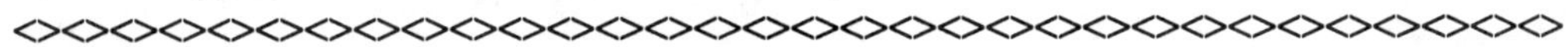

Example 1

Let $\quad \mathbf{A} := \begin{pmatrix} 7 & 5 & 7 \\ 3 & 4 & 3 \\ 2 & 1 & 4 \end{pmatrix}$

Compute $\mathbf{A}^{-1}$ by computing the RREF of $\quad \mathbf{M} := \begin{pmatrix} 7 & 5 & 7 & 1 & 0 & 0 \\ 3 & 4 & 3 & 0 & 1 & 0 \\ 2 & 1 & 4 & 0 & 0 & 1 \end{pmatrix}$

Step 1: Clean out the first column:

$$\mathbf{M1} := \begin{bmatrix} \frac{1}{7} & 0 & 0 \\ -\frac{3}{7} & 1 & 0 \\ -\frac{2}{7} & 0 & 1 \end{bmatrix} \cdot \mathbf{M} \qquad \mathbf{M1} = \begin{pmatrix} 1 & 0.714 & 1 & 0.143 & 0 & 0 \\ 0 & 1.857 & 0 & -0.429 & 1 & 0 \\ 0 & -0.429 & 2 & -0.286 & 0 & 1 \end{pmatrix}$$

Step 2: Clean out the second column:

$$\mathbf{M2} := \begin{bmatrix} 1 & -\frac{.714}{1.857} & 0 \\ 0 & \frac{1}{1.857} & 0 \\ 0 & \frac{.429}{1.857} & 1 \end{bmatrix} \cdot \mathbf{M1} \qquad \mathbf{M2} = \begin{pmatrix} 1 & 2.308 \cdot 10^{-4} & 1 & 0.308 & -0.384 & 0 \\ 0 & 1 & 0 & -0.231 & 0.539 & 0 \\ 0 & 4.616 \cdot 10^{-4} & 2 & -0.385 & 0.231 & 1 \end{pmatrix}$$

Step 3: Clean out the third column:

$$\mathbf{M3} := \begin{bmatrix} 1 & 0 & -\frac{1}{2} \\ 0 & 1 & 0 \\ 0 & 0 & \frac{1}{2} \end{bmatrix} \cdot \mathbf{M2} \qquad \mathbf{M3} = \begin{pmatrix} 1 & 0 & 0 & 0.5 & -0.5 & -0.5 \\ 0 & 1 & 0 & -0.231 & 0.539 & 0 \\ 0 & 2.308 \cdot 10^{-4} & 1 & -0.192 & 0.116 & 0.5 \end{pmatrix}$$

Step 4: Select the last three columns of **M3** and call them **B**.

$$\mathbf{B} := \mathbf{M3}\cdot\begin{bmatrix} 0 & 0 & 0 \\ 0 & 0 & 0 \\ 0 & 0 & 0 \\ 1 & 0 & 0 \\ 0 & 1 & 0 \\ 0 & 0 & 1 \end{bmatrix} \qquad \mathbf{B} = \begin{pmatrix} 0.5 & -0.5 & -0.5 \\ -0.231 & 0.539 & 0 \\ -0.192 & 0.116 & 0.5 \end{pmatrix}$$

Check to see if **B** is $\mathbf{A}^{-1}$. We have displayed ten digits of accuracy.

$$\mathbf{A}\cdot\mathbf{B} = \begin{pmatrix} 0.999538426 & 0.0010770059 & 0 \\ -0.000230787 & 1.000538503 & 0 \\ -0.000230787 & 0.000538503 & 1 \end{pmatrix}$$

$$\mathbf{B}\cdot\mathbf{A} = \begin{pmatrix} 1 & 0 & 0 \\ 0 & 1.000076929 & 0 \\ 0 & 0.000230787 & 1 \end{pmatrix}$$

Obviously round off error has affected the calculation as we can see by comparing **B** with $\mathbf{A}^{-1}$.

$$\mathbf{B} = \begin{pmatrix} 0.5 & -0.5 & -0.5 \\ -0.2307869836 & 0.5385029618 & 0 \\ -0.1923609508 & 0.1155088853 & 0.5 \end{pmatrix}$$

$$\mathbf{A}^{-1} = \begin{pmatrix} 0.5 & -0.5 & -0.5 \\ -0.2307692308 & 0.5384615385 & 0 \\ -0.1923076923 & 0.1153846154 & 0.5 \end{pmatrix}$$

$$\mathbf{A}^{-1} - \mathbf{B} = \begin{pmatrix} 0 & 0 & 0 \\ 0.0000177528 & -0.0000414233 & 0 \\ 0.0000532585 & -0.0001242699 & 0 \end{pmatrix}$$

Note: To display 10 digits, box the quantity you wish to change, pull down the **Math** menu, select **Numerical Format** and change **Displayed Precision** and **Exponential Threshold** to 10.

<><><><><><><><><><><><><><><><><><><><><><><><><><><><><><><><><><><><><>

Example 2

You could try to use the full accuracy of Mathcad by using the appropriate matrix entries in the calculations.

Let $\mathbf{A} := \begin{pmatrix} 7 & 5 & 7 \\ 3 & 4 & 3 \\ 2 & 1 & 4 \end{pmatrix}$ Compute $\mathbf{A}^{-1}$ by computing the RREF of $\mathbf{M} := \begin{pmatrix} 7 & 5 & 7 & 1 & 0 & 0 \\ 3 & 4 & 3 & 0 & 1 & 0 \\ 2 & 1 & 4 & 0 & 0 & 1 \end{pmatrix}$

Step 1: Clean out the first column:

$$\mathbf{M1} := \begin{bmatrix} \frac{1}{M_{1,1}} & 0 & 0 \\ -\frac{M_{2,1}}{M_{1,1}} & 1 & 0 \\ -\frac{M_{3,1}}{M_{1,1}} & 0 & 1 \end{bmatrix} \cdot \mathbf{M} \qquad \mathbf{M1} = \begin{pmatrix} 1 & 0.714 & 1 & 0.143 & 0 & 0 \\ 0 & 1.857 & 0 & -0.429 & 1 & 0 \\ 0 & -0.429 & 2 & -0.286 & 0 & 1 \end{pmatrix}$$

Step 2: Clean out the second column:

$$\mathbf{M2} := \begin{bmatrix} 1 & -\frac{M1_{1,2}}{M1_{2,2}} & 0 \\ 0 & \frac{1}{M1_{2,2}} & 0 \\ 0 & \frac{-M1_{3,2}}{M1_{2,2}} & 1 \end{bmatrix} \cdot \mathbf{M1} \qquad \mathbf{M2} = \begin{pmatrix} 1 & 0 & 1 & 0.308 & -0.385 & 0 \\ 0 & 1 & 0 & -0.231 & 0.538 & 0 \\ 0 & 0 & 2 & -0.385 & 0.231 & 1 \end{pmatrix}$$

Step 3: Clean out the third column:

$$\mathbf{M3} := \begin{bmatrix} 1 & 0 & \frac{-M2_{1,3}}{M2_{3,3}} \\ 0 & 1 & 0 \\ 0 & 0 & \frac{1}{M2_{3,3}} \end{bmatrix} \cdot \mathbf{M2} \qquad \mathbf{M3} = \begin{pmatrix} 1 & 0 & 0 & 0.5 & -0.5 & -0.5 \\ 0 & 1 & 0 & -0.231 & 0.538 & 0 \\ 0 & 0 & 1 & -0.192 & 0.115 & 0.5 \end{pmatrix}$$

Step 4: Select the last three columns of **M3** and call them **B**.

$$B := M3 \cdot \begin{bmatrix} 0 & 0 & 0 \\ 0 & 0 & 0 \\ 0 & 0 & 0 \\ 1 & 0 & 0 \\ 0 & 1 & 0 \\ 0 & 0 & 1 \end{bmatrix} \qquad B = \begin{pmatrix} 0.5 & -0.5 & -0.5 \\ -0.231 & 0.538 & 0 \\ -0.192 & 0.115 & 0.5 \end{pmatrix}$$

Check to see if **B** is $\mathbf{A}^{-1}$. We have displayed ten digits of accuracy.

$$A \cdot B = \begin{pmatrix} 1 & 0 & 0 \\ 0 & 1 & 0 \\ 0 & 0 & 1 \end{pmatrix}$$

$$B \cdot A = \begin{pmatrix} 1 & 0 & 0 \\ 0 & 1 & 0 \\ 0 & 0 & 1 \end{pmatrix}$$

$$A^{-1} - B = \begin{pmatrix} 0 & 0 & 0 \\ 0 & 0 & 0 \\ 0 & 0 & 0 \end{pmatrix}$$

This solved the problem but involved a bit more typing.

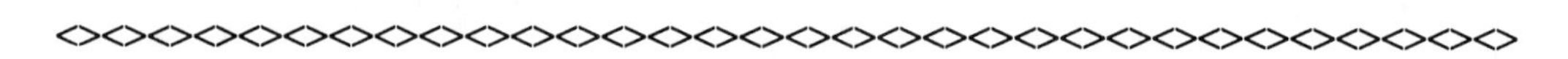

Example 3

There is also another way to achieve the same goal and that is to use the symbolic processor. The symbolic processor gives us exact answers that we can use.

Let $\quad A := \begin{pmatrix} 7 & 5 & 7 \\ 3 & 4 & 3 \\ 2 & 1 & 4 \end{pmatrix}$

Compute $\mathbf{A}^{-1}$ by finding the RREF of $\quad M := \begin{pmatrix} 7 & 5 & 7 & 1 & 0 & 0 \\ 3 & 4 & 3 & 0 & 1 & 0 \\ 2 & 1 & 4 & 0 & 0 & 1 \end{pmatrix}$

Step 1: Clean out the first column by evaluating symbolically.

$$\begin{bmatrix} \frac{1}{7} & 0 & 0 \\ -\frac{3}{7} & 1 & 0 \\ -\frac{2}{7} & 0 & 1 \end{bmatrix} \cdot \begin{pmatrix} 7 & 5 & 7 & 1 & 0 & 0 \\ 3 & 4 & 3 & 0 & 1 & 0 \\ 2 & 1 & 4 & 0 & 0 & 1 \end{pmatrix} = \begin{bmatrix} 1 & \frac{5}{7} & 1 & \frac{1}{7} & 0 & 0 \\ 0 & \frac{13}{7} & 0 & \frac{-3}{7} & 1 & 0 \\ 0 & \frac{-3}{7} & 2 & \frac{-2}{7} & 0 & 1 \end{bmatrix}$$

Step 2: Clean out the second column by evaluating symbolically.

$$\begin{bmatrix} 1 & -\frac{5}{13} & 0 \\ 0 & \frac{7}{13} & 0 \\ 0 & \frac{3}{13} & 1 \end{bmatrix} \cdot \begin{bmatrix} 1 & \frac{5}{7} & 1 & \frac{1}{7} & 0 & 0 \\ 0 & \frac{13}{7} & 0 & \frac{-3}{7} & 1 & 0 \\ 0 & \frac{-3}{7} & 2 & \frac{-2}{7} & 0 & 1 \end{bmatrix} = \begin{bmatrix} 1 & 0 & 1 & \frac{4}{13} & \frac{-5}{13} & 0 \\ 0 & 1 & 0 & \frac{-3}{13} & \frac{7}{13} & 0 \\ 0 & 0 & 2 & \frac{-5}{13} & \frac{3}{13} & 1 \end{bmatrix}$$

Step 3: Clean out the third column by evaluating symbolically.

$$\begin{bmatrix} 1 & 0 & -\frac{1}{2} \\ 0 & 1 & 0 \\ 0 & 0 & \frac{1}{2} \end{bmatrix} \cdot \begin{bmatrix} 1 & 0 & 1 & \frac{4}{13} & \frac{-5}{13} & 0 \\ 0 & 1 & 0 & \frac{-3}{13} & \frac{7}{13} & 0 \\ 0 & 0 & 2 & \frac{-5}{13} & \frac{3}{13} & 1 \end{bmatrix} = \begin{bmatrix} 1 & 0 & 0 & \frac{1}{2} & \frac{-1}{2} & \frac{-1}{2} \\ 0 & 1 & 0 & \frac{-3}{13} & \frac{7}{13} & 0 \\ 0 & 0 & 1 & \frac{-5}{26} & \frac{3}{26} & \frac{1}{2} \end{bmatrix}$$

Step 4: Delete the first three columns of the above result by using the Matrix dialog box and evaluate the product.

$$\begin{bmatrix} \frac{1}{2} & \frac{-1}{2} & \frac{-1}{2} \\ \frac{-3}{13} & \frac{7}{13} & 0 \\ \frac{-5}{26} & \frac{3}{26} & \frac{1}{2} \end{bmatrix} \cdot \begin{pmatrix} 7 & 5 & 7 \\ 3 & 4 & 3 \\ 2 & 1 & 4 \end{pmatrix} = \begin{pmatrix} 1 & 0 & 0 \\ 0 & 1 & 0 \\ 0 & 0 & 1 \end{pmatrix}$$

Example 4

$$\begin{pmatrix} 7 & 5 & 7 \\ 3 & 4 & 3 \\ 2 & 1 & 4 \end{pmatrix}$$

If we box the matrix on the left and choose invert matrix from the Symbolic Menu we get

$$\begin{bmatrix} \frac{1}{2} & \frac{-1}{2} & \frac{-1}{2} \\ \frac{-3}{13} & \frac{7}{13} & 0 \\ \frac{-5}{26} & \frac{3}{26} & \frac{1}{2} \end{bmatrix}$$

As you can see this is an exact answer, However if we set this matrix = **B**, Mathcad immediately transforms it to decimal (floating point) representation.

$$B := \begin{bmatrix} \frac{1}{2} & \frac{-1}{2} & \frac{-1}{2} \\ \frac{-3}{13} & \frac{7}{13} & 0 \\ \frac{-5}{26} & \frac{3}{26} & \frac{1}{2} \end{bmatrix} \qquad B = \begin{pmatrix} 0.5 & -0.5 & -0.5 \\ -0.231 & 0.538 & 0 \\ -0.192 & 0.115 & 0.5 \end{pmatrix}$$

The matrix **B** has much greater precision than is displayed as we can see by changing the Numerical format.

$$B = \begin{pmatrix} 0.5 & -0.5 & -0.5 \\ -0.230769230769231 & 0.538461538461538 & 0 \\ -0.192307692307692 & 0.115384615384615 & 0.5 \end{pmatrix}$$

If we **cut and paste** the numerical matrix above we lose all nondisplayed precision. To see this we do the following:

Click on the right matrix bracket and then click on the copy icon (or click on **Copy** in the **Edit** menu.

$$C := \begin{pmatrix} 0.5 & -0.5 & -0.5 \\ -0.231 & 0.538 & 0 \\ -0.192 & 0.115 & 0.5 \end{pmatrix}$$

Now paste into the right side of the expression above by boxing the right side and clicking on the paste icon.

$$\mathbf{C} = \begin{pmatrix} 0.5 & -0.5 & -0.5 \\ -0.231 & 0.538 & 0 \\ -0.192 & 0.115 & 0.5 \end{pmatrix}$$

If we increase the precision of **C** after pasting by clicking on **Numerical Format** on the **Math** menu, we still get.

$$\mathbf{C} = \begin{pmatrix} 0.5 & -0.5 & -0.5 \\ -0.231 & 0.538 & 0 \\ -0.192 & 0.115 & 0.5 \end{pmatrix}$$

Pasting or clicking and dragging are Windows operations.

What you get after these operations is what you see.

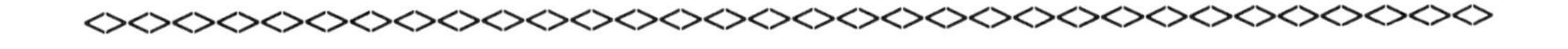

CHAPTER 3

DETERMINANTS AND THEIR APPLICATIONS

$$\left|\begin{pmatrix} 2 & 12 & 2 \\ -3 & 4 & -5 \\ 0 & 5 & 1 \end{pmatrix}\right| = 64$$

Section 3-1 The Determinant Function

Name:________________

Score:____________

Overview

The determinant function has as ***input*** a square matrix and as ***output*** a scalar. In this section you will investigate the properties of the determinant and its relationship to other concepts of linear algebra.

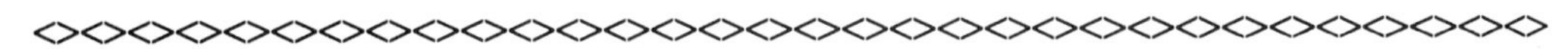

The determinant function is a mapping from the set of square matrices to the set of scalars. If the entries of the matrix are real, then the output scalar is real and if the entries are complex the output scalar is complex. Pictorially we think of the determinant function as

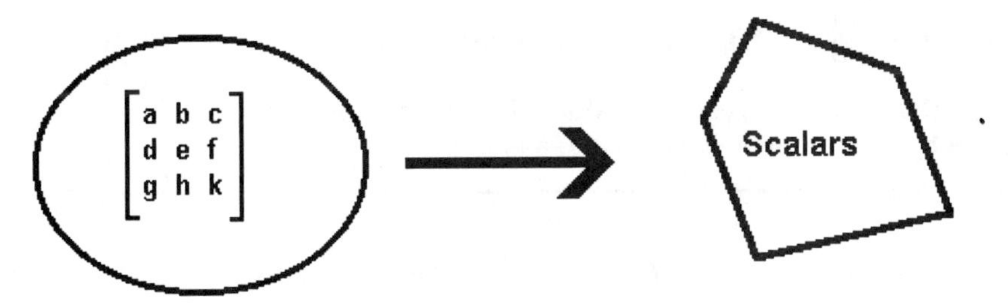

The notation for a determinant is similar to that for other functions we have discussed. More precisely, the absolute value function | | of Mathcad operates on several different types of variables.

On real numbers it is the usual absolute value function: $|x| = \sqrt{x^2}$

$|2| = 2$ $\quad |-3| = 3$ <== *Press F9.*

On vectors, | | is the length function: $|v| = \sqrt{v \cdot v}$

$u := \begin{pmatrix} 3 \\ -2 \\ 4 \end{pmatrix}$ $\quad |u| = 5.385$ <== *Press F9.*

Evaluate the following expression symbolically.

$\left| \begin{pmatrix} a \\ b \\ c \\ d \end{pmatrix} \right|$

Remember that for Mathcad vector means column vector so if this operation is applied to a row vector the answer might be different.

$\left|u^T\right| =$ *<== Press F9.*

The Mathcad 'absolute value' function also is defined on certain matrices. This function is called the ***determinant***. We investigate its properties.

$$M := \begin{pmatrix} 2 & 2 & 4 & -1 \\ 4 & 3 & 3 & 0 \\ -1 & 0 & 2 & 1 \end{pmatrix}$$

Calculate the determinant of **M**, |**M**|.

| | can be entered either from the icon on the calculator palette

(palette in Mathcad 5.0) or by using the | key.

The determinant can also be evaluated from the Symbolic menu by clicking on Determinant of a matrix.

When writing about determinants the symbols |M| and det(M) are used to denote the determinant of matrix M.

What do you conclude about the set of matrices for which the determinant function is defined?
ANSWER:

<><><><><><><><><><><><><><><><><><><><><><><><><><><><><><><><><><><><><><><>

Given matrices

$$B := \begin{pmatrix} 2 & -2 & 0 & 3 \\ 0 & 2 & -3 & 1 \\ 0 & 2 & -2 & 1 \\ 1 & -1 & 2 & 0 \end{pmatrix}$$

compute |**A**|, |**B**|, |**A** + **B**| and |**A**·**B**|. Display the results below.
ANSWER:

Even though we are using the | | symbol, the output from the determinant function can be negative.

Let $\quad \mathbf{C} := \begin{pmatrix} 2 & 1 & 1 \\ 0 & 0 & 3 \\ 1 & 2 & 4 \end{pmatrix} \qquad \mathbf{D} := \begin{pmatrix} 5 & 3 & 0 \\ 1 & 1 & -3 \\ -2 & 4 & 2 \end{pmatrix}$

Compute $|\mathbf{C}|$, $|\mathbf{D}|$, $|\mathbf{C} + \mathbf{D}|$ and $|\mathbf{C} \cdot \mathbf{D}|$. Display the results below.

ANSWER:

Try the same experiment for a pair of 2 by 2 and a pair of 5 by 5 matrices. Based on the observations from these experiments form a conjecture as a response to the following:

If **P** and **Q** are square matrices of the same size, then how are $|\mathbf{P}|$, $|\mathbf{Q}|$, and $|\mathbf{P} + \mathbf{Q}|$ related?
ANSWER:

If **P** and **Q** are square matrices of the same size, then how are $|\mathbf{P}|$, $|\mathbf{Q}|$, and $|\mathbf{P} \cdot \mathbf{Q}|$ related?
ANSWER:

If **P** and **Q** are square matrices of the same size, carry out experiments to form a conjecture about the relationship between $|\mathbf{P} \cdot \mathbf{Q}|$ and $|\mathbf{Q} \cdot \mathbf{P}|$.

ANSWER:

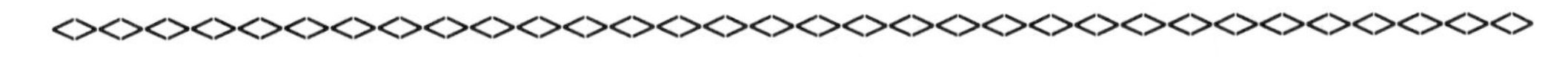

Compute $|\mathbf{P}^T|$ and $|\mathbf{P}|$ for some test matrices **P**. How are $|\mathbf{P}^T|$ and $|\mathbf{P}|$ related?
ANSWER:

<><><><><><><><><><><><><><><><><><><><><><><><><><><><><><><>

Let **I** be the n by n identity matrix. What can you conclude about the determinant of **I** from the fact that **I·A** = **A** for all n by n matrices **A**? (Hint: Use the result for the determinant of a product of matrices developed earlier.)

ANSWER:

Suppose that matrix **P** has an inverse. How is the determinant of $|\mathbf{P}^{-1}|$ related to $|\mathbf{P}|$? (Hint: What is $|\mathbf{P}\cdot\mathbf{P}^{-1}|$? Then use the result for the determinant of a product of matrices developed earlier.)

ANSWER:

What can you conclude if **det(P)** = 0?

ANSWER:

The determinant of a diagonal matrix

Compute the determinant of each of the following.

$$I := \begin{pmatrix} 1 & 0 & 0 & 0 \\ 0 & 1 & 0 & 0 \\ 0 & 0 & 1 & 0 \\ 0 & 0 & 0 & 1 \end{pmatrix} \qquad D1 := \begin{pmatrix} 3 & 0 & 0 & 0 \\ 0 & -1 & 0 & 0 \\ 0 & 0 & 2 & 0 \\ 0 & 0 & 0 & 1 \end{pmatrix} \qquad D2 := \begin{pmatrix} 2 & 0 & 0 \\ 0 & 7 & 0 \\ 0 & 0 & 3 \end{pmatrix}$$

det(I) = ______ **det(D1)** = ______ **det(D2)** = ______

Use the **Symbolic** menu to compute the determinant of each of the following.

$$\begin{bmatrix} a & 0 & 0 & 0 & 0 \\ 0 & b & 0 & 0 & 0 \\ 0 & 0 & c & 0 & 0 \\ 0 & 0 & 0 & d & 0 \\ 0 & 0 & 0 & 0 & e \end{bmatrix} \qquad \begin{pmatrix} a & 0 & 0 & 0 \\ 0 & b & 0 & 0 \\ 0 & 0 & c & 0 \\ 0 & 0 & 0 & d \end{pmatrix} \qquad \begin{pmatrix} a & 0 & 0 \\ 0 & b & 0 \\ 0 & 0 & c \end{pmatrix} \qquad \begin{pmatrix} a & 0 \\ 0 & b \end{pmatrix}$$

Complete the following conjecture:

The determinant of a diagonal matrix is ______________________________.

Check your conjecture on the following matrix.

$$\begin{pmatrix} 3 & 0 & 0 & 0 \\ 0 & w & 0 & 0 \\ 0 & 0 & -4 & 0 \\ 0 & 0 & 0 & z \end{pmatrix}$$

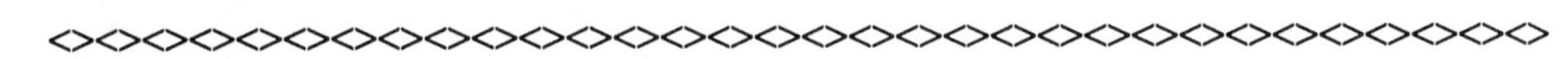

The determinant of a triangular matrix

Compute the determinant of each of the following matrices and record your answer below the matrix.

$$\begin{pmatrix} 1 & 0 & 0 \\ 2 & 6 & 0 \\ -3 & 5 & -3 \end{pmatrix} \qquad \begin{pmatrix} a & 0 & 0 \\ b & d & 0 \\ c & e & f \end{pmatrix} \qquad \begin{pmatrix} a & 9 & 0 \\ 0 & 7 & e \\ 0 & 0 & f \end{pmatrix} \qquad \begin{pmatrix} a & b & c & g \\ 0 & d & e & h \\ 0 & 0 & f & i \\ 0 & 0 & 0 & j \end{pmatrix}$$

Complete the following conjecture:

The determinant of a triangular matrix is ______________________________.

The determinant of a 2 by 2 matrix

From the above, if the value of the (1,1) entry, a, is not zero, we know the determinant of two of the three matrices below. Use the fact that $|\mathbf{P}\cdot\mathbf{Q}| = |\mathbf{P}|\cdot|\mathbf{Q}|$ to compute the determinant of the matrix:

$$\begin{pmatrix} a & b \\ c & d \end{pmatrix}$$

$$\begin{pmatrix} \frac{1}{a} & 0 \\ -\frac{c}{a} & 1 \end{pmatrix} \cdot \begin{pmatrix} a & b \\ c & d \end{pmatrix} = \begin{pmatrix} 1 & \frac{1}{a}\cdot b \\ 0 & \frac{-c}{a}\cdot b + d \end{pmatrix}$$

ANSWER: The determinant of first matrix on the left is _____ .

The determinant of matrix on the right is ______ .

What is the determinant of the matrix $\begin{pmatrix} a & b \\ c & d \end{pmatrix}$? ________

Check your answer to the previous problem by using the **Determinant** function on the **Symbolic** menu.

You should have concluded that ***the determinant of a 2 by 2 matrix is the product of the diagonal entries minus the product of the backward diagonal entries.***

The following diagram shows the *2 by 2 trick* for computing the determinant of a 2 by 2 matrix.

$$\det(A) = \det\begin{pmatrix} a & c \\ b & d \end{pmatrix} = ad - bc$$

- b·c

+a·d

If $\mathbf{A} := \begin{pmatrix} 5 & -4 \\ 3 & 8 \end{pmatrix}$ then **det(A)** = (5)(8) - (3)(-4) = 52.

<><><><><><><><><><><><><><><><><><><><><><><><><><><><><><><><><><><><>

The determinant of a 3 by 3 matrix

The determinant of a 3 by 3 matrix is more complicated. It is computed as a linear combination of 6 products of three entries each with the entries selected in a particular fashion. Let **A** be the following matrix

$$\mathbf{A} = \begin{pmatrix} a_{1,1} & a_{1,2} & a_{1,3} \\ a_{2,1} & a_{2,2} & a_{2,3} \\ a_{3,1} & a_{3,2} & a_{3,3} \end{pmatrix}$$

Compute the determinant of **A** by boxing the expression below and pressing Shift F9.

$$\left|\begin{pmatrix} a_{1,1} & a_{1,2} & a_{1,3} \\ a_{2,1} & a_{2,2} & a_{2,3} \\ a_{3,1} & a_{3,2} & a_{3,3} \end{pmatrix}\right|$$

A careful inspection of this expression shows that if we form the diagram

$$\begin{pmatrix} a_{1,1} & a_{1,2} & a_{1,3} \\ a_{2,1} & a_{2,2} & a_{2,3} \\ a_{3,1} & a_{3,2} & a_{3,3} \end{pmatrix} \begin{matrix} a_{1,1} & a_{1,2} \\ a_{2,1} & a_{2,2} \\ a_{3,1} & a_{3,2} \end{matrix}$$

then each term of the determinant is the product of terms on a diagonal of the diagram. Moreover the terms with a positive sign are all on forward diagonals and the terms with a negative sign are on backward diagonals.

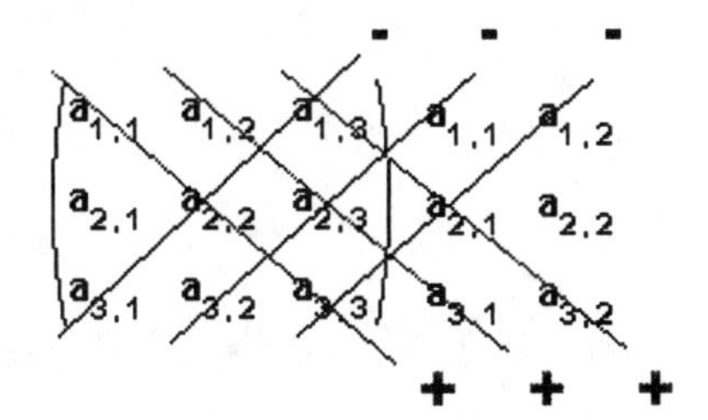

The expression that we obtain is the same as the general formula for the determinant of a 3 by 3 matrix obtained symbolically above. We call this the *3 by 3 trick* for evaluating a determinant. The *2 by 2* and *3 by 3 tricks* are useful when calculating determinants by hand.

We illustrate the diagonal device for the *3 by 3 trick* on a numerical matrix next. Compute the determinant of the following matrix **A** using the *3 by 3 trick*.

$$\mathbf{A} := \begin{pmatrix} 4 & 2 & -1 \\ 0 & 5 & 8 \\ 3 & -2 & 7 \end{pmatrix}$$

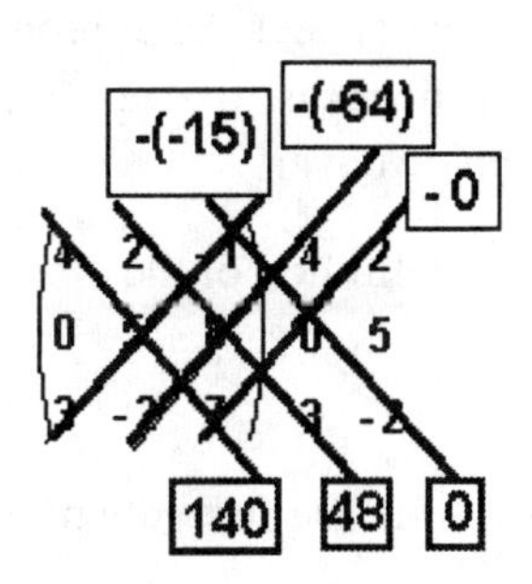

$$|\mathbf{A}| = 140 + 48 + 0 - (-15) - (-64) - 0 = 267$$

Verify the answer by using Mathcad to evaluate the determinant of **A**.

ANSWER:

<><><><><><><><><><><><><><><><><><><><><><><><><><><><><><><><><><>

Determinant of n by n matrices, n > 3

The determinant of an n by n matrix for n > 3 is much more complicated.

The determinant of $\begin{bmatrix} a_{1,1} & a_{1,2} & a_{1,3} & a_{1,4} \\ a_{2,1} & a_{2,2} & a_{2,3} & a_{2,4} \\ a_{3,1} & a_{3,2} & a_{3,3} & a_{3,4} \\ a_{4,1} & a_{4,2} & a_{4,3} & a_{4,4} \end{bmatrix}$ is given below.

$$a_{1,1}\cdot a_{2,2}\cdot a_{3,3}\cdot a_{4,4} - a_{1,1}\cdot a_{2,2}\cdot a_{3,4}\cdot a_{4,3} - a_{1,1}\cdot a_{3,2}\cdot a_{2,3}\cdot a_{4,4} \cdots$$

$$+ a_{1,1}\cdot a_{3,2}\cdot a_{2,4}\cdot a_{4,3} + a_{1,1}\cdot a_{4,2}\cdot a_{2,3}\cdot a_{3,4} - a_{1,1}\cdot a_{4,2}\cdot a_{2,4}\cdot a_{3,3} \cdots$$

$$- a_{2,1}\cdot a_{1,2}\cdot a_{3,3}\cdot a_{4,4} + a_{2,1}\cdot a_{1,2}\cdot a_{3,4}\cdot a_{4,3} + a_{2,1}\cdot a_{3,2}\cdot a_{1,3}\cdot a_{4,4} \cdots$$

$$- a_{2,1}\cdot a_{3,2}\cdot a_{1,4}\cdot a_{4,3} - a_{2,1}\cdot a_{4,2}\cdot a_{1,3}\cdot a_{3,4} + a_{2,1}\cdot a_{4,2}\cdot a_{1,4}\cdot a_{3,3} \cdots$$

$$+ a_{3,1}\cdot a_{1,2}\cdot a_{2,3}\cdot a_{4,4} - a_{3,1}\cdot a_{1,2}\cdot a_{2,4}\cdot a_{4,3} - a_{3,1}\cdot a_{2,2}\cdot a_{1,3}\cdot a_{4,4} \cdots$$

$$+ a_{3,1}\cdot a_{2,2}\cdot a_{1,4}\cdot a_{4,3} + a_{3,1}\cdot a_{4,2}\cdot a_{1,3}\cdot a_{2,4} - a_{3,1}\cdot a_{4,2}\cdot a_{1,4}\cdot a_{2,3} \cdots$$

$$- a_{4,1}\cdot a_{1,2}\cdot a_{2,3}\cdot a_{3,4} + a_{4,1}\cdot a_{1,2}\cdot a_{2,4}\cdot a_{3,3} + a_{4,1}\cdot a_{2,2}\cdot a_{1,3}\cdot a_{3,4} \cdots$$

$$- a_{4,1}\cdot a_{2,2}\cdot a_{1,4}\cdot a_{3,3} - a_{4,1}\cdot a_{3,2}\cdot a_{1,3}\cdot a_{2,4} + a_{4,1}\cdot a_{3,2}\cdot a_{1,4}\cdot a_{2,3}$$

There really is a pattern to the terms that appear in the products, the sign that precedes them, and even the subscripts. The pattern is not easy to determine by inspection; however, there are aspects of the pattern that you can discover.

How many summands are there in the definition of the 4 by 4 determinant?

ANSWER:

How many summands include a term from the first row of the matrix; i.e., a term of the form $a_{1,*}$?

ANSWER:

How many summands include two terms from the first row of the matrix?

ANSWER:

How many summands include $a_{1,1}$? $a_{1,2}$? any given element from the first row?
ANSWER:

Would your answers above change if you looked at some other row of the matrix?
ANSWER:

Would your answers change if you looked at columns instead of rows?
ANSWER:

What can you deduce from the answers to the preceding questions about the terms that appear in each summand of the expression for the determinant of a 4 by 4 matrix?
ANSWER:

What portion of the entries in the expression for the determinant of the 4 by 4 matrix are preceded by a + sign?
ANSWER:

How many summands are there in the expression for the determinant of an n by n matrix?
ANSWER:

What fraction of the entries in the expression for the determinant of an n by n matrix are preceded by a + sign?
ANSWER:

Figure out a rule for listing the various terms. Don't worry about the signs.
ANSWER:

WARNING: The ***3 by 3 TRICK*** for the computation of the determinant does not work for a 4 by 4 or larger matrix.

How many summands are there in the expression for the determinant of a 4 by 4 matrix?

ANSWER:

How many terms would there be using the *diagonal trick*?

ANSWER:

Compare your responses to the last two questions, then explain why there is no *4 by 4 trick*.

ANSWER:

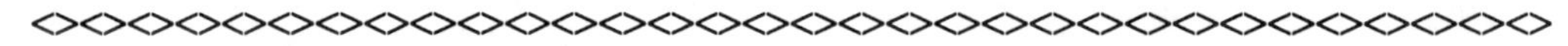

There are several computational procedures for computing determinants. The determinant of an n by n matrix can be expressed as a linear combination of determinants of matrices of size (n -1) by (n -1). This is called Lagrange expansion or cofactor expansion. It is discussed in Section 4 of this chapter.

A more useful way of computing determinants is based upon the following three facts.

1) The determinant of an upper triangular matrix is the product of the diagonal elements.

2) The determinant of a product of matrices is the product of the determinants of the matrices.

3) There are elementary matrices $\mathbf{E}_1, \ldots, \mathbf{E}_k$ such that $\mathbf{E}_k \cdot \mathbf{E}_{k-1} \cdot \ldots \cdot \mathbf{E}_1 \cdot \mathbf{A} = \mathbf{U}$ where $\mathbf{U}$ is upper triangular.

Assume we know how to evaluate the determinant of elementary matrices. Express the determinant of **A** in terms of the determinant of **U** and the determinants of the elementary matrices.

ANSWER:

Since the determinant of the diagonal matrix is the product of the diagonal entries, we know how to compute the determinant of the elementary matrix used to multiply a row by a constant.

Thus it is enough to figure out the determinant for

(1) the elementary matrices that interchange rows, and

(2) the elementary matrices that add a multiple of one row to another.

Since we will show below that we can express the elementary matrix that interchanges rows as a product of the other two types of elementary matrices, it is enough to compute the determinant of the elementary matrices that add a multiple of one row to another.

For each of the following matrices tell what elementary operation corresponds to the matrix and compute the determinant.

$$\begin{pmatrix} 1 & 0 & 0 & 0 \\ 0 & 1 & 0 & 0 \\ 3 & 0 & 1 & 0 \\ 0 & 0 & 0 & 1 \end{pmatrix} \quad \begin{pmatrix} 1 & 0 & 0 & 0 \\ 0 & 1 & 0 & 0 \\ 0 & 0 & 1 & 0 \\ 0 & 0 & -5 & 1 \end{pmatrix} \quad \begin{pmatrix} 1 & 0 & 2 & 0 \\ 0 & 1 & 0 & 0 \\ 0 & 0 & 1 & 0 \\ 0 & 0 & 0 & 1 \end{pmatrix} \quad \begin{pmatrix} 1 & 0 & 0 & 0 \\ 0 & 1 & 0 & -3 \\ 0 & 0 & 1 & 0 \\ 0 & 0 & 0 & 1 \end{pmatrix}$$

ANSWER:

Each of these matrices is either lower or upper triangular. What can you conclude about the determinant of the matrix that corresponds to the elementary operation of adding a multiple of one row to another?

ANSWER:

Consider the matrix that interchanges row 2 with row 4 (for example).

$$\begin{pmatrix} 1 & 0 & 0 & 0 \\ 0 & 0 & 0 & 1 \\ 0 & 0 & 1 & 0 \\ 0 & 1 & 0 & 0 \end{pmatrix}$$

Start with the identity matrix

add row 2 to row 4

add -1 times row 4 to row 2

add row 2 to row 4

multiply row 2 by -1

Thus interchange of rows can be done by adding multiples of one row to another and multiplying a row by -1. The calculation done here can easily be generalized to interchange any two rows.

$$\begin{pmatrix} 1 & 0 & 0 & 0 \\ 0 & 1 & 0 & 0 \\ 0 & 0 & 1 & 0 \\ 0 & 0 & 0 & 1 \end{pmatrix} \Longrightarrow \begin{pmatrix} 1 & 0 & 0 & 0 \\ 0 & 1 & 0 & 0 \\ 0 & 0 & 1 & 0 \\ 0 & 1 & 0 & 1 \end{pmatrix} \Longrightarrow \begin{pmatrix} 1 & 0 & 0 & 0 \\ 0 & 0 & 0 & -1 \\ 0 & 0 & 1 & 0 \\ 0 & 1 & 0 & 1 \end{pmatrix}$$

$$\Longrightarrow \begin{pmatrix} 1 & 0 & 0 & 0 \\ 0 & 0 & 0 & -1 \\ 0 & 0 & 1 & 0 \\ 0 & 1 & 0 & 0 \end{pmatrix} \Longrightarrow \begin{pmatrix} 1 & 0 & 0 & 0 \\ 0 & 0 & 0 & 1 \\ 0 & 0 & 1 & 0 \\ 0 & 1 & 0 & 0 \end{pmatrix}$$

What can you conclude about the determinant of the matrix of the elementary matrix that interchanges two rows?

ANSWER:

Summary

Let the elementary matrix corresponding to a row operation be denoted by the identity matrix **I** with the row operation as a subscript. Complete each of the following:

If $\mathbf{B} = (\mathbf{I}_{\text{k·row i ==> row i}}) \cdot \mathbf{A}$, then det(**B**) = _____ det(**A**).

If $\mathbf{C} = (\mathbf{I}_{\text{k·row i + row j ==> row i}}) \cdot \mathbf{A}$, then det(**C**) = ____ det(**A**).

If $\mathbf{D} = (\mathbf{I}_{\text{row i <==> row j}}) \cdot \mathbf{A}$, then det(**D**) = _____ det(**A**).

<><><><><><><><><><><><><><><><><><><><><><><><><><><><><><><><><><>

Exercises

1. Use elementary matrices to compute the determinant of each of the following matrices.

$$A := \begin{pmatrix} 3 & 1 & 2 \\ 4 & 6 & 8 \\ 5 & 7 & 9 \end{pmatrix} \qquad B := \begin{pmatrix} 1 & 2 & 5 & 0 \\ 3 & 5 & 3 & 0 \\ 7 & 8 & 1 & 2 \\ 9 & 1 & 9 & 1 \end{pmatrix}$$

ANSWERS:

In Chapter 2 we introduced the elementary row operations **rswap** and **rcomb** that have been added to Mathcad. If the file rowops.dll is not contained in the directory efi, then insert the file rowops.mcd below.

2. Use **rswap** and **rcomb** to reduce each of the following matrices to triangular form. The determinant is then the product of the diagonal elements of the reduced matrix times $(-1)^n$ where n is the number of times **rswap** was used.

$$C := \begin{pmatrix} 0 & 3 & 6 \\ 1 & 4 & 7 \\ 2 & 5 & 8 \end{pmatrix} \qquad D := \begin{bmatrix} 0 & 2 & 3 & 2 & 0 \\ 0 & 6 & 7 & 0 & 0 \\ 3 & 8 & 1 & 1 & 5 \\ 0 & 1 & 0 & 0 & 0 \\ 1 & 0 & 1 & 0 & 1 \end{bmatrix}$$

ANSWERS:

<><><><><><><><><><><><><><><><><><><><><><><><><><><><><><><>

3. Let k be a constant and **A** a 3 by 3 matrix. What is $|k \cdot \mathbf{A}|$?
Same question for a 4 by 4 matrix? n by n? Explain your reasoning.
ANSWER:

<><><><><><><><><><><><><><><><><><><><><><><><><><><><><><><>

4. If **det(A)** = 6, what is $\mathbf{det}(\mathbf{A}^5)$?
If **det(A)** = -5, what is $\mathbf{det}(\mathbf{A}^3)$?
Write a general formula relating $\mathbf{det}(\mathbf{A}^p)$ and **det(A)**.
ANSWER:

<><><><><><><><><><><><><><><><><><><><><><><><><><><><><><><>

5. Let **A** be an n by n matrix with two equal rows. What is **det(A)**? Explain in terms of elementary matrices.
ANSWER:

<><><><><><><><><><><><><><><><><><><><><><><><><><><><><><><>

6. Explore the relationship between the determinant of a matrix and whether the matrix is singular on nonsingular.

Below each matrix put the following information:

> On the first line put whether it is singular or nonsingular. (Use the **rref** function to determine this.)

> On the second line put the value of its determinant.

$$A := \begin{pmatrix} 3 & 1 \\ 4 & 2 \end{pmatrix} \quad B := \begin{pmatrix} 1 & 1 \\ 5 & 5 \end{pmatrix} \quad C := \begin{pmatrix} 1 & 2 & 7 \\ 0 & 4 & 2 \\ 0 & 1 & 2 \end{pmatrix} \quad D := \begin{pmatrix} 2 & 1 & 0 & 0 \\ 2 & 2 & 0 & 0 \\ 0 & 0 & 5 & 2 \\ 0 & 0 & 2 & 1 \end{pmatrix} \quad E := \begin{pmatrix} 2 & 4 & 6 \\ 3 & -2 & 1 \\ -1 & 0 & -1 \end{pmatrix}$$

________ ________ ________ ________ ________

________ ________ ________ ________ ________

Complete the following conjectures:

If a matrix is singular then its determinant is ____________ .

If a matrix is nonsingular then its determinant is __________ .

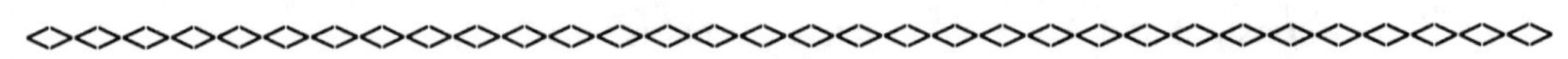

7. (Individual) Write a summary of the MATHEMATICS you learned in this section. Specifically make a list of the properties of the determinant function.

Section 3-2 Volume and the Determinant

Name:_______________

Score:_____________

Overview

Let (a,b) and (c,d) be vectors in $\mathbf{R}^2$. The area of the parallelogram whose sides are (a,b) and (c,d) equals:

$$\left|\begin{vmatrix} a & b \\ c & d \end{vmatrix}\right| \quad = \text{the absolute value of the determinant.}$$

More generally, if $\mathbf{u}_1, \mathbf{u}_2, \ldots, \mathbf{u}_n$ are n vectors in $\mathbf{R}^n$, the volume of the parallelepiped whose sides are the **u**'s is the absolute value of the determinant of the matrix whose rows (or columns) are the vectors $\mathbf{u}_1, \mathbf{u}_2, \ldots, \mathbf{u}_n$.

The relationship in $\mathbf{R}^2$ will follow as the result of explicit calculations below. The result in higher dimensions is stated without proof.

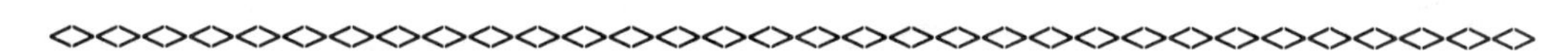

Cubes and Parallelepipeds

Let $\mathbf{t} = \begin{bmatrix} t_1 \\ t_2 \\ \cdots \\ t_n \end{bmatrix}$ be a vector in $\mathbf{R}^n$

The **unit cube** in $\mathbf{R}^n$ is defined to be the set of all vectors **t** with $0 \le t_1, t_2, \ldots, t_n \le 1$. In the plane this is the unit square and in $\mathbf{R}^3$ it is the 'usual' cube. What do you think the volume of the unit cube in $\mathbf{R}^n$ should be?

ANSWER:

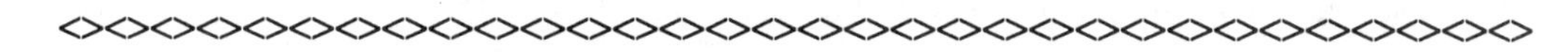

Let **u** and **v** be vectors in $\mathbf{R}^2$ and let **A** be the matrix whose first column is **u** and whose second column is **v**. As we saw in Section 1-7, the matrix **A** defines a map from the plane to itself by matrix multiplication. If **t** is the vector given above, then $\mathbf{A}\cdot\mathbf{t} = t_1\cdot\mathbf{u} + t_2\cdot\mathbf{v}$. We illustrate this map as follows.

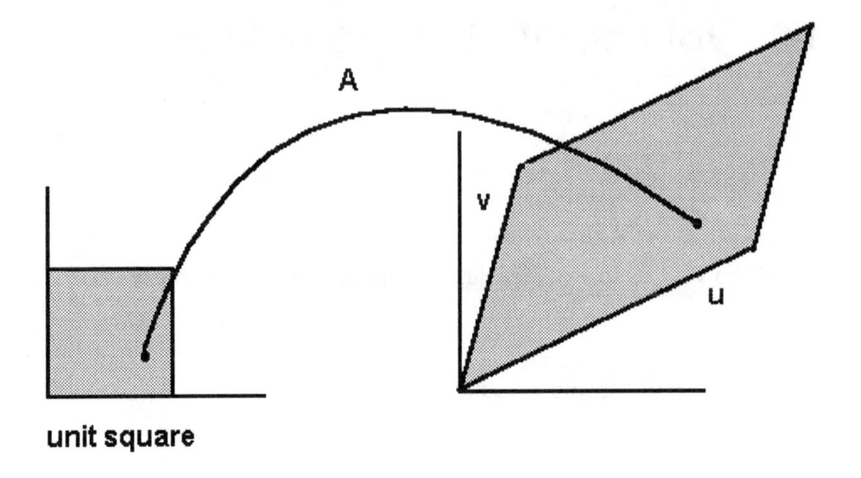

The image of the unit square is the parallelogram with sides **u** and **v**.

There is a similar picture in $\mathbf{R}^3$.

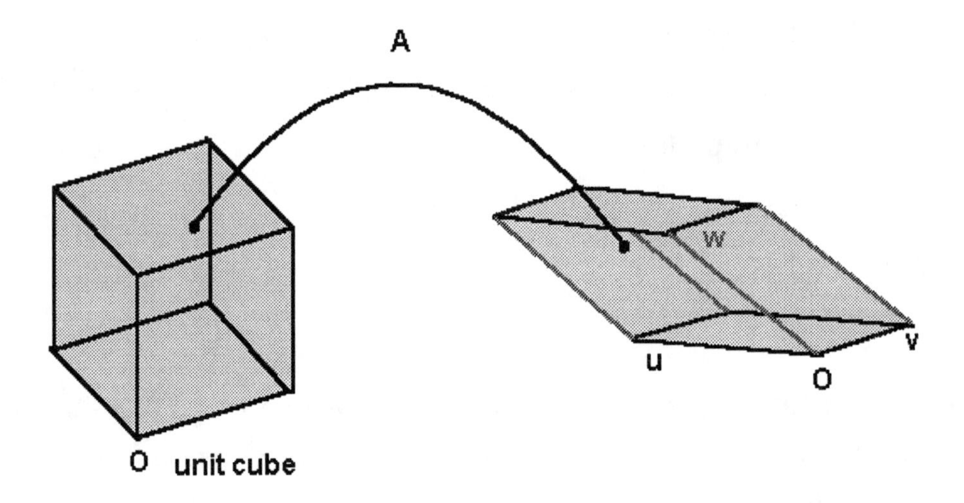

The image of the unit cube in $\mathbf{R}^3$ is called the **parallelepiped** with sides **u**, **v**, and **w**.

More generally the image of the n-cube under a matrix mapping is called a **parallelepiped**.
As we saw above a parallelepiped is an *n-dimensional parallelogram.*
If $\mathbf{u}_1, \mathbf{u}_2, \ldots, \mathbf{u}_n$ are vectors in $\mathbf{R}^n$ then the set of all points of the form

$$t_1 \cdot u_1 + t_2 \cdot u_2 + \ldots + t_n \cdot u_n$$

where $0 \le t_j \le 1$ for j = 1 to n is called the **parallelepiped generated by** $\mathbf{u}_1, \mathbf{u}_2, \ldots, \mathbf{u}_n$.

We also refer to this as the parallelepiped whose sides are $\mathbf{u}_1, \mathbf{u}_2, \ldots, \mathbf{u}_n$.

Area of a parallelogram

We consider the parallelogram whose sides are the vectors (a,b) and (c,d).

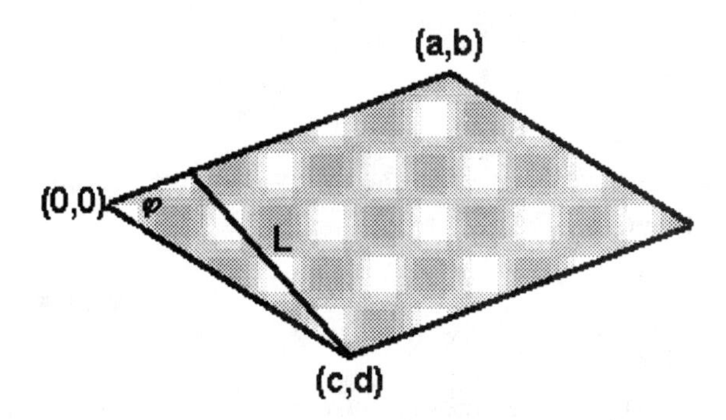

We wish to calculate the area of the above parallelogram. As we can easily see, the area is equal to the length of the line L times the length of the line from (0,0) to (a,b). The length of L is sin(φ) times the length of the line from (0,0) to (c,d).

Write an expression for the area of the parallelogram in terms of a, b, c, d and sin(φ).

ANSWER:

Complete the expression for angle, φ, below in terms of a, b, c and d.

So that we may use Maple to simplify this expression, **do not use | |** in the expression.

ANSWER:

$$\phi := \mathbf{acos}(\blacksquare)$$

The area of the parallelogram is equal to

$$\sqrt{a^2+b^2}\cdot\sqrt{c^2+d^2}\cdot\sin(\phi)$$

Substitute below the expression that you have determined for φ and simplify the result using Maple.

ANSWER:

$$\sqrt{a^2+b^2}\cdot\sqrt{c^2+d^2}\cdot\sin(\blacksquare)$$

Express the result in words: The area of the parallelogram is ______________.

There is a good chance that Maple has made an 'error' in the above calculation. To see if this is the case set a = b = 1 and c = 5, d = 2. Calculate the answer before and after Maple has simplified. Do you get the same answer? What error do you think Maple made and what should the correct answer be.

ANSWER:

Test the expression that you have derived on the following data. If you do not get the correct answer, check your work above.

(a,b)=(4,0) (c,d)=(7,5)	**AREA=20**	(a,b)=(5,3) (c,d)=(7,-2)	**AREA=31**

If your work above is correct you should have shown that the area of the region generated by the vectors (a,b) and (c,d) in the plane is the absolute value of the determinant of the matrix:

$$\begin{pmatrix} a & b \\ c & d \end{pmatrix}$$

Compute the area of the parallelogram generated by each of the following pairs of vectors.

a) (1,7) and (3,2)

b) (0,5) and (3,3)

c) (0,5) and (3, 25)

d) (2, 5) and (6,15)

ANSWERS:

a) b) c) d)

<><><><><><><><><><><><><><><><><><><><><><><><><><><><><><><><><><><><>

In $\mathbf{R}^3$ consider the parallelepiped generated by the three vectors

$$\mathbf{u} = (u_1, u_2, u_3),\ \mathbf{v} = (v_1, v_2, v_3), \text{ and } \mathbf{w} = (w_1, w_2, w_3).$$

The volume of the parallelepiped is the absolute value of the determinant whose rows are the vectors: **u**, **v** and **w**.

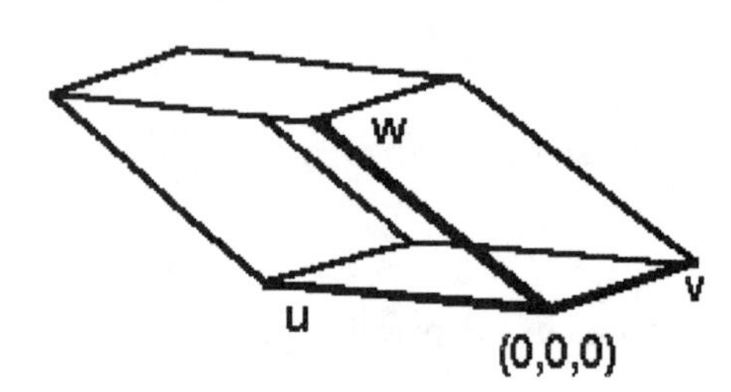

Find the volume of the parallelepiped generated by the following triples of vectors.

a) $\mathbf{u} = (2, 0, 0)$, $\mathbf{v} = (0, 3, 0)$, $\mathbf{w} = (0, 0, 5)$

b) $\mathbf{u} = (1, 4, 2)$, $\mathbf{v} = (3, -2, 2)$, $\mathbf{w} = (-2, 5, 0)$

c) $\mathbf{u} = (1, 3, 5)$, $\mathbf{v} = (0, 3, 5)$, $\mathbf{w} = (0, 0, 5)$

ANSWERS:

<><><><><><><><><><><><><><><><><><><><><><><><><><><><><><><><><><><><><>

What do we mean by the **VOLUME OF A PARALLELEPIPED** in more than three dimensions? Discuss. You might begin by thinking how you would explain three dimensional volume by using two dimensional volume (area) and how you could extend those ideas to explain four dimensional volume using three dimensional volume.

ANSWER:

Compute the volume generated by the following vectors in $\mathbf{R}^5$.

$$\begin{bmatrix} 2 \\ 0 \\ -3 \\ 8 \\ 1 \end{bmatrix} \quad \begin{bmatrix} 0 \\ 3 \\ 1 \\ 1 \\ 0 \end{bmatrix} \quad \begin{bmatrix} 7 \\ -8 \\ 2 \\ 11 \\ 0 \end{bmatrix} \quad \begin{bmatrix} 1 \\ 1 \\ 0 \\ -1 \\ -1 \end{bmatrix} \quad \begin{bmatrix} 2 \\ -5 \\ 1 \\ 0 \\ 4 \end{bmatrix}$$

ANSWER:

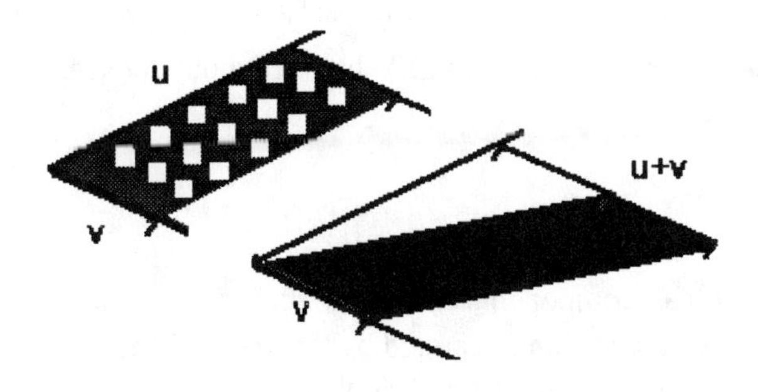

Which of the regions, patterned or solid, is the parallelogram generated by **u** and **v**?

ANSWER:

Which of the regions, patterned or solid, is the region generated by **u** + **v** and **v**?

ANSWER:

Are the areas of the patterned and solid regions related?

ANSWER:

Explain geometrically why the area of the patterned and solid regions are equal. What fact about determinants corresponds to the geometric fact that the area of the region generated by **u** + c·**v** and **v** is equal to the area of the region generated by **u** and **v**?

ANSWER:

Explain geometrically why the area of the region generated by the vectors **u** and **v** is the same as the area of the region generated by **v** and **u**. What fact about determinants corresponds to this geometric fact?

ANSWER:

Explain geometrically the relationship between the area of the region generated by the vectors **u** and **v** and the area of the region generated by c·**u** and **v**. What fact about determinants corresponds to this geometric fact?

ANSWER:

Let $\mathbf{u}_1, \mathbf{u}_2, ..., \mathbf{u}_n$ be n vectors in $\mathbf{R}^n$. Suppose that for some i and j, $i \neq j$, $\mathbf{u}_i = \mathbf{u}_j$. What is the volume of the region generated by the vectors? What fact about determinants corresponds to this geometric fact?

ANSWER:

Let **ej** be the vector in $\mathbf{R}^n$ with all components except the jth zero and with the jth component equal 1. What is the volume of the region generated by the vectors: **e1, e2**, ..., **en**? What fact about determinants corresponds to this geometric fact?

ANSWER:

Let T be the triangle shown at the right. Suppose we know the coordinates of its three vertices.

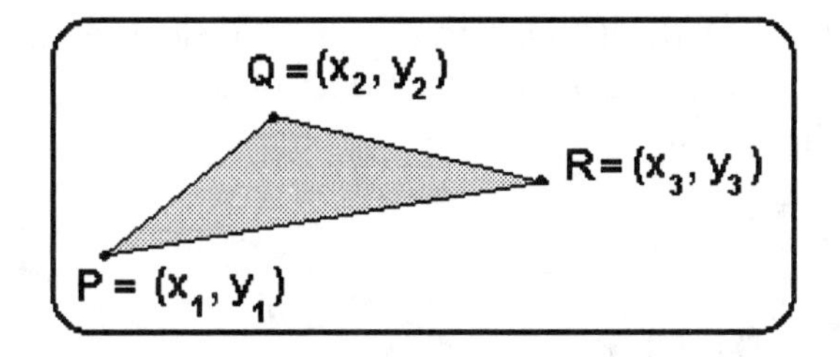

Express the area of triangle T symbolically using a determinant of a 2 by 2 matrix involving the points given above. (Hint: The triangle is half a parallelogram.)

ANSWER:

Use Maple to expand the determinant that gives the area of triangle T.

ANSWER:

Expand the following expression symbolically. Then explain why the absolute value of the determinant of this matrix is equal to the area of the triangle given above.

$$\frac{1}{2}\cdot\left|\left|\begin{pmatrix} x_1 & y_1 & 1 \\ x_2 & y_2 & 1 \\ x_3 & y_3 & 1 \end{pmatrix}\right|\right|$$

ANSWER:

<><><><><><><><><><><><><><><><><><><><><><><><><><><><><><>

Compute the area of the triangles with vertices P, Q, and R by using a determinant given in terms of the coordinates of the points P, Q, and R.

a) P = (2, 3), Q = (-4, 7), R = (1, 0)

ANSWER:

b) P = (-12, 3), Q = (5, 7), R = (0, 2)

ANSWER:

<><><><><><><><><><><><><><><><><><><><><><><><><><><><><><>

Let points P, Q, R, and S determine the parallelepiped given below. Extend the previous result to derive a matrix such that the absolute value of its determinant gives the volume of the parallelepiped?

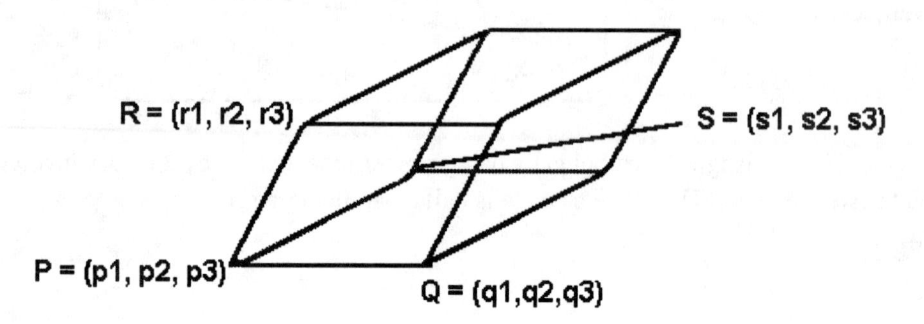

ANSWER:

<><><><><><><><><><><><><><><><><><><><><><><><><><><><><><><><><><><><><><><><>

A ***polygonal region*** is one enclosed by a boundary of straight line segments. For the polygonal region depicted below, assume we know the coordinates of the vertices A, B, C, D, and E. Explain how to determine the area of this region.

ANSWER:

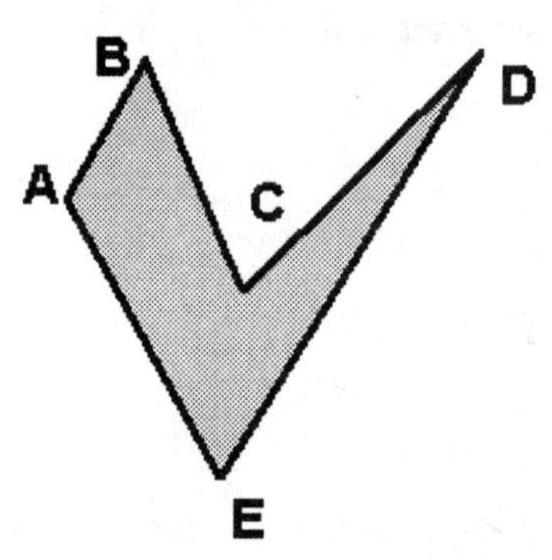

<><><><><><><><><><><><><><><><><><><><><><><><><><><><><><><><><><><><><><><><>

Exercises

1. Find the area of the trapezoid with vertices A = (-2,0), B = (-2,3), C = (4,7), and D = (4,0). To see the trapezoid, press F9.

$j := 1..5 \quad X := \begin{pmatrix} -2 & -2 & 4 & 4 & -2 \\ 0 & 3 & 7 & 0 & 0 \end{pmatrix}$

ANSWER:

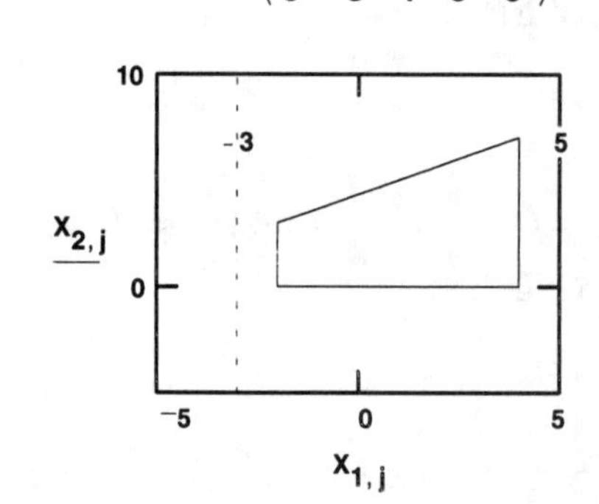

2. A farmer's triangular field is to be cut by a road. The state must pay the farmer $1500 per sq. unit of land taken by the road. Determine the amount the farmer gets for his land. To see the diagram of the land (red) and road (blue) press F9.

$j := 1..4 \quad i := 1..2$ $\quad L1 := \begin{pmatrix} -3 & 7 \\ -1 & -1 \end{pmatrix} \quad L2 := \begin{pmatrix} -3 & 7 \\ 3 & 3 \end{pmatrix}$

$$T := \begin{pmatrix} \frac{1}{3} & 5 & 5 & \frac{1}{3} \\ -1 & 6 & -8 & -1 \end{pmatrix}$$

ANSWER:

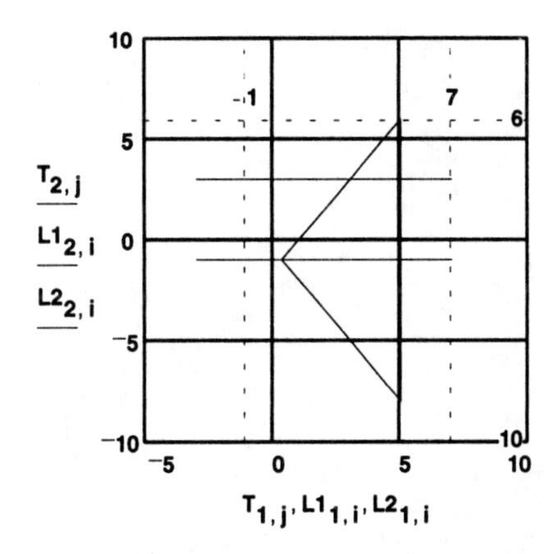

<><><><><><><><><><><><><><><><><><><><><><><><><><><><><><><><><><><><>

3. (Individual) Write a summary of the MATHEMATICS that you learned in this section.

Section 3-3 Determinants and Linear Independence

Name:________________

Score:_____________

Overview

The goal of this section is to find a condition in terms of a determinant that distinguishes a linearly independent set of vectors in $\mathbf{R}^n$ from a linearly dependent set.

<><><><><><><><><><><><><><><><><><><><><><><><><><><><><><><>

Let **M** be an n by n matrix. From Section 3-1 we know

M has an inverse if and only if det(M) is not zero.

From Section 2-8 we know

M has an inverse if and only if rref(M) = I.

From Section 2-7 we know that forming the **rref(M)** eliminates linearly dependent rows of **M** and that the nonzero rows of the **rref(M)** are a basis for the row space of **M** (denoted **Row(M)**). Complete the following:

M has an inverse if and only if dim(Row(M)) = ________ .

M has an inverse if and only if the rows of M are linearly ____________ .

From Section 2-7 we also know that

dim(Row(M)) + dim(Solution(M)) = n.

(**Solution(M)** is the solution space of the homogeneous system $\mathbf{M}\cdot\mathbf{x} = \mathbf{0}$.)

Complete the following:

M has an inverse if and only if dim(Solution(M)) = ________ .

If **M** has an inverse, the only solution of $\mathbf{M}\cdot\mathbf{x} = \mathbf{0}$ is ________.

The matrix-vector product $\mathbf{M}\cdot\mathbf{x}$ is expressible as a linear combination of the columns of **M** as

$$\mathbf{M}\cdot\mathbf{x} = \mathbf{x}_1\cdot\mathbf{col}_1(\mathbf{M}) + \mathbf{x}_2\cdot\mathbf{col}_2(\mathbf{M}) + \ldots + \mathbf{x}_n\cdot\mathbf{col}_n(\mathbf{M})$$

Thus if **M** is invertible a linear combination of columns of **M** is the zero vector only if $\mathbf{x}_i =$ ____ for each i. Complete the following:

M has an inverse if and only if the columns of M are linearly ____________ .

<><><><><><><><><><><><><><><><><><><><><><><><><><><><><><><>

Summary

For an n by n matrix **M**, the following are equivalent:

(1) **M** has an inverse.

(2) The rows of **M** are __________________________.

(3) The columns of **M** are ___________________________.

(4) det(**M**) = __________________________.

(5) rref(**M**) = ________.

(6) The solution space of $\mathbf{M}\cdot\mathbf{x} = \mathbf{0}$ is __________________.

(7) dim(Row(**M**)) = ________ .

<><><><><><><><><><><><><><><><><><><><><><><><><><><><><><><>

In Section 3-2 we stated that the absolute value of the determinant of a matrix, **M**, whose rows are n vectors in $\mathbf{R}^n$ was the volume of the parallelepiped generated by the vectors. Suppose that the determinant of **M** equals 0. What does this mean *geometrically* about the parallelepiped generated by the rows of **M**?

ANSWER:

<><><><><><><><><><><><><><><><><><><><><><><><><><><><><><><>

For each of the following matrices determine for what value(s) of k the matrix has an inverse. Outline your procedure, then carry out the computations.

a) $\begin{pmatrix} 1 & 7 & 0 & 0 \\ k & -5 & 2 & 3 \\ 3 & 4 & k & 5 \\ 2 & 1 & 1 & 8 \end{pmatrix}$ b) $\begin{pmatrix} 1 & 0 & 2 & 0 \\ -4 & k & -8 & k \\ 3 & 5 & 1 & 10 \\ 2 & 6 & 1 & 9 \end{pmatrix}$ c) $\begin{pmatrix} 1 & 1 & k & 0 \\ -2 & 2 & 1 & 2 \\ 6 & 0 & 3 & -6 \\ -4 & 1 & k & 5 \end{pmatrix}$

ANSWERS:

<><><><><><><><><><><><><><><><><><><><><><><><><><><><><><><>

Let $\mathbf{M} := \begin{pmatrix} 1 & 1 & 3 \\ -2 & 1 & 5 \\ 3 & 0 & -6 \end{pmatrix}$

We may rewrite the equation $\mathbf{M}\cdot\mathbf{x} = k\cdot\mathbf{x}$ as $\mathbf{M}\cdot\mathbf{x} = k\cdot\mathbf{I}\cdot\mathbf{x}$, where **I** is the identity matrix. This may then be rewritten as $(\mathbf{M} - k\cdot\mathbf{I})\cdot\mathbf{x} = \mathbf{0}$.

Use the latter form of this equation and determinants to find the value(s) of k for which the equation, $\mathbf{M}\cdot\mathbf{x} = k\cdot\mathbf{x}$, has a nonzero solution.

ANSWER:

Generalize the approach used above: Let **M** be any square matrix. Give a procedure that determines for which values of k is there a solution to $\mathbf{M}\cdot\mathbf{x} = k\cdot\mathbf{x}$.

ANSWER:

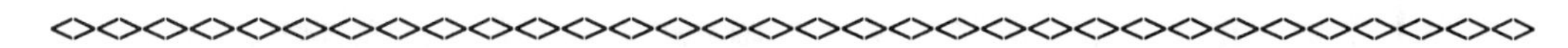

The use of determinants provides an easy method for checking when n vectors in $\mathbf{R}^m$ are linearly independent. We are now interested in answering a more general question:

When are n vectors in R^m linearly independent?

Let **M** be the matrix whose ***columns*** are the given vectors. **M** is an m by n matrix. The columns are linearly dependent if and only if there is a nonzero n-vector $\mathbf{x} = [x_1, \ldots, x_n]^T$ such that $\mathbf{M}\cdot\mathbf{x} = \mathbf{0}$; that is,

$$M\cdot x = x_1\cdot M^{<1>} + x_2\cdot M^{<2>} + \ldots + x_n\cdot M^{<n>} = 0$$

(Note: $\mathbf{M}^{<j>}$ is the jth column of **M**.)

The case where n > m

If n>m then any n vectors in R^m are linearly dependent.

There are more columns than rows. This means that in the rref at least one column can not be a pivot column and must correspond to a free variable. Thus there is a nonzero solution to $\mathbf{M}\cdot\mathbf{x} = \mathbf{0}$.

The case where n < m

Claim: $\mathbf{M}\cdot\mathbf{x} = \mathbf{0}$ has the unique solution $\mathbf{x} = \mathbf{0}$ if and only if $\mathbf{M}^T\cdot\mathbf{M}\cdot\mathbf{x} = \mathbf{0}$ has the unique solution $\mathbf{x} = \mathbf{0}$. (We must show that $\mathbf{M}^T\cdot\mathbf{M}\cdot\mathbf{x} = \mathbf{0}$ if and only if $\mathbf{M}\cdot\mathbf{x} = \mathbf{0}$).

Suppose that **x** is a solution of $\mathbf{M}^T\cdot\mathbf{M}\cdot\mathbf{x} = \mathbf{0}$.

Then $\mathbf{x}^T\cdot\mathbf{M}^T\cdot\mathbf{M}\cdot\mathbf{x} = 0$ but $\mathbf{x}^T\cdot\mathbf{M}^T\cdot\mathbf{M}\cdot\mathbf{x} = (\mathbf{M}\cdot\mathbf{x})^T\cdot(\mathbf{M}\cdot\mathbf{x}) = |\mathbf{M}\cdot\mathbf{x}|^2$

Thus $|\mathbf{M}\cdot\mathbf{x}| = 0$ and **x** is a solution of $\mathbf{M}\cdot\mathbf{x} = \mathbf{0}$.

Obviously any solution of $\mathbf{M}\cdot\mathbf{x} = \mathbf{0}$ is also a solution of $\mathbf{M}^T\cdot\mathbf{M}\cdot\mathbf{x} = \mathbf{0}$. Since the two sets of solutions coincide, if **0** is the unique solution for one equation, it must also be the unique solution for the other.

$\mathbf{M}^T \cdot \mathbf{M}$ is a ***square*** matrix (n by n). Thus $\mathbf{0}$ is the unique solution to $\mathbf{M}^T \cdot \mathbf{M} \cdot \mathbf{x} = \mathbf{0}$ if and only if $\mathbf{M}^T \cdot \mathbf{M}$ is invertible if and only if the determinant of $\mathbf{M}^T \cdot \mathbf{M}$ is nonzero.

Conclusion: **Let $M^{<1>}, \ldots, M^{<n>}$ be vectors in R^m with $n < m$. Then $M^{<1>}, \ldots, M^{<n>}$ are linearly independent if and only if $|M^T \cdot M|$ is not zero.**

<><><><><><><><><><><><><><><><><><><><><><><><><><><><><><><><><><><>

Example

$$M^{<1>} := \begin{bmatrix} 2 \\ 0 \\ -2 \\ 1 \\ 3 \end{bmatrix} \qquad M^{<2>} := \begin{bmatrix} 0 \\ 2 \\ 1 \\ 0 \\ 2 \end{bmatrix} \qquad M^{<3>} := \begin{bmatrix} -1 \\ 2 \\ -1 \\ 1 \\ 0 \end{bmatrix}$$

$$\left| M^T \cdot M \right| = 875$$

$$M^{<4>} := 2 \cdot M^{<1>} - M^{<2>} + M^{<3>}$$

$$\left| M^T \cdot M \right| = 0$$

Be careful!! Because Mathcad generally does not perform exact arithmetic, the determinant may appear to be nonzero even when it is zero. This is even true when all the matrix entries are integers. One way of avoiding this problem is given below.

$$M_{5,3} := 5$$

$$M^{<4>} := 2 \cdot M^{<1>} - M^{<2>} + M^{<3>}$$

$$\left| M^T \cdot M \right| = -3.517186542 \cdot 10^{-12}$$

Calculate $\mathbf{M}^T \cdot \mathbf{M}$ and then use Maple to evaluate the determinant.

ANSWER:

<><><><><><><><><><><><><><><><><><><><><><><><><><><><><><><><><><><>

Exercises

1. For each of the following sets of vectors, determine if the vectors are linearly independent.

a) $\begin{pmatrix} \frac{1}{2} \\ -3 \\ 4 \end{pmatrix}$ $\begin{pmatrix} 2 \\ \frac{1}{3} \\ 0 \end{pmatrix}$ $\begin{pmatrix} 5 \\ \frac{1}{4} \\ 3 \end{pmatrix}$ $\begin{pmatrix} -\frac{1}{5} \\ 2 \\ 1 \end{pmatrix}$ ***ANSWER***:

b) $\begin{pmatrix} \frac{1}{2} \\ -3 \\ 4 \end{pmatrix}$ $\begin{pmatrix} 2 \\ \frac{1}{3} \\ 0 \end{pmatrix}$ $\begin{pmatrix} 5 \\ \frac{1}{4} \\ 3 \end{pmatrix}$ ***ANSWER***:

c) $\begin{pmatrix} 2 \\ -5 \\ 11 \\ 17 \end{pmatrix}$ $\begin{pmatrix} 19 \\ 31 \\ -17 \\ 3 \end{pmatrix}$ $\begin{pmatrix} 181 \\ 254 \\ -98 \\ 112 \end{pmatrix}$ ***ANSWER***:

<><><><><><><><><><><><><><><><><><><><><><><><><><><><>

2. Determine if $\begin{pmatrix} 9 \\ 45 \\ -35 \\ -14 \end{pmatrix}$ is in the span of $\begin{pmatrix} 3 \\ 8 \\ -6 \\ 2 \end{pmatrix}$ $\begin{pmatrix} 2 \\ -5 \\ 10 \\ 8 \end{pmatrix}$ $\begin{pmatrix} 13 \\ 14 \\ 17 \\ 22 \end{pmatrix}$

ANSWER:

<><><><><><><><><><><><><><><><><><><><><><><><><><><><>

3. Determine if the columns of the matrix **M** given below are linearly independent. Do this first by computing rref($\mathbf{M}^T$) and then by computing $|\mathbf{M}^T \cdot \mathbf{M}|$. Compute $|\mathbf{M}^T \cdot \mathbf{M}|$ using Mathcad and then using the symbolic processor. Explain the results.

$$M := \begin{bmatrix} \frac{1}{3} & \frac{1}{5} & \frac{8}{3} \\ 1 & \frac{1}{5} & 26 \\ 1 & 7 & -280 \\ 3 & 21 & -840 \\ 2 & \frac{36}{5} & -254 \end{bmatrix}$$

ANSWER:

$$\mathrm{rref}\left(M^T\right) =$$

$$\left|M^T \cdot M\right| =$$

$$\left|\begin{bmatrix} \frac{1}{3} & 1 & 1 & 3 & 2 \\ \frac{1}{5} & \frac{1}{5} & 7 & 21 & \frac{36}{5} \\ \frac{8}{3} & 26 & -280 & -840 & -254 \end{bmatrix} \cdot \begin{bmatrix} \frac{1}{3} & \frac{1}{5} & \frac{8}{3} \\ 1 & \frac{1}{5} & 26 \\ 1 & 7 & -280 \\ 3 & 21 & -840 \\ 2 & \frac{36}{5} & -254 \end{bmatrix}\right|$$

<== ***Evaluate symbolically.***

<><><><><><><><><><><><><><><><><><><><><><><><><><><><><><><>

4. (Individual) Write a summary of the MATHEMATICS that you learned in this section.

Section 3-4 Cofactor Expansion of a Determinant

Name:________________

Score:_____________

Overview

The cofactor or Lagrange expansion of a determinant expresses the determinant of an n by n matrix in terms of the determinants of certain (n-1) by (n-1) submatrices. The expansion is motivated by an exploration of some 3 by 3 matrices. The cofactor expansion is used in Section 3-5 to derive Cramer's rule for the solution of systems of linear equations in terms of determinants.

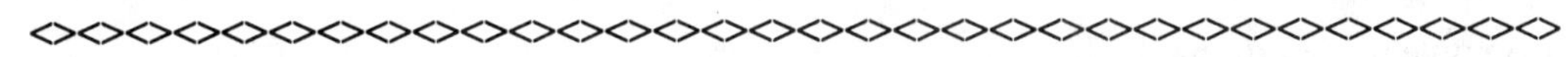

Compute the determinant of each of the following matrices and record the value below the matrix.

$$\begin{pmatrix} a & b & c \\ 3 & 5 & 2 \\ 2 & 7 & 1 \end{pmatrix} \quad \begin{pmatrix} a & 0 & 0 \\ 3 & 5 & 2 \\ 2 & 7 & 1 \end{pmatrix} \quad \begin{pmatrix} 0 & b & 0 \\ 3 & 5 & 2 \\ 2 & 7 & 1 \end{pmatrix} \quad \begin{pmatrix} 0 & 0 & c \\ 3 & 5 & 2 \\ 2 & 7 & 1 \end{pmatrix}$$

________ ________ ________ ________

Guess a relationship between the determinant of the first matrix and the determinants of the other three matrices.
ANSWER:

Same question.

$$\begin{pmatrix} 3 & 5 & 2 \\ a & b & c \\ 2 & 7 & 1 \end{pmatrix} \quad \begin{pmatrix} 3 & 5 & 2 \\ a & 0 & 0 \\ 2 & 7 & 1 \end{pmatrix} \quad \begin{pmatrix} 3 & 5 & 2 \\ 0 & b & 0 \\ 2 & 7 & 1 \end{pmatrix} \quad \begin{pmatrix} 3 & 5 & 2 \\ 0 & 0 & c \\ 2 & 7 & 1 \end{pmatrix}$$

________ ________ ________ ________

Did we have to compute the determinants to answer the second question or could we have deduced the result from above? Explain your answer.
ANSWER:

If you did the above calculations correctly you should have concluded that

$$\left|\begin{pmatrix} a & b & c \\ 3 & 5 & 2 \\ 2 & 7 & 1 \end{pmatrix}\right| = \left[\left|\begin{pmatrix} a & 0 & 0 \\ 3 & 5 & 2 \\ 2 & 7 & 1 \end{pmatrix}\right| + \left|\begin{pmatrix} 0 & b & 0 \\ 3 & 5 & 2 \\ 2 & 7 & 1 \end{pmatrix}\right|\right] + \left|\begin{pmatrix} 0 & 0 & c \\ 3 & 5 & 2 \\ 2 & 7 & 1 \end{pmatrix}\right|$$

To compute the determinant on the left we need to understand how to compute each of the determinants on the right. We begin by examining the determinant of

$$\begin{pmatrix} a & 0 & 0 \\ 3 & 5 & 2 \\ 2 & 7 & 1 \end{pmatrix}$$

Compute its determinant.

ANSWER:

Replace the 3 and 2 in the first column by x and y and recompute the determinant.

$$\begin{pmatrix} a & 0 & 0 \\ x & 5 & 2 \\ y & 7 & 1 \end{pmatrix}$$

ANSWER:

How is the determinant of $\begin{pmatrix} a & 0 & 0 \\ x & 5 & 2 \\ y & 7 & 1 \end{pmatrix}$ related to the determinant of $\begin{pmatrix} 5 & 2 \\ 7 & 1 \end{pmatrix}$?

ANSWER:

More generally, what is the relationship between the determinants of the following matrices?

$$\begin{pmatrix} a & b & c \\ 3 & 5 & 2 \\ 2 & 7 & 1 \end{pmatrix} \quad \begin{pmatrix} a & 0 & 0 \\ 3 & 5 & 2 \\ 2 & 7 & 1 \end{pmatrix} \quad \begin{pmatrix} 0 & b & 0 \\ 3 & 5 & 2 \\ 2 & 7 & 1 \end{pmatrix} \quad \begin{pmatrix} 0 & 0 & c \\ 3 & 5 & 2 \\ 2 & 7 & 1 \end{pmatrix} \quad \begin{pmatrix} 5 & 2 \\ 7 & 1 \end{pmatrix} \quad \begin{pmatrix} 3 & 2 \\ 2 & 1 \end{pmatrix} \quad \begin{pmatrix} 3 & 5 \\ 2 & 7 \end{pmatrix}$$

ANSWER:

What would happen above if a, b, and c were in another row? What if they were in a column?

ANSWER:

<><><><><><><><><><><><><><><><><><><><><><><><><><><><><><>

Let **A** be an n by n matrix. The ***(i,j)th minor*** of **A** is the (n-1) by (n-1) matrix obtained by deleting the ith row and the jth column. This is denoted by **A(i,j)**.

The ***(i,j)th cofactor*** is defined by: $(-1)^{(i+j)}\,|\mathbf{A(i,j)}|$.

We discovered above that

$$\left|\begin{pmatrix} a & b & c \\ 3 & 5 & 2 \\ 2 & 7 & 1 \end{pmatrix}\right| = \left[\left|\begin{pmatrix} a & 0 & 0 \\ 3 & 5 & 2 \\ 2 & 7 & 1 \end{pmatrix}\right| + \left|\begin{pmatrix} 0 & b & 0 \\ 3 & 5 & 2 \\ 2 & 7 & 1 \end{pmatrix}\right|\right] + \left|\begin{pmatrix} 0 & 0 & c \\ 3 & 5 & 2 \\ 2 & 7 & 1 \end{pmatrix}\right|$$

and that
$$\left|\begin{pmatrix} a & 0 & 0 \\ 3 & 5 & 2 \\ 2 & 7 & 1 \end{pmatrix}\right| = a\cdot\left|\begin{pmatrix} 5 & 2 \\ 7 & 1 \end{pmatrix}\right|$$

with similar formulas for the second and third terms on the right hand side above.

Express the formula (or formulas) for the determinant of **A** (the left hand side above) in words. Your answer should use the word cofactor or cofactors.

ANSWER:

Test your formula on the following matrix.

$$A := \begin{pmatrix} 3 & 2 & -3 \\ 7 & 5 & 2 \\ -1 & 0 & 1 \end{pmatrix}$$

$$A11 := \begin{pmatrix} 5 & 2 \\ 0 & 1 \end{pmatrix} \quad A12 := \begin{pmatrix} 7 & 2 \\ -1 & 1 \end{pmatrix} \quad A13 := \begin{pmatrix} 7 & 5 \\ -1 & 0 \end{pmatrix}$$

$$A21 := \begin{pmatrix} 2 & -3 \\ 0 & 1 \end{pmatrix} \quad A22 := \begin{pmatrix} 3 & -3 \\ -1 & 1 \end{pmatrix} \quad A23 := \begin{pmatrix} 3 & 2 \\ -1 & 0 \end{pmatrix}$$

$$A31 := \begin{pmatrix} 2 & -3 \\ 5 & 2 \end{pmatrix} \quad A32 := \begin{pmatrix} 3 & -3 \\ 7 & 2 \end{pmatrix} \quad A33 := \begin{pmatrix} 3 & 2 \\ 7 & 5 \end{pmatrix}$$

ANSWER:

a) Use row 1.

b) Use column 1.

c) Use row 3.

The procedure that you have developed above is called ***cofactor expansion***. It is particularly useful when a matrix has many zero terms. If we use row 1 in the expansion we say that we ***expand on row 1***. Similarly we can expand on any row or any column. We illustrate with the following computation:

step 1: expand on the 6th column

$$\left|\begin{bmatrix} 0 & 8 & 0 & 23 & 0 & 0 & 3 \\ 1 & 3 & -2 & 5 & 7 & 4 & 1 \\ 1 & 5 & 0 & 11 & 2 & 0 & 6 \\ 0 & 0 & 0 & 2 & 0 & 0 & 0 \\ 5 & 2 & 3 & -9 & -1 & 0 & 7 \\ 1 & 1 & 0 & 8 & 0 & 0 & 3 \\ 0 & 0 & 0 & 2 & 0 & 0 & 1 \end{bmatrix}\right| \qquad 4\cdot|A(2,6)| = 4\cdot\left|\begin{bmatrix} 0 & 8 & 0 & 23 & 0 & 3 \\ 1 & 5 & 0 & 11 & 2 & 6 \\ 0 & 0 & 0 & 2 & 0 & 0 \\ 5 & 2 & 3 & -9 & -1 & 7 \\ 1 & 1 & 0 & 8 & 0 & 3 \\ 0 & 0 & 0 & 2 & 0 & 1 \end{bmatrix}\right|$$

step 2: expand on the 3rd row

$$4\cdot(-2)\cdot\begin{bmatrix} 0 & 8 & 0 & 0 & 3 \\ 1 & 5 & 0 & 2 & 6 \\ 5 & 2 & 3 & -1 & 7 \\ 1 & 1 & 0 & 0 & 3 \\ 0 & 0 & 0 & 0 & 1 \end{bmatrix}$$

step 3 : expand on the fifth row

$$1\cdot\left[4\cdot(-2)\cdot\begin{pmatrix} 0 & 8 & 0 & 0 \\ 1 & 5 & 0 & 2 \\ 5 & 2 & 3 & -1 \\ 1 & 1 & 0 & 0 \end{pmatrix}\right]$$

step 4: expand on the 3rd column

$$3\cdot\left[1\cdot\left[4\cdot(-2)\cdot\begin{pmatrix} 0 & 8 & 0 \\ 1 & 5 & 2 \\ 1 & 1 & 0 \end{pmatrix}\right]\right]$$

step 5: expand on the 3rd column

$$(-2)\cdot\left[3\cdot\left[1\cdot\left[4\cdot(-2)\cdot\begin{pmatrix} 0 & 8 \\ 1 & 1 \end{pmatrix}\right]\right]\right]$$

Compute:

$$(-8)\cdot((-2)\cdot(3\cdot(1\cdot(4\cdot(-2))))) = -384$$

SUMMARY: Let **A** be an n by n matrix, then

$$|A| = \sum_i (-1)^{i+j}\cdot A_{i,j}\,|A(i,j)| = \sum_j (-1)^{i+j} A_{i,j}\,|A(i,j)|$$

where the sum is taken for any j in the first sum or for any i in the second sum.

Explain the formula $\sum_{i} (-1)^{i+j} \cdot A_{i,j} \cdot |A(i,j)|$ in words.

ANSWER:

Explain the formula $\sum_{j} (-1)^{i+j} \cdot A_{i,j} \cdot |A(i,j)|$ in words.

ANSWER:

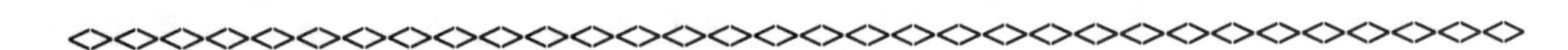

Exercises

1. Use cofactors to compute the determinant of the following matrices.

$$\begin{bmatrix} 0 & 0 & 1 & 3 & 5 \\ 5 & 0 & 2 & 0 & 1 \\ 0 & 4 & 0 & 0 & 0 \\ 2 & 6 & 7 & 0 & -7 \\ 0 & 3 & 0 & 2 & 0 \end{bmatrix} \qquad \begin{pmatrix} a_{1.1} & a_{1,2} & a_{1,3} \\ a_{2,1} & a_{2,2} & a_{2,3} \\ a_{3,1} & a_{3,2} & a_{3,3} \end{pmatrix}$$

ANSWER: *ANSWER*:

2. Let the vector (x,y,z) be defined by setting x = coefficient of i, y = coefficient of j and z = coefficient of k in the expression for the determinant of the following matrix.

$$\begin{pmatrix} 2 & 0 & -3 \\ 1 & 4 & 5 \\ i & j & k \end{pmatrix}$$

Let (u,v,w) be any vector in $\mathbf{R}^3$. Explain why (using cofactor expansion) the dot product of (x,y,z) and (u,v,w) is the determinant of

$$\begin{pmatrix} 2 & 0 & -3 \\ 1 & 4 & 5 \\ u & v & w \end{pmatrix}$$

ANSWER:

3. Let (x,y,z) be the vector defined in the previous exercise. Using the properties of determinants explain why

$$\begin{pmatrix} 2 \\ 0 \\ -3 \end{pmatrix} \cdot \begin{pmatrix} x \\ y \\ z \end{pmatrix} = 0 \qquad \text{and} \qquad \begin{pmatrix} 1 \\ 4 \\ 5 \end{pmatrix} \cdot \begin{pmatrix} x \\ y \\ z \end{pmatrix} = 0$$

ANSWER:

<><><><><><><><><><><><><><><><><><><><><><><><><><><><><><><><><><><><><><><><><>

4. (Individual) Write a summary of the MATHEMATICS that you learned in this section.

Section 3-5
The Adjoint Matrix, Inverses and Cramer's Rule

Name:________________

Score:______________

Overview

For any matrix, **M**, we define the adjoint matrix, **adj(M)**. A formula for the inverse of **M** is given in terms of the adjoint. Using this, a solution for the system of equations **M·x = b**, can be given in terms of the determinants of **M** and some related matrices. The formula for this solution is often called Cramer's Rule.

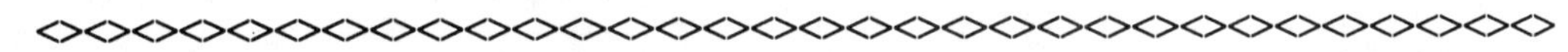

For any matrix **M** define the ***adjoint matrix*** of **M**, **adj(M)**, by setting the (i,j)th element of **adj(M)** equal to the (j,i)th cofactor.

For example, if **A** is a 3 by 3 matrix, then

$$\mathbf{adj}(\mathbf{A}) := \begin{pmatrix} |\mathbf{A}(1,1)| & -|\mathbf{A}(2,1)| & |\mathbf{A}(3,1)| \\ -|\mathbf{A}(1,2)| & |\mathbf{A}(2,2)| & -|\mathbf{A}(3,2)| \\ |\mathbf{A}(1,3)| & -|\mathbf{A}(2,3)| & |\mathbf{A}(3,3)| \end{pmatrix}$$

<><><><><><><><><><><><><><><><><><><><><><><><><><><><><><><><>

Let $$\mathbf{A} = \begin{pmatrix} a_{1,1} & a_{1,2} & a_{1,3} \\ a_{2,1} & a_{2,2} & a_{2,3} \\ a_{3,1} & a_{3,2} & a_{3,3} \end{pmatrix}$$

The (i,j)th element of the product, **A·adj(A)** is

$$a_{i,1}\cdot(-1)^{j+1}\cdot|\mathbf{A}(j,1)| + a_{i,2}\cdot(-1)^{j+2}\cdot|\mathbf{A}(j,2)| + a_{i,3}\cdot(-1)^{j+3}\cdot|\mathbf{A}(j,3)| \qquad (1)$$

If i = j, expression (1) is the cofactor expansion for |**A**|.

Let $$\mathbf{B} = \begin{pmatrix} a_{1,1} & a_{1,2} & a_{1,3} \\ a_{1,1} & a_{1,2} & a_{1,3} \\ a_{3,1} & a_{3,2} & a_{3,3} \end{pmatrix}$$

Suppose that i = 1 and j = 2. Explain why the (1,2) element of $\mathbf{A \cdot adj(A)}$ (see expression (1) above) is the same as the (2,2) element of $\mathbf{B \cdot adj(B)}$.

ANSWER:

Explain why the (2,2) element of $\mathbf{B \cdot adj(B)}$ equals 0.

ANSWER:

Explain how you would show that if $i \neq j$, the (i,j)th term of $\mathbf{A \cdot adj(A)}$ is 0.

ANSWER:

Complete the following sentences:

if $i = j$ $(\mathbf{A \cdot adj(A)})_{i,j}$ equals ____________ .

if $i \neq j$ $(\mathbf{A \cdot adj(A)})_{i,j}$ equals ____________ .

Verify your answer above by computing $\mathbf{adj(A)}$ and $\mathbf{A \cdot adj(A)}$ for the matrix $\mathbf{A}$ given below.

$$A := \begin{pmatrix} 2 & 1 & 10 \\ 7 & 0 & 2 \\ -5 & 6 & -3 \end{pmatrix}$$

ANSWER:

Compute: $A \cdot \left(\frac{adj(A)}{|A|} \right)$

ANSWER:

Explain why the following is true.

Let M be a matrix with $|M| \neq 0$. Then $M^{-1} = \frac{adj(M)}{|M|}$.

ANSWER:

Cramer's Rule (general case)

Let $\mathbf{A}\cdot\mathbf{x} = \mathbf{b}$ be a system of n equations in n unknowns and assume that $\det(\mathbf{A}) \neq 0$.

Let $\mathbf{A}_i$ be the matrix obtained from $\mathbf{A}$ by substituting $\mathbf{b}$ in the ith column of $\mathbf{A}$.

Then $x_i = \dfrac{|A_i|}{|A|}$ $\quad$ ($\mathbf{x}_i$ is the ith component of the solution $\mathbf{x}$)

We show that this follows from the equation for the inverse given above.

1. First note that $\mathbf{x} = \mathbf{A}^{-1}\cdot\mathbf{b}$ and therefore $\mathbf{x}_i$ equals the dot product of $\mathbf{b}$ with the ith row of $\mathbf{A}^{-1}$.

2. Since the ith row of $\mathbf{A}^{-1}$ is the ith row of **adj(A)** divided by $|\mathbf{A}|$ it follows that $\mathbf{x}_i$ is $\mathbf{b}$ times the cofactors coming from the cofactor expansion of the ith column of $\mathbf{A}$. But this is just the determinant of the matrix that arises by replacing the ith column of $\mathbf{A}$ with the vector $\mathbf{b}$.

Cramer's Rule (2 by 2 case)

Solve $\mathbf{A}\cdot\mathbf{x} = \mathbf{b}$, where $A := \begin{pmatrix} 3 & -1 \\ 5 & 2 \end{pmatrix}$ and $b := \begin{pmatrix} 19 \\ 6 \end{pmatrix}$

ANSWER:

$$x_1 := \frac{\left|\begin{pmatrix} 19 & -1 \\ 6 & 2 \end{pmatrix}\right|}{\left|\begin{pmatrix} 3 & -1 \\ 5 & 2 \end{pmatrix}\right|} \qquad x_2 := \frac{\left|\begin{pmatrix} 3 & 19 \\ 5 & 6 \end{pmatrix}\right|}{\left|\begin{pmatrix} 3 & -1 \\ 5 & 2 \end{pmatrix}\right|}$$

Check: $A\cdot x = \begin{pmatrix} 19 \\ 6 \end{pmatrix}$

<><><><><><><><><><><><><><><><><><><><><><><><><><><><><><>

Exercises

Sometimes it is useful to concatenate two matrices or a vector and a matrix to form a new matrix. (For example, one can form the matrix needed for Cramer's Rule by concatenation.) The Mathcad function augment performs this function. An example follows:

$$w := \begin{pmatrix} 1 & 3 \\ 0 & -1 \\ 7 & 5 \end{pmatrix} \qquad v := \text{augment}\left[w, \begin{pmatrix} 11 \\ 5 \\ -1 \end{pmatrix} \right] \qquad v = \begin{pmatrix} 1 & 3 & 11 \\ 0 & -1 & 5 \\ 7 & 5 & -1 \end{pmatrix}$$

The augment function only accepts two arguments. To concatenate more than two objects, repeat the use of augment, e.g., augment(augment(u , v) , w).

1. Use Cramer's Rule to compute the first component, $\mathbf{x}_1$, of the solution to $\mathbf{M} \cdot \mathbf{x} = \mathbf{b}$.

$$M := \begin{bmatrix} .12 & .73 & .35 & 0 & -1.2 & 1 & 3 \\ 5 & 1 & -3 & 2.3 & 2 & 3 & 2 \\ 3 & .002 & 12 & .87 & 3 & 2 & 0 \\ -1 & .3 & 4 & 1 & 2.71 & 9.81 & 0 \\ 0 & 7.1 & .731 & 0 & -6 & .021 & 1.7 \\ 2 & 1.2 & -3.21 & 0 & 8 & 2 & -2.135 \\ 5 & 3 & .5 & 1 & 2 & 1 & 7 \end{bmatrix} \qquad b := \begin{bmatrix} 1 \\ 0 \\ 1 \\ 0 \\ 12 \\ -3 \\ 4 \end{bmatrix}$$

ANSWER:

<><><><><><><><><><><><><><><><><><><><><><><><><><><><><><><><><><><><><><><>

2. Use the adjoint matrix to compute the inverse of the following matrix.

$$\begin{pmatrix} 1 & 0 & 1 \\ 3 & 4 & 0 \\ 0 & 5 & -2 \end{pmatrix}$$

ANSWER:

<><><><><><><><><><><><><><><><><><><><><><><><><><><><><><><><><><><><><><><>

3. (Individual) Write a summary of the MATHEMATICS that you learned in this section.

CHAPTER 4

LINES
AND
PLANES

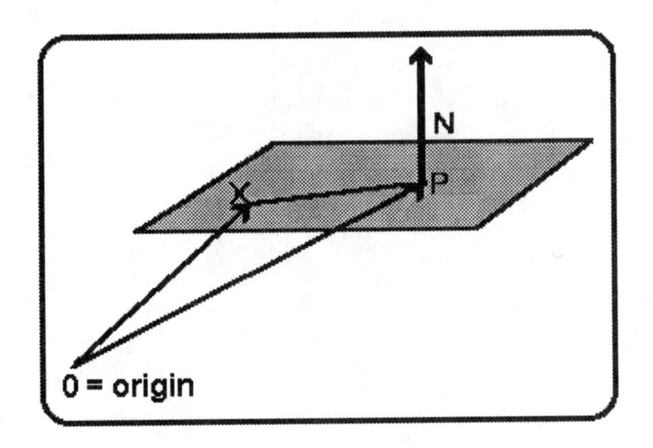

Section 4-1 Lines in R^n

Name:________________

Score:_____________

Overview

Linear Algebra and Geometry are two sides of the same coin.

The goal of this section is to describe lines in $\mathbf{R}^2$ and generalize that description to $\mathbf{R}^n$. Familiar concepts such as parallelism and orthogonality are then investigated.

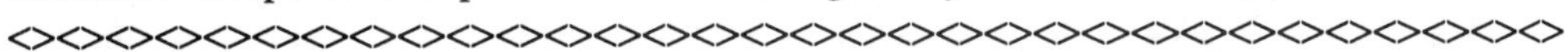

We begin by looking at the description of a line in terms of equations involving the coordinates of points on the line.

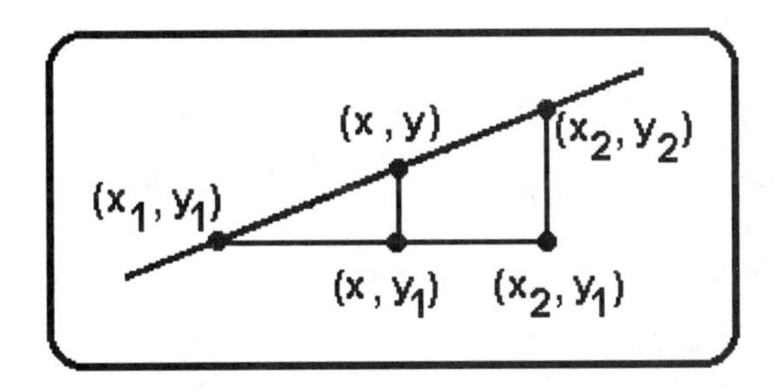

Use similar triangles, to give a condition that shows that the point (x,y) is on the line from (x_1,y_1) to (x_2,y_2).

ANSWER:

Write your equation in the form: **Ax + By = C.**

ANSWER:

This representation is called the ***general form of an equation of a line in the plane***. The form you are most familiar with, y = mx + b, leaves out the important line x = c, where c is a constant.

Are A, B and C unique or can there be several equations that describe the same line?

ANSWER:

Find two points on the line 3x + 7y = 1.

ANSWER:

Is the point (7,-3) on the line $3x + 7y = 1$?

ANSWER:

Use the fact that a line has an equation of the form, $Ax + By = C$ to set up two simultaneous linear equations that indicate that the points (2,5) and (-1,4) lie on a given line. Solve these equations to find the equation of the line.

ANSWER:

Is the equation unique? Can we really speak about '*the equation of a line*' or are there many equations for a given line? Explain.

ANSWER:

We can extend this approach to $\mathbf{R}^3$. To do this consider three points

$$P_1 = \begin{pmatrix} x_2 \\ y_2 \\ z_1 \end{pmatrix} \quad P_2 = \begin{pmatrix} x_2 \\ y_2 \\ z_2 \end{pmatrix} \quad \text{and} \quad \begin{pmatrix} x \\ y \\ z \end{pmatrix}$$

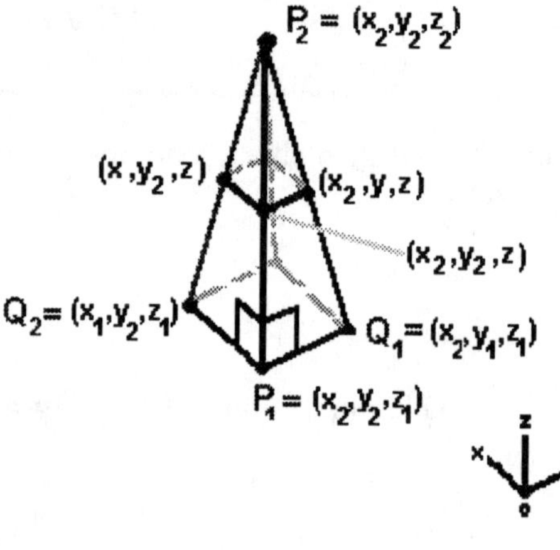

$P_2 = (x_2, y_2, z_2)$

(x, y_2, z)

(x_2, y, z)

(x_2, y_2, z)

$Q_2 = (x_1, y_2, z_1)$

$Q_1 = (x_2, y_1, z_1)$

$P_1 = (x_2, y_2, z_1)$

z x y o

The picture on the right is analogous to the one at the beginning of this section. **All the triangles in this picture are right triangles.** We are interested in

$\Delta P_1 Q_2 P_2$ and $\Delta P_1 Q_1 P_2$

Construct vectors P_2P_1 and Q_1P_1 and show that angle $P_2P_1Q_1$ is a right angle.

ANSWER:

Construct vectors P_1P_2 and Q_2P_1 and show that angle $Q_2P_1P_2$ is a right angle.

ANSWER:

The two triangles of interest are shown below.

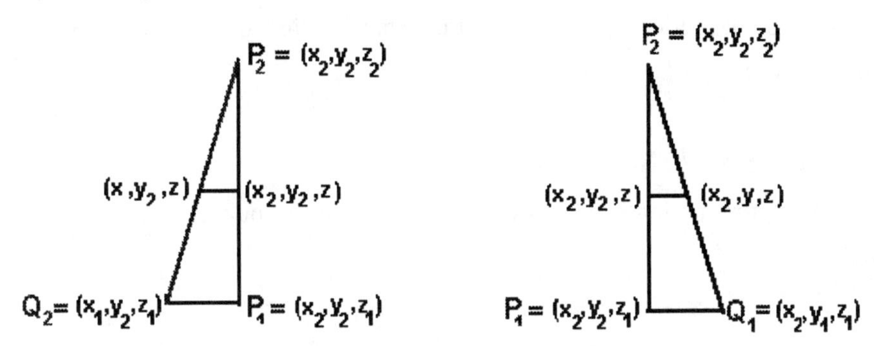

Complete the following using similar triangles.

$$\frac{y_2 - y}{y_2 - y_1} = \blacksquare$$ **Use the triangle on the right.**

$$\frac{x_2 - x}{x_2 - x_1} = \blacksquare$$ **Use the triangle on the left.**

If you transform these by cross multiplying and collecting terms you should get two equations in the following form:

$$\mathbf{Ax + Cz = D}$$
$$\mathbf{Ry + Sz = T}$$

where D and T are expressions for constants.

As we have seen, a line in $\mathbf{R}^3$ is given by a system of two equations. We shall see later that each of these equations represents a plane and therefore a line in $\mathbf{R}^3$ is given as the intersection of two planes.

Find the representation of the line through the points, **p** and **q**, (given below) as the intersection of two planes (i.e., a system of two equations). (Hint: Use the result derived above from similar triangles.)

$$\mathbf{p} := \begin{pmatrix} 2 \\ 7 \\ -1 \end{pmatrix} \qquad \mathbf{q} := \begin{pmatrix} -1 \\ 0 \\ 2 \end{pmatrix}$$

ANSWER:

<><><><><><><><><><><><><><><><><><><><><><><><><><><><><><><><>

Explain how this representation would extend to $\mathbf{R}^4$ for a pair of points (p_1, p_2, p_3, p_4) and (q_1, q_2, q_3, q_4). (Do not produce a formula, just indicate the kind and number of equations that need to be solved.)

ANSWER:

Explain how this representation would extend to $\mathbf{R}^n$ for a pair of points.

ANSWER:

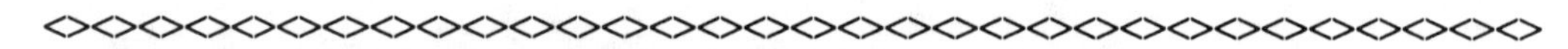

Vector representations of a line in the plane

There are other ways to think about a line in $\mathbf{R}^2$. We saw in Section 1-3 in our discussion of convex linear combinations that given two points $\mathbf{p}$ and $\mathbf{q}$ in $\mathbf{R}^2$ the line segment between them was described as the set of points $t \cdot \mathbf{p} + (1 - t) \cdot \mathbf{q}$ where $0 \le t \le 1$ (see picture below).

$t := 0, .05 .. 1$ $\quad\quad$ $p := \begin{pmatrix} 3 \\ 5 \end{pmatrix}$ $\quad\quad$ $q := \begin{pmatrix} -1 \\ 1 \end{pmatrix}$

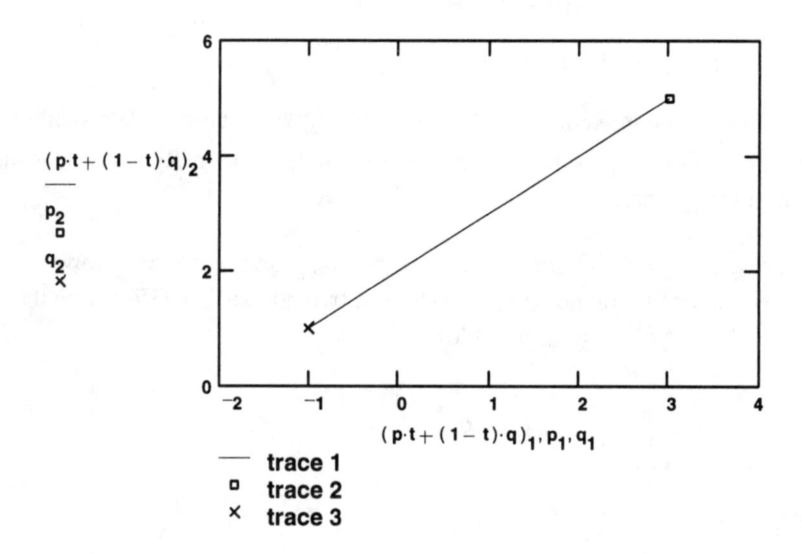

What happens if we allow a larger range for t ? $t := 0, .1 .. 5$ $s := -5, -4.9 .. 0$

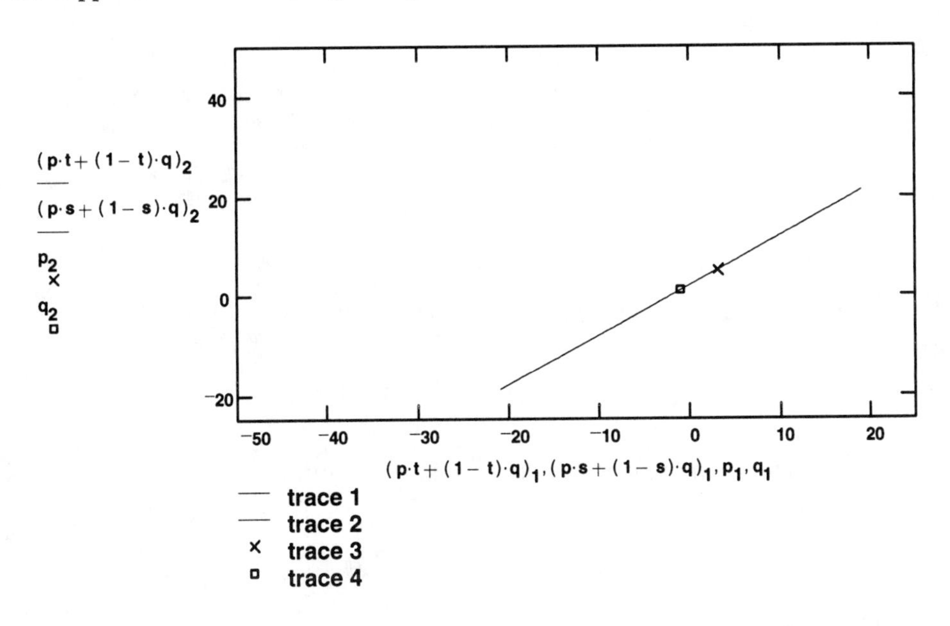

How has the graph changed? Describe in terms of the line between the points p and q.

ANSWER:

Describe the set when $-\infty < t < \infty$; i.e., t can take on any real value.

ANSWER:

The equation $t \cdot \mathbf{p} + (1 - t) \cdot \mathbf{q}$ is called the ***vector equation for the line through p and q***.

Write the vector equation for the line, L, passing through the points (2,5) and (-1,4).

ANSWER:

Which of the following points (if any) lie on the line L?

a) $\begin{pmatrix} 13 \\ 0 \end{pmatrix}$ b) $\begin{pmatrix} 11 \\ 8 \end{pmatrix}$ c) $\begin{pmatrix} -4 \\ 3 \end{pmatrix}$

ANSWER:

◇◇◇◇◇◇◇◇◇◇◇◇◇◇◇◇◇◇◇◇◇◇◇◇◇◇◇◇◇◇◇◇◇◇◇◇

Write the equation of the line through the **p** and the **q** in the above vector form for each of the following points **p**.

$\mathbf{q} := \begin{pmatrix} 1 \\ 2 \end{pmatrix}$ i) $\begin{pmatrix} 6 \\ 10 \end{pmatrix}$ ii) $\begin{pmatrix} -2 \\ 7 \end{pmatrix}$ iii) $\begin{pmatrix} 0 \\ 3 \end{pmatrix}$ iv) $\begin{pmatrix} 5 \\ 0 \end{pmatrix}$

ANSWERS:

Write an equation for each of the preceding lines in the form $Ax + By = C$.

ANSWERS:

Write an equation for each of the following lines in vector form.

a) $5x + 3y = 0$

b) $3x - y = 7$

c) $x = 13$

d) $y = -7$

ANSWERS:

◇◇◇◇◇◇◇◇◇◇◇◇◇◇◇◇◇◇◇◇◇◇◇◇◇◇◇◇◇◇◇

Consider the line determined by **p** and **q**.
The vector equation for this line is: $t \cdot \mathbf{p} + (1 - t) \cdot \mathbf{q}$
If we collect on t we get the alternate form: $\mathbf{q} + t \cdot (\mathbf{p} - \mathbf{q})$

The vector $(\mathbf{p} - \mathbf{q})$ is called a ***direction vector*** for the line.

From this we can see that a line is determined by a direction vector and a point on the line.

Let **v** be a vector. Describe the set $t \cdot \mathbf{v}$ where $-\infty < t < \infty$. (Hint: Think about drawing this in $\mathbf{R}^2$ or $\mathbf{R}^3$.)

ANSWER:

Let **p** be another vector. How does adding **p** to t·**v** change the set of points?

(Hint: Think about drawing this in $\mathbf{R}^2$ or $\mathbf{R}^3$.)

ANSWER:

Consider vectors **p** and **v** in $\mathbf{R}^n$. Describe the set of points **p** + t·**v** where $-\infty < t < \infty$.

ANSWER:

Other than the fact that the points are in $\mathbf{R}^n$ rather than in $\mathbf{R}^2$ does the description differ from that in $\mathbf{R}^2$? Explain.

ANSWER:

You should be able to conclude:

> A **LINE** in $\mathbf{R}^n$ is determined by
> (1) a **DIRECTION VECTOR**, v, and
> (2) a **POINT**, **p**.

Give an equation of the line, L, in $\mathbf{R}^3$ passing through the points (1,2,1) and (5,-1,0) in the vector form described above.

ANSWER:

Find a third point on the line, L.

ANSWER:

<><><><><><><><><><><><><><><><><><><><><><><><><><><><><><><><>

Which, if any, of the following lines are parallel?

a) $\begin{pmatrix}1\\0\\2\end{pmatrix} + t\cdot\begin{pmatrix}1\\3\\-2\end{pmatrix}$ b) $\begin{pmatrix}11\\9\\-5\end{pmatrix} + t\cdot\begin{pmatrix}2\\-3\\1\end{pmatrix}$ c) $\begin{pmatrix}0\\0\\17\end{pmatrix} + t\cdot\begin{pmatrix}4\\12\\-8\end{pmatrix}$

d) $\begin{pmatrix}2\\15\\-117\end{pmatrix} + t\cdot\begin{pmatrix}3\\0\\-1\end{pmatrix}$ e) $\begin{pmatrix}4\\13\\2\end{pmatrix} + 3\cdot t\cdot\begin{pmatrix}2\\6\\-4\end{pmatrix}$ f) $\begin{bmatrix}\pi\\ \frac{\pi}{2}\\ -\pi\end{bmatrix} + \frac{t}{3}\cdot\begin{pmatrix}10\\-15\\5\end{pmatrix}$

ANSWER:

In general can you tell if two lines are parallel? Explain. (Hint: Think about lines in $\mathbf{R}^3$.)

ANSWER:

Find an equation for the line through the points (1,-2,4) and (3,0,7).

ANSWER:

Is the point (7,4,13) on this line?

ANSWER:

Find an equation for the line parallel to the line given above and through the point (5,4,3).

ANSWER:

<><><><><><><><><><><><><><><><><><><><><><><><><><><><><><><><><><><><>

Do the following lines intersect?

$\begin{pmatrix}1\\0\\2\end{pmatrix} + t\cdot\begin{pmatrix}-2\\3\\1\end{pmatrix}$ $\begin{pmatrix}-9\\0\\4\end{pmatrix} + t\cdot\begin{pmatrix}0\\5\\1\end{pmatrix}$

ANSWER:

By merely inspecting the equations of two lines in $\mathbf{R}^3$ can you tell if they intersect? Explain.

ANSWER:

What criteria is used to determine if two lines in $\mathbf{R}^3$ intersect?

ANSWER:

How can you tell if two lines in $\mathbf{R}^n$ intersect?

ANSWER:

<><><><><><><><><><><><><><><><><><><><><><><><><><><><><><><><><><><><><><>

Describe in vector form the line through the points

$$\mathbf{p} := \begin{bmatrix} 2 \\ 0 \\ -1 \\ 3 \\ 11 \\ 7 \end{bmatrix} \qquad \mathbf{q} := \begin{bmatrix} 0 \\ 2 \\ 2 \\ 3 \\ -5 \\ 4 \end{bmatrix}$$

ANSWER:

<><><><><><><><><><><><><><><><><><><><><><><><><><><><><><><><><><><><><><>

Let a curve be given in 4-space parametrically by

$$\mathbf{p(t)} := \begin{bmatrix} t^2 - 1 \\ \sin(\pi \cdot t) \\ \sqrt{t^4 + 4} \\ t \end{bmatrix}$$

What is p(0)? p(1)?

ANSWER:

The line tangent to this curve at a point on the curve, $\mathbf{p}(t)$, is the line through $\mathbf{p}(t)$ with slope $\mathbf{p}'(t)$. Interpret this statement and use it to calculate the tangent line to $\mathbf{p}(t)$ when $t = 1$.
ANSWER:

<><><><><><><><><><><><><><><><><><><><><><><><><><><><><><><><><><><><><><><>

Recall from Section 1-5 that we can use the dot product to define the angle between two vectors. In particular, if $\mathbf{u}$ and $\mathbf{v}$ are vectors then the angle t between them is defined by

$$\cos(t) = \frac{\mathbf{u} \cdot \mathbf{v}}{|\mathbf{u}| \cdot |\mathbf{v}|}$$

We can use this definition to define the angle between two lines.

Let L1 be the line $(1,3) + t \cdot (2,-2)$ and L2 be the line $(1,3) + t \cdot (3,1)$

Compute the angle between these lines.
ANSWER:

Does it make sense to speak about the angle between two lines in $\mathbf{R}^n$ if the lines do not intersect?
ANSWER:

What is the angle between the two lines in $\mathbf{R}^5$ given below?

L1: $\begin{bmatrix} 1 \\ 3 \\ 5 \\ 0 \\ 2 \end{bmatrix} + t \cdot \begin{bmatrix} 2 \\ 0 \\ 5 \\ 1 \\ 1 \end{bmatrix}$ L2: $\begin{bmatrix} 1 \\ -3 \\ 0 \\ 1 \\ -1 \end{bmatrix} + t \cdot \begin{bmatrix} 5 \\ 2 \\ 1 \\ 1 \\ -5 \end{bmatrix}$

ANSWER:

Two vectors, $\mathbf{u}$ and $\mathbf{v}$, are called ***orthogonal*** or ***perpendicular*** if the angle between them is $\pi/2$. This is the case if and only if the dot product $\mathbf{u} \cdot \mathbf{v} = 0$.

Find a nonzero vector perpendicular to the vector (a,b).

ANSWER:

Write an equation in vector form of the line through (5,3) that is perpendicular to the line through (3,1) with direction vector (4,1).

ANSWER:

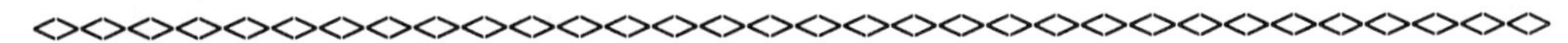

Let's return to the general form of an equation for a line: $Ax + By = C$. Next we interpret what the coefficients A, B and C mean.

1) Find an equation for all vectors (x,y) that are orthogonal (perpendicular) to a given vector (A,B).

2) Describe the set of all such (x,y) geometrically.

3) Finally, give the equation of all points (x,y) such that the vector from (p,q) to (x,y) is perpendicular to a given vector (A,B). (See picture below).

The vector (A,B) is perpendicular to the vector from (p,q) to (x,y). Write down the equation that corresponds to this fact.

ANSWER:

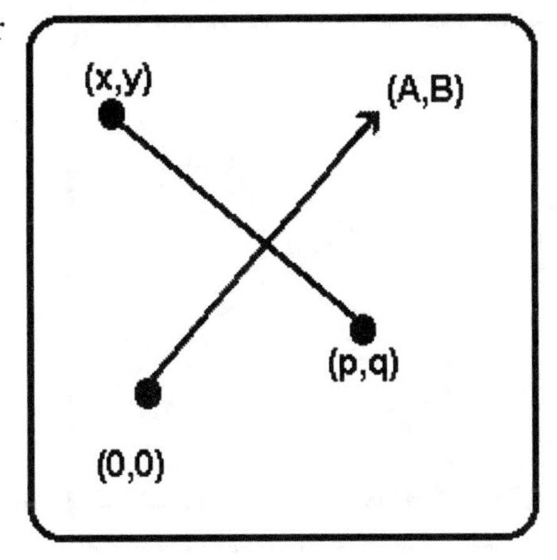

Why is the vector (A,B) perpendicular to the line $Ax + By = C$?

(Vector (A,B) is called the ***normal vector*** to line $Ax + By = C$.)

ANSWER:

What is the slope of the line $Ax + By = C$?

ANSWER:

What is the slope of the vector (A,B)? (Note the fact that slope is a single number is an accident of two-space. If a vector has only two components then its direction can be determined by the ratio of the coordinates.)

ANSWER:

Compute the equation of the line through (3,1) with normal (1,5).

ANSWER:

<><><><><><><><><><><><><><><><><><><><><><><><><><><><><><><><><><><><><><><><><>

Our next objective is to find a formula for the distance from the line through the point (p,q) with normal vector (A,B) to the origin. (As usual distance means perpendicular distance).

From the picture it is clear that the desired distance is $|(p,q)| \cos(t)$. Explain.

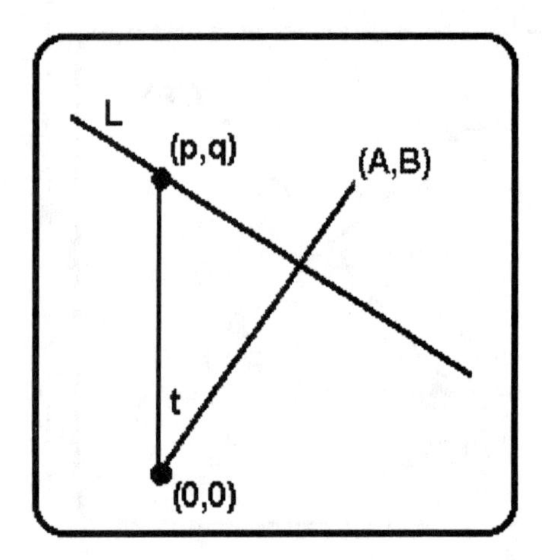

Use the formula for the angle between two vectors to compute the cosine between the vectors (p,q) and (A,B).

ANSWER:

Substitute this into the previous formula to derive an expression for the distance from the origin to the line $Ax + By = C$. Use the fact that $Ap + Bq = C$ to simplify the equation.

ANSWER:

As we have seen, '*the*' equation for a line $Ax + By = C$ is only determined up to a constant multiple i.e., $3x + 2y = 1$ and $6x + 4y = 2$ are the same line. If we insist that the vector (A,B) be a unit vector ($A^2 + B^2 = 1$) then A and B are determined up to their sign. If we further insist that the constant C be non-negative the constants A, B, and C are uniquely determined for each line except when $C = 0$. We call this the ***standard form of the equation of the line***.

Write the standard form for the equation of each of the following lines.

***ANSWERS*:**

a) $3x + 2y = 7$

b) $-x + 5y = 3$

c) $Ax + By = C \quad C > 0$

If the equation of the line $Ax + By = C$ is in standard form what can you conclude about the meaning of the constant C?

***ANSWER*:**

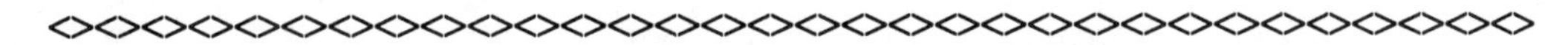

Exercise: (Individual) Write a summary of the MATHEMATICS that you learned in this section.

Section 4-2 Planes in R^n

Name:________________

Score:_____________

Overview

In the previous section we related the geometry of lines in $\mathbf{R}^n$ and linear algebra. In this section we study planes in $\mathbf{R}^3$ in terms of our knowledge of linear algebra and extend the description to planes and hyperplanes in $\mathbf{R}^n$.

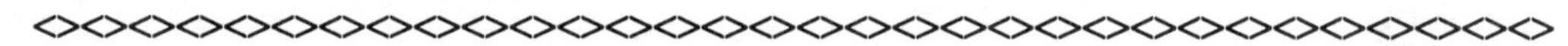

All vectors (x,y,z) orthogonal to a given vector (A,B,C) satisfy the equation Ax + By + Cz = 0. The set of such vectors form a **plane** through the origin. (A,B,C) is called the **normal vector** to the plane.

Find an equation of the plane through the origin that contains the vectors

$$\mathbf{u} := \begin{pmatrix} 1 \\ 5 \\ 0 \end{pmatrix} \quad \text{and} \quad \mathbf{v} := \begin{pmatrix} -2 \\ 3 \\ 5 \end{pmatrix} \quad \text{Call the normal to the plane} \quad \mathbf{N} := \begin{pmatrix} A \\ B \\ C \end{pmatrix}$$

Use the equations $\mathbf{N}\cdot\mathbf{u} = \mathbf{0}$ and $\mathbf{N}\cdot\mathbf{v} = \mathbf{0}$ to set up and solve two equations in three unknowns.

ANSWER:

This is similar to the situation we had in the previous section with the equation of a line. Why is there more than one equation for a plane? Is the normal vector **N** unique?

ANSWER:

Now that we understand how to represent a plane through the origin, we can ask the same question about planes through an arbitrary point P.

$$\mathbf{P} := \begin{pmatrix} p_1 \\ p_2 \\ p_3 \end{pmatrix} \qquad \mathbf{X} := \begin{pmatrix} x \\ y \\ z \end{pmatrix}$$

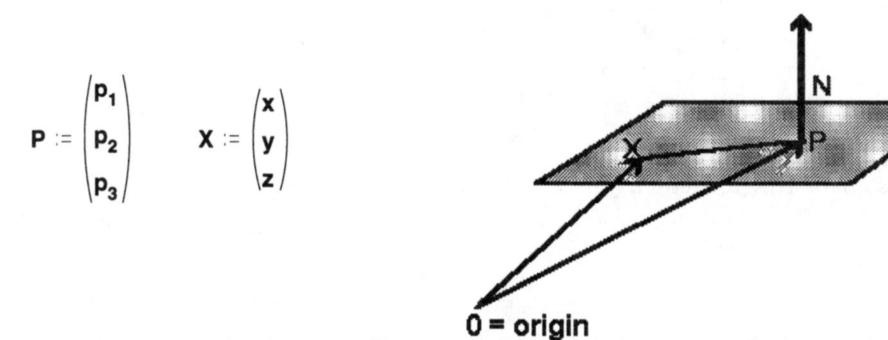

This generalizes the situation of the equation of the line in the plane with normal vector (A,B) through a point **P**.

N must be perpendicular to the vector **X** - **P**. Write this as an equation (do not expand the equation).
ANSWER:

Find an equation of the plane with normal vector, **N** = (2,-1,3) through the point, **P** = (0,5,1).
ANSWER:

How can you tell when two planes are parallel? (Hint: Draw a picture of the planes together with their normal vectors.)
ANSWER:

Are the planes: 3x - 2y + z = 2 and 9x - 6y + 3x = 5 parallel?
ANSWER:

Somewhere you learned that *three points determine a plane.* (This explains why 3 legged stools are more stable than four legged chairs.) There are several ways of determining the equation of a plane given three points. Remember, once again, that there are many equations for a given plane.

Find an equation of the plane containing the points **u**, **v** and **w** given below.

$$\mathbf{u} := \begin{pmatrix} 3 \\ 7 \\ -2 \end{pmatrix} \qquad \mathbf{v} := \begin{pmatrix} 1 \\ 1 \\ 0 \end{pmatrix} \qquad \mathbf{w} := \begin{pmatrix} 0 \\ 2 \\ -5 \end{pmatrix}$$

Assume that the plane has the equation: Ax + By + Cz = D. Then each point yields an equation. So we have three equations in four unknowns.

$$\begin{pmatrix} 3 & 7 & -2 & -1 \\ 1 & 1 & 0 & -1 \\ 0 & 2 & -5 & -1 \end{pmatrix} \cdot \begin{pmatrix} A \\ B \\ C \\ D \end{pmatrix} = \begin{pmatrix} 0 \\ 0 \\ 0 \end{pmatrix}$$

Solve the system by row-reduction. (Hint: Use rref.)
ANSWER:

<><><><><><><><><><><><><><><><><><><><><><><><><><><><><><><><><><><><><>

Use the same procedure to find the equation of the plane containing the points

$$u := \begin{pmatrix} 1 \\ 8 \\ -3 \end{pmatrix} \qquad v := \begin{pmatrix} 0 \\ 2 \\ 3 \end{pmatrix} \qquad w := \begin{pmatrix} 3 \\ 18 \\ -18 \end{pmatrix}$$

ANSWER:

In the first example above we could solve for the equation of the plane in terms of the constant D. We can't do this in the second example. Explain why.
ANSWER:

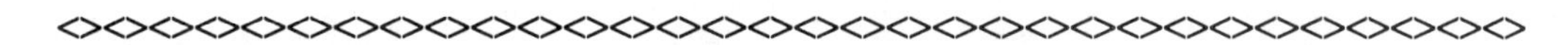

Another way to find the equation is to consider the vectors

$$uv := u - v \qquad uv = \begin{pmatrix} 1 \\ 6 \\ -6 \end{pmatrix} \qquad \text{and} \qquad uw := u - w \qquad uw = \begin{pmatrix} -2 \\ -10 \\ 15 \end{pmatrix}$$

The normal, (A,B,C), must be perpendicular to each of these vectors. This yields two equations in three unknowns and enables us to compute the normal vector. We can then calculate D by substitution. Perform this calculation for points

$$u := \begin{pmatrix} 1 \\ 8 \\ -3 \end{pmatrix} \qquad v := \begin{pmatrix} 0 \\ 2 \\ 3 \end{pmatrix} \qquad w := \begin{pmatrix} 3 \\ 18 \\ -18 \end{pmatrix}$$

ANSWER:

In the equation $Ax + By + Cz = D$ what does D represent? This is a similar question to the one we asked in the previous section for the equation of a line. It should not be surprising to find that the answer is similar to the previous case. What is the (perpendicular) distance of the plane $Ax + By + Cz = D$ to the origin?

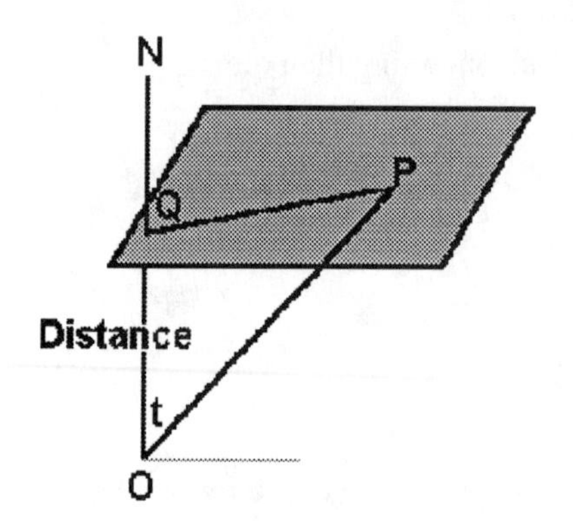

Triangle (**QPO**) is a right triangle. Why?

The length of side (**OQ**) is the perpendicular distance from the origin. The length of side (**OP**) is |**P**|.

Distance = |**P**|·cos(t)

Express cos(t) in terms of the dot product of **P** and **N** and derive a formula for the distance of the plane from the origin.

ANSWERS:

Let Ax + By + Cz = D be a plane. Assume that (A,B,C) is a unit vector. What is its distance from the origin?

ANSWER:

Calculate the distance of the plane, 3x + 5y - z = 7 from the origin.

ANSWER:

What is the equation of the plane with normal vector (0,2,1) whose distance from the origin is 7?

ANSWER:

You know that two planes either

a) are parallel

b) coincide, or

c) intersect in a line.

How can one tell from the equations of the two planes if the planes are parallel?

ANSWER:

How can one tell from the equations of the two planes if the planes are the same?

ANSWER:

Find the line of intersection of the planes

$2 \cdot x + y - z = 3$ and $x + z = 2$

i.e., find all solutions to the system of equations

$$2 \cdot x + y - z = 3$$
$$x + z = 2$$

This is a nonhomogeneous system of two linear equations in three unknowns. Assume that the planes are distinct and not parallel. What is the general form of the solution set? Compare this to the equation of a line in three space.

ANSWER:

<><><><><><><><><><><><><><><><><><><><><><><><><><><><><><><>

Systems of equations and the geometry of lines, planes, etc. are two sides of the same coin. For each statement in one language there is a corresponding statement in the other, for example:

Geometry

Find the line of intersection of the planes

$3x + 2y - z = 2$

$4x + y + z = 3$

Find the equation of the plane containing the points

(2,3,0) **(1,0,5)** **(7,-2,1)**

Algebra

Solve: $3x + 2y - z = 2$

$4x + y + z = 3$

Solve: $2A + 3B = D$

$A + 5B = D$

$7A - 2B + C = D$

<><><><><><><><><><><><><><><><><><><><><><><><><><><><><><><>

Consider two vectors, **p** and **q**, in the plane.

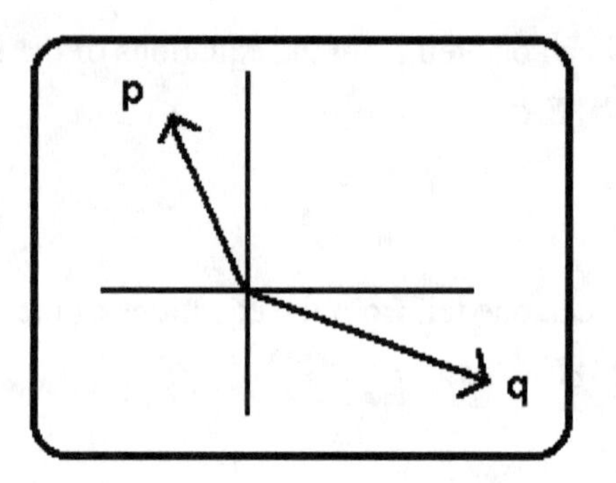

Which other vectors **x** can be written as a linear combination of **p** and **q**?
The algebraic formulation of this question is:
Find a and b so that $a \cdot \mathbf{p} + b \cdot \mathbf{q} = \mathbf{x}$

$$a \cdot \begin{pmatrix} p_1 \\ p_2 \end{pmatrix} + b \cdot \begin{pmatrix} q_1 \\ q_2 \end{pmatrix} = \begin{pmatrix} x_1 \\ x_2 \end{pmatrix}$$

or looking at coordinate-wise equality
$$\begin{aligned} a \cdot p_1 + b \cdot q_1 &= x_1 \\ a \cdot p_2 + b \cdot q_2 &= x_2 \end{aligned}$$

When does this have a solution?

ANSWER:

Consider the same question for 3 vectors. Given two vectors, can a third be expressed in terms of the first two?

ANSWER:

Let **p**, **q**, and **x** be vectors in $\mathbf{R}^3$. When are there scalars, a and b, such that $a \cdot \mathbf{p} + b \cdot \mathbf{q} = \mathbf{x}$? Write this as a system of equations.

ANSWER:

Solve: $x - 2y + z = 7$.

What does this mean? After all the solution set is simply the set of all (x,y,z) that satisfy the equation. However, we do have a procedure for solving such a system, even though in this case it seems obvious; namely,

$$x = 7 + 2 \cdot y - z$$

Below is the general solution in vector notation

$$\begin{pmatrix} 7 \\ 0 \\ 0 \end{pmatrix} + y \cdot \begin{pmatrix} 2 \\ 1 \\ 0 \end{pmatrix} + z \cdot \begin{pmatrix} -1 \\ 0 \\ 1 \end{pmatrix}$$

The last two vectors above are solutions to the associated homogeneous system; i.e. the plane through the origin. Thus another way to view a plane is as a point and the related plane through the origin. This formulation is very similar to the definition of a line. A line was a point and a direction vector. A plane is a point and two **direction** vectors. In particular we see that given two non-collinear vectors in 3 space, they determine a unique plane.

Find the equation of the plane (through the origin) which contains the vectors ===============> by calculating a vector orthogonal to each of the vectors.

$$\begin{pmatrix} 1 \\ 3 \\ -2 \end{pmatrix} \qquad \begin{pmatrix} 4 \\ 0 \\ 2 \end{pmatrix}$$

ANSWER:

<><><><><><><><><><><><><><><><><><><><><><><><><><><><><><><>

Planes in R^n

We saw in the previous section that a line in $\mathbf{R}^n$ was given by a direction vector and a point. This generalized the situation in the plane. In an analogous manner a plane in $\mathbf{R}^n$ is given by two direction vectors and a point. In linear algebra terms a ***plane in R^n through the origin*** is a two dimensional subspace; i.e., all linear combinations of two non-collinear vectors. As we have seen earlier subspaces arise as solutions of a homogeneous system of equations. More generally, a plane in $\mathbf{R}^n$ that does not contain the origin is the solution of a nonhomogeneous system of equations; i.e., one with two free variables.

Example

Consider the system $\mathbf{A}\cdot\mathbf{x} = \mathbf{b}$ where $\quad \mathbf{A} := \begin{pmatrix} 1 & 0 & 2 & 0 & 2 \\ 5 & 7 & 10 & 0 & 3 \\ 2 & 4 & 4 & 1 & 1 \end{pmatrix} \quad$ and $\quad \mathbf{b} := \begin{pmatrix} 1 \\ 0 \\ 1 \end{pmatrix}$

$$\text{rref}\begin{pmatrix} 1 & 0 & 2 & 0 & 2 & 3 \\ 5 & 7 & 10 & 0 & 3 & 29 \\ 2 & 4 & 4 & 1 & 1 & 13 \end{pmatrix} = \begin{pmatrix} 1 & 0 & 2 & 0 & 2 & 3 \\ 0 & 1 & 0 & 0 & -1 & 2 \\ 0 & 0 & 0 & 1 & 1 & -1 \end{pmatrix}$$

The solution is: $\quad \begin{bmatrix} 3 \\ 2 \\ 0 \\ -1 \\ 0 \end{bmatrix} + s\cdot\begin{bmatrix} -2 \\ 0 \\ 1 \\ 0 \\ 0 \end{bmatrix} + t\cdot\begin{bmatrix} -2 \\ 1 \\ 0 \\ -1 \\ 1 \end{bmatrix}$

This is a plane in $\mathbf{R}^5$.

<><><><><><><><><><><><><><><><><><><><><><><><><><><><><><><>

Hyperplanes

Consider the set given in $\mathbf{R}^7$ by $\mathbf{p} + r \cdot \mathbf{u} + s \cdot \mathbf{v} + t \cdot \mathbf{w}$ where **p**, **u**, **v**, and **w** are vectors and r, s, and t are scalars. We call this a 3-dimensional hyperplane. The word **hyperplane** is chosen since we can not choose a different word for each $\mathbf{R}^n$. Each hyperplane in $\mathbf{R}^n$ is the solution of a system of linear equations in n variables. If the system is homogeneous the vector **p** is zero. In particular the solution to the equation

$$a_1 \cdot x_1 + a_2 \cdot x_2 + \dots + a_n \cdot x_n = b$$

is an (n-1)-dimensional hyperplane in $\mathbf{R}^n$. The vector $(a_1, \dots, a_n)$ is the **normal** to this hyperplane.

<><><><><><><><><><><><><><><><><><><><><><><><><><><><><><><><><><><><><><><><>

Consider the system of k linear equations:

$$a_{11} \cdot x_1 + a_{12} \cdot x_2 + \dots + a_{1n} \cdot x_n = b_1$$

$$a_{21} \cdot x_1 + a_{22} \cdot x_2 + \dots + a_{2n} \cdot x_n = b_2$$

....

$$a_{k1} \cdot x_1 + a_{k2} \cdot x_2 + \dots + a_{kn} \cdot x_n = b_k$$

Write in your own words (a) what each equation represents geometrically and (b) what the solution set of the system represents geometrically. What does it mean geometrically when the system is inconsistent?

ANSWER:

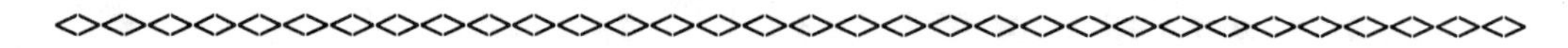

Exercise: (Individual) Write a summary of the MATHEMATICS that you learned in this section.

Section 4-3 The Cross Product

Name:_______________

Score:____________

Overview

Given two vectors in $\mathbf{R}^3$, their cross product is a vector in $\mathbf{R}^3$ that is orthogonal to each of the vectors. Use of the cross product simplifies computations involving planes in $\mathbf{R}^3$. In this section we explore the properties of the cross product.

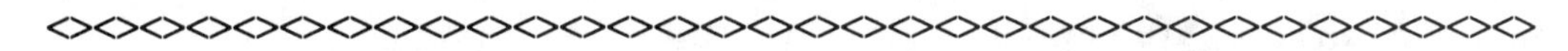

Before you begin this section, you should do Section 3-4 if you skipped it earlier.

<><><><><><><><><><><><><><><><><><><><><><><><><><><><><><><><><><>

Let **u**, **v**, and **w** be vectors in $\mathbf{R}^3$.

$$u=\begin{pmatrix} u_1 \\ u_2 \\ u_3 \end{pmatrix} \qquad v=\begin{pmatrix} v_1 \\ v_2 \\ v_3 \end{pmatrix} \qquad w=\begin{pmatrix} w_1 \\ w_2 \\ w_3 \end{pmatrix}$$

Form the matrix, **M**, whose rows are **u**, **v** and **w**.

$$M=\begin{pmatrix} u_1 & u_2 & u_3 \\ v_1 & v_2 & v_3 \\ w_1 & w_2 & w_3 \end{pmatrix}$$

Box $\begin{pmatrix} u_1 & u_2 & u_3 \\ v_1 & v_2 & v_3 \\ w_1 & w_2 & w_3 \end{pmatrix}$ and select **Determinant of Matrix** from the **Symbolic** menu.

Box w_1 in the resulting expression and select **Collect on Subexpression** from the **Symbolic** menu. Repeat for w_2 and w_3.

You can recognize the resulting expression as the dot product of **w** with a vector that we denote (for the moment) as $\mathbf{X}(\mathbf{u},\mathbf{v})$ to indicate that it is a function of **u** and **v**.

To summarize:

$$\left|\begin{pmatrix} u_1 & u_2 & u_3 \\ v_1 & v_2 & v_3 \\ w_1 & w_2 & w_3 \end{pmatrix}\right| = X(u,v)\cdot w$$

Use the properties of the determinant to explain why:

1) $\mathbf{X}(c\mathbf{u},\mathbf{v}) = c\mathbf{X}(\mathbf{u},\mathbf{v})$ and $\mathbf{X}(\mathbf{u},c\mathbf{v}) = c\mathbf{X}(\mathbf{u},\mathbf{v})$ where c is a scalar.

2) $\mathbf{X}(\mathbf{u}+\mathbf{z},\mathbf{v}) = \mathbf{X}(\mathbf{u},\mathbf{v}) + \mathbf{X}(\mathbf{z},\mathbf{v})$ and $\mathbf{X}(\mathbf{u},\mathbf{v}+\mathbf{z}) = \mathbf{X}(\mathbf{u},\mathbf{v}) + \mathbf{X}(\mathbf{u},\mathbf{z})$

   where **z** is a vector in $\mathbf{R}^3$

3) $\mathbf{X}(\mathbf{u},\mathbf{v}) = -\mathbf{X}(\mathbf{v},\mathbf{u})$

4) $\mathbf{X}(\mathbf{u},\mathbf{v})\cdot\mathbf{w} = -\mathbf{X}(\mathbf{u},\mathbf{w})\cdot\mathbf{v}$ and $\mathbf{X}(\mathbf{u},\mathbf{v})\cdot\mathbf{w} = -\mathbf{X}(\mathbf{w},\mathbf{v})\cdot\mathbf{u}$

5) $\mathbf{X}(\mathbf{u},\mathbf{v})\cdot\mathbf{u} = \mathbf{X}(\mathbf{u},\mathbf{v})\cdot\mathbf{v} = 0$

6) $\mathbf{X}(\mathbf{u},\mathbf{u}) = 0$

ANSWERS:

The function $\mathbf{X}(\mathbf{u},\mathbf{v})$ is called the **cross product** of **u** and **v** and is usually denoted as $\mathbf{u} \times \mathbf{v}$.

In Mathcad the cross-product is entered by using the [x×y] button
on the matrix palette (palette [2] in Mathcad 5.0).
This produces the template ▪ × ▪ in which the placeholders can be filled.

Evaluate the following symbolically and compare with your computation above.

$$\begin{pmatrix} u_1 \\ u_2 \\ u_3 \end{pmatrix} \times \begin{pmatrix} v_1 \\ v_2 \\ v_3 \end{pmatrix}$$

Properties 1 and 2 above are linearity. Property 3 says that $\mathbf{u} \times \mathbf{v} = -\mathbf{v} \times \mathbf{u}$ and Property 4 relates the cross and dot products; i.e., $(\mathbf{u} \times \mathbf{v})\cdot\mathbf{w} = -(\mathbf{u} \times \mathbf{w})\cdot\mathbf{v}$. Property 5 says that the cross product of **u** and **v** is orthogonal to both **u** and **v**. If **u** and **v** are linearly independent the orthogonality property fixes the direction of $\mathbf{u} \times \mathbf{v}$. We now explore the length of $\mathbf{u} \times \mathbf{v}$.

As we saw above:

$$(u \times v)\cdot w = \left| \begin{pmatrix} u_1 & u_2 & u_3 \\ v_1 & v_2 & v_3 \\ w_1 & w_2 & w_3 \end{pmatrix} \right|$$

But the above determinant is the volume of the parallelepiped generated by **u**, **v**, and **w**. In particular, if

$$\mathbf{w} = \mathbf{u} \times \mathbf{v}$$

then

$$(\mathbf{u} \times \mathbf{v}) \cdot (\mathbf{u} \times \mathbf{v}) = (|\mathbf{u} \times \mathbf{v}|)^2$$

but it is also the volume of the parallelepiped.

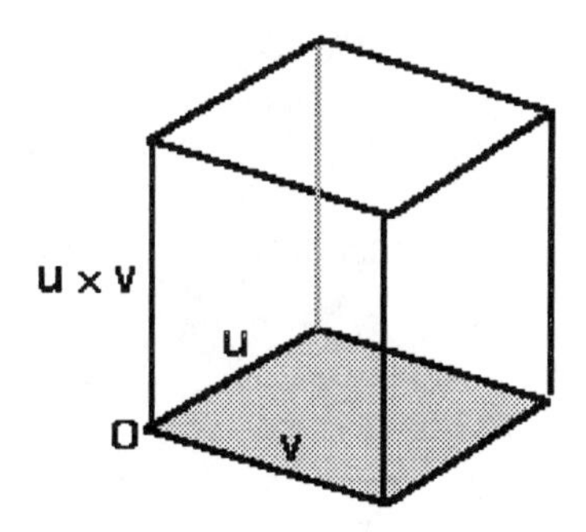

Since $\mathbf{u} \times \mathbf{v}$ is perpendicular to the base, the volume equals $|\mathbf{u} \times \mathbf{v}| \cdot$ BaseArea.

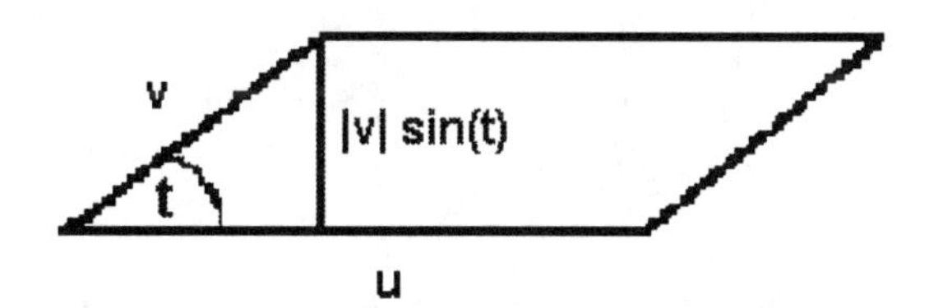

Write an expression for the area of the base involving the angle t.

ANSWER:

Substitute this into $(|\mathbf{u} \times \mathbf{v}|)^2 = |\mathbf{u} \times \mathbf{v}| \cdot \mathbf{BaseArea}$

and divide both sides by $|\mathbf{u} \times \mathbf{v}|$

to yield an expression for the length of $\mathbf{u} \times \mathbf{v}$ in terms of the length of **u**, the length of **v** and the angle between them.

ANSWER:

You have now determined the magnitude of the cross product of **u** and **v** in terms of $|\mathbf{u}|$, $|\mathbf{v}|$ and the angle between them. In addition we know that the cross product is perpendicular to **u** and **v**. Is this enough to uniquely determine the cross product?

ANSWER:

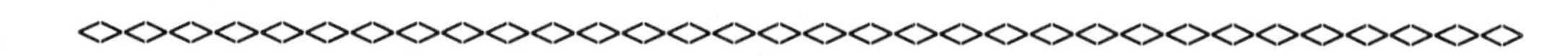

Representations of the cross product

If $\quad \mathbf{u} = \begin{pmatrix} u1 \\ u2 \\ u3 \end{pmatrix} \quad$ and $\quad \mathbf{v} = \begin{pmatrix} v1 \\ v2 \\ v3 \end{pmatrix}$

their cross product $\mathbf{u} \times \mathbf{v}$ is often written as the determinant

$$\left| \begin{pmatrix} u1 & u2 & u3 \\ v1 & v2 & v3 \\ i & j & k \end{pmatrix} \right|$$

where **i**, **j** and **k** are used to denote the standard basis in $\mathbf{R}^3$.

$$i := \begin{pmatrix} 1 \\ 0 \\ 0 \end{pmatrix} \qquad j := \begin{pmatrix} 0 \\ 1 \\ 0 \end{pmatrix} \qquad k := \begin{pmatrix} 0 \\ 0 \\ 1 \end{pmatrix}$$

This is not a formal definition since the determinant of a matrix with vector entries has not been defined. To evaluate this *determinant* perform the following steps.

(1) Use Maple to compute the determinant.
(2) Collect on the terms **i**, **j**, and **k**.
(3) Delete symbol **i** and substitute the vector value of **i**.
 Repeat for symbols **j** and **k**.
(4) Evaluate symbolically.

If you have successfully completed these steps above you will get

$$\begin{pmatrix} u2 \cdot v3 - u3 \cdot v2 \\ -u1 \cdot v3 + v1 \cdot u3 \\ -v1 \cdot u2 + u1 \cdot v2 \end{pmatrix}$$

Check your work by computing the cross product on the right symbolically.

$$\begin{pmatrix} u1 \\ u2 \\ u3 \end{pmatrix} \times \begin{pmatrix} v1 \\ v2 \\ v3 \end{pmatrix}$$

You may also verify that the first coordinate of the cross product is the (3,1) cofactor of the matrix given above; the second term is the (3,2) cofactor, and the third term is the (3,3) cofactor.

$$\left|\begin{array}{c} \left|\begin{pmatrix} u2 & u3 \\ v2 & v3 \end{pmatrix}\right| \\ -\left|\begin{pmatrix} u1 & u3 \\ v1 & v3 \end{pmatrix}\right| \\ \left|\begin{pmatrix} u1 & u2 \\ v1 & v2 \end{pmatrix}\right| \end{array}\right|$$

Do not try to save a Mathcad document that contains an expression similar to the one above. You will be unable to load the file once it has been saved. The vector above is a picture of the object and not a Mathcad expression.

<><><><><><><><><><><><><><><><><><><><><><><><><><><><><><>

Applications of the cross product

Find an equation of the form Ax + By + Cz = D for the plane containing the points:

$$\begin{pmatrix} 1 \\ 0 \\ 2 \end{pmatrix} \qquad \begin{pmatrix} 3 \\ -2 \\ 4 \end{pmatrix} \qquad \begin{pmatrix} 7 \\ 12 \\ 1 \end{pmatrix}$$

Solution 1:

Set up and solve a system of linear equations for A, B, C and D.

ANSWER:

Solution 2:

Call the points, **P**, **Q** and **R** (for the moment). Then the normal must be orthogonal to the vectors **P** - **R** and **P** - **Q**. But the cross product is normal to these vectors as shown above. Thus the normal is

$$\mathbf{N}=(\mathbf{P}-\mathbf{R})\times(\mathbf{Q}-\mathbf{R})$$

$$N := \left[\begin{pmatrix} 1 \\ 0 \\ 2 \end{pmatrix} - \begin{pmatrix} 7 \\ 12 \\ 1 \end{pmatrix}\right] \times \left[\begin{pmatrix} 3 \\ -2 \\ 4 \end{pmatrix} - \begin{pmatrix} 7 \\ 12 \\ 1 \end{pmatrix}\right] \qquad N = \begin{pmatrix} -22 \\ 14 \\ 36 \end{pmatrix}$$

What part of equation Ax + By + Cz = D does the normal determine?

ANSWER:

To find D substitute one of the points below into the equation Ax + By + Cz = D.

$$\begin{pmatrix}1\\0\\2\end{pmatrix} \qquad \begin{pmatrix}3\\-2\\4\end{pmatrix} \qquad \begin{pmatrix}7\\12\\1\end{pmatrix}$$

This is easily done by evaluating symbolically the expression on the right to get the equation of the plane.

$$\begin{pmatrix}-22\\14\\36\end{pmatrix} \cdot \begin{pmatrix}x\\y\\z\end{pmatrix} = \begin{pmatrix}-22\\14\\36\end{pmatrix} \cdot \begin{pmatrix}1\\0\\2\end{pmatrix}$$

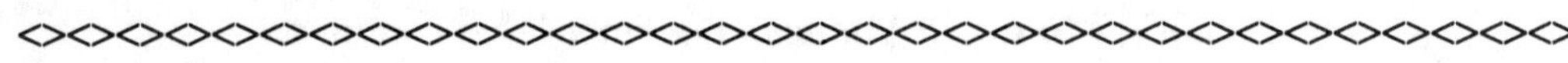

Exercises

1. Compute the equation of the plane that contains the two lines

$$\begin{pmatrix}3\\7\\-2\end{pmatrix} + t\cdot\begin{pmatrix}5\\-2\\1\end{pmatrix} \qquad \text{and} \qquad \begin{pmatrix}3\\7\\-2\end{pmatrix} + t\cdot\begin{pmatrix}11\\6\\-4\end{pmatrix}$$

ANSWER:

2. Compute the equation of the plane that contains the three points

$$\begin{pmatrix}1\\0\\-3\end{pmatrix} \qquad \begin{pmatrix}2\\5\\7\end{pmatrix} \qquad \begin{pmatrix}0\\11\\13\end{pmatrix}$$

ANSWER:

3. The cross product of two vectors in $\mathbf{R}^3$ is again a vector in $\mathbf{R}^3$. Explain why it would be unrealistic to expect that if **u** and **v** are vectors in $\mathbf{R}^n$ (n > 3) we could construct a cross product $\mathbf{u} \times \mathbf{v}$ with the property that $\mathbf{u} \times \mathbf{v}$ is orthogonal to each of **u** and **v**.
ANSWER:

Let $\quad \mathbf{x} := \begin{bmatrix} x_1 \\ x_2 \\ x_3 \\ x_4 \end{bmatrix} \quad \mathbf{y} := \begin{bmatrix} y_1 \\ y_2 \\ y_3 \\ y_4 \end{bmatrix} \quad \mathbf{z} := \begin{bmatrix} z_1 \\ z_2 \\ z_3 \\ z_4 \end{bmatrix}$

Find a vector $\mathbf{u} := \begin{bmatrix} u_1 \\ u_2 \\ u_3 \\ u_4 \end{bmatrix}$ such that **u** is orthogonal to each of **x**, **y**, and **z**.

Solution 1: Set up a system of linear equations that will allow you to solve for **u**.

ANSWER:

Solution 2: Generalize the cross product definition. **HOW?**

Consider the following matrix:

$$\begin{bmatrix} x_1 & x_2 & x_3 & x_4 \\ y_1 & y_2 & y_3 & y_4 \\ z_1 & z_2 & z_3 & z_4 \\ u_1 & u_2 & u_3 & u_4 \end{bmatrix}$$

We can mimic the construction of the cross product given above except now the cross product will be a function of ***three*** vectors, **x**, **y**, and *z*, instead of two as in the previous case. Call this function $\mathbf{X}(\mathbf{x},\mathbf{y},\mathbf{z})$. $\mathbf{X}(\mathbf{x},\mathbf{y},\mathbf{z})$ will then be determined by setting the determinant of the above matrix equal to $\mathbf{X}(\mathbf{x},\mathbf{y},\mathbf{z})\cdot\mathbf{u}$. Generalize the properties of the 3 dimensional cross product given above.

ANSWER:

Define $\mathbf{X(x,y,z)}$ in terms of the cofactors of $\begin{bmatrix} x_1 & x_2 & x_3 & x_4 \\ y_1 & y_2 & y_3 & y_4 \\ z_1 & z_2 & z_3 & z_4 \\ u_1 & u_2 & u_3 & u_4 \end{bmatrix}$

ANSWER:

Define a 'cross product' in $\mathbf{R}^n$. How many vectors will be needed to determine this 'product'?

ANSWER:

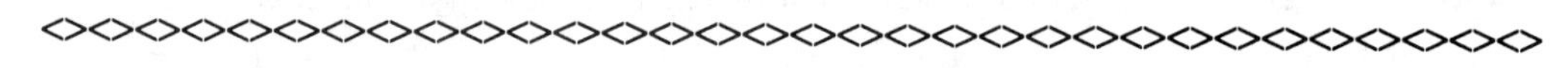

Find a vector in $\mathbf{R}^4$ that is orthogonal to the following three vectors.

$$\begin{pmatrix} 2 \\ -5 \\ 6 \\ 1 \end{pmatrix} \qquad \begin{pmatrix} 3 \\ 0 \\ 8 \\ -2 \end{pmatrix} \qquad \begin{pmatrix} 1 \\ 3 \\ 0 \\ -5 \end{pmatrix}$$

ANSWER:

In $\mathbf{R}^5$ find a vector that is orthogonal to the following four vectors.

$$\begin{bmatrix} 1 \\ 5 \\ 0 \\ 2 \\ 1 \end{bmatrix} \qquad \begin{bmatrix} -3 \\ 0 \\ 5 \\ .3 \\ 1.06 \end{bmatrix} \qquad \begin{bmatrix} 1 \\ -3.2 \\ 0 \\ 2 \\ 4 \end{bmatrix} \qquad \begin{bmatrix} 2.1 \\ 0.7 \\ 3 \\ 3 \\ 3 \end{bmatrix}$$

ANSWER:

<><><><><><><><><><><><><><><><><><><><><><><><><><><><><><><><><><><><><><><>

Problem

Solve $\mathbf{u} \times \mathbf{z} = \mathbf{v}$ for **z**.

a) First solve the equation when

$$\mathbf{u} := \begin{pmatrix} 5 \\ 3 \\ -4 \end{pmatrix} \qquad \mathbf{v} := \begin{pmatrix} 3 \\ -1 \\ 3 \end{pmatrix}$$

ANSWER:

b) Then solve for

$$\mathbf{u} := \begin{pmatrix} 5 \\ 3 \\ -4 \end{pmatrix} \qquad \mathbf{v} := \begin{pmatrix} 3 \\ 1 \\ 3 \end{pmatrix}$$

ANSWER:

c) Under what conditions will there be a solution?

ANSWER:

d) Is there ever a unique solution?

ANSWER:

e) Compute the solution in terms of **u** and **v**?

ANSWER:

<><><><><><><><><><><><><><><><><><><><><><><><><><><><><><><>

A complete solution of the above problem appears in the file 4_3prob.mcd.

<><><><><><><><><><><><><><><><><><><><><><><><><><><><><><><>

Exercise: (Individual) Write a summary of the MATHEMATICS that you learned in this section.

CHAPTER 5

APPLICATIONS
OF THE
DOT PRODUCT

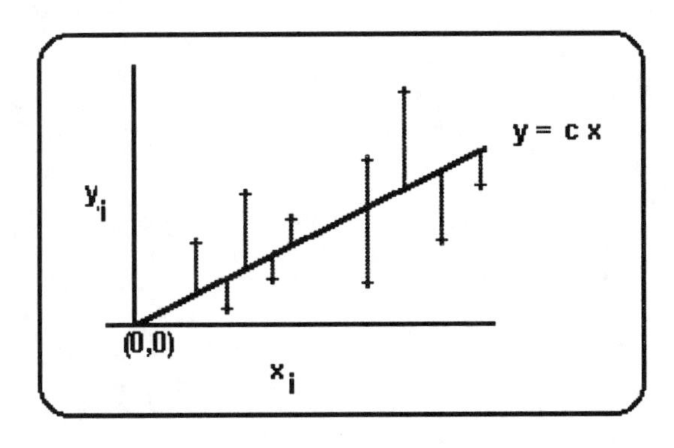

Section 5-1 Projections in R^n

Name:________________

Score:_____________

Overview

Given two vectors **a** and **b** in $\mathbf{R}^n$ the dot product can be used to express **b** as a linear combination of a scalar multiple of **a** and a vector orthogonal to **a**. The scalar multiple of **a** is called the *component of **b** in the direction of **a*** or the *projection* of **b** in the **a** direction. The projection of one vector onto another is applied to compute the distance from a point to a plane, the distance from a point to a line, and to simple regression.

<><><><><><><><><><><><><><><><><><><><><><><><><><><><><><>

Intuitively the projection **p** of a vector **b** onto another vector **a** is the shadow cast by **b** in the direction of **a**.

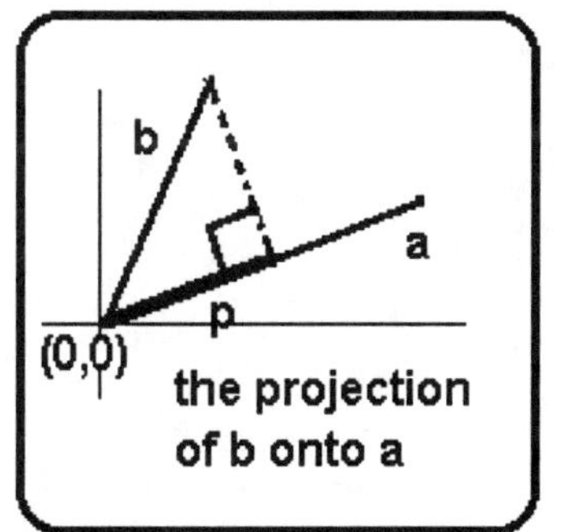

OR

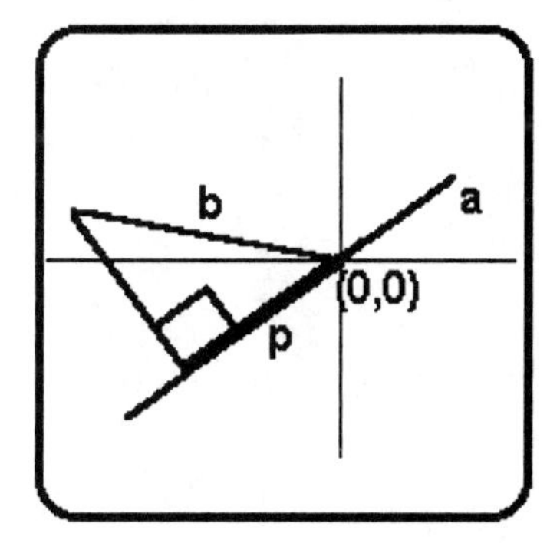

Geometrically the projection **p** is a scalar multiple of the vector **a**. The easiest way to express **p** is as a scalar k times a unit vector in the direction of **a**.

Given a nonzero vector **a**, what is the formula for a unit vector in the **a** direction?

ANSWER:

It is easily seen that the length of the projection is $|\mathbf{b}|\cos(\theta)$ as indicated in the diagram. Note that this is a scalar quantity. The actual ***projection of b in the a direction*** is a vector in the direction of **a** with length $|\mathbf{b}|\cos(\theta)$.

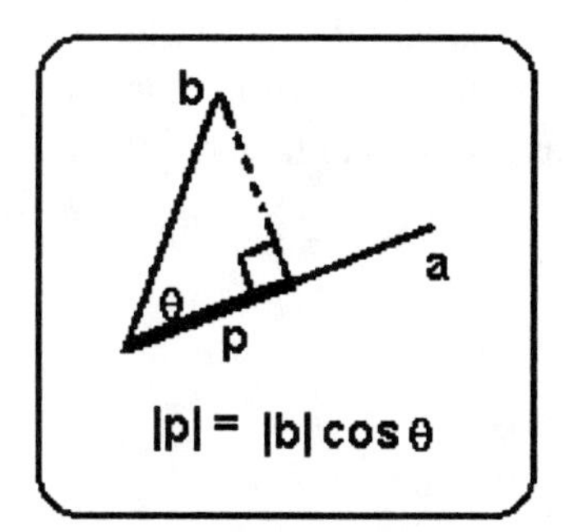

Give a formula for the length of the projection of **b** in the **a** direction in terms of the vectors **a** and **b**. (Hint: Express cos(θ) in terms of the vectors **a** and **b**.)

ANSWER:

Give a formula for the projection, **p,** of **b** in the **a** direction in terms of the vectors **a** and **b**. (**This is a vector!**)

ANSWER:

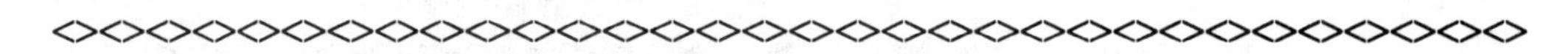

Check your answer on the following example.

Example:

$$\mathbf{a} := \begin{pmatrix} 1 \\ 6 \\ 7 \\ 3 \end{pmatrix} \qquad \mathbf{b} := \begin{pmatrix} -2 \\ 5 \\ 0 \\ 1 \end{pmatrix}$$

The projection of **b** in the **a** direction is $\begin{bmatrix} \frac{31}{95} \\ \frac{186}{95} \\ \frac{217}{95} \\ \frac{93}{95} \end{bmatrix}$

<><><><><><><><><><><><><><><><><><><><><><><><><><><><><><><><><><><><>

Given vectors **a** and **b** as shown in the adjacent diagram, draw the picture that corresponds to the projection of **a** in the **b** direction. (Hint: Copy the diagram into *paintbrush*, complete it, and then copy it back.)

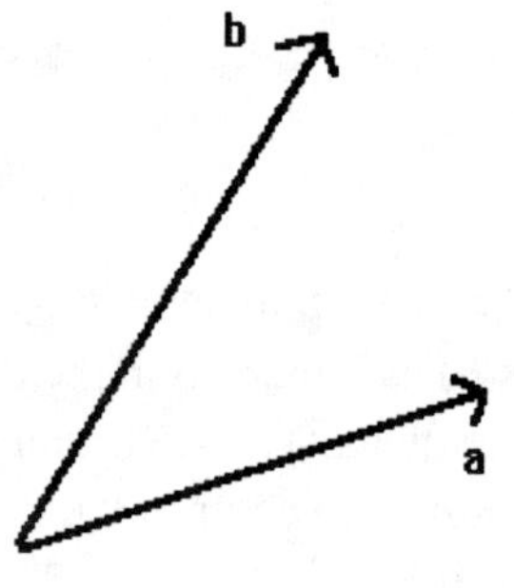

Give the formula for the projection, **q**, of **a** in the direction of **b**.

ANSWER:

Check your answer on the following example.

Example:

$$\mathbf{a} := \begin{pmatrix} 1 \\ 6 \\ 7 \\ 3 \end{pmatrix} \qquad \mathbf{b} := \begin{pmatrix} -2 \\ 5 \\ 0 \\ 1 \end{pmatrix}$$

The projection of **a** in the direction of **b** is $\quad \mathbf{q} = \begin{pmatrix} -2.067 \\ 5.167 \\ 0 \\ 1.033 \end{pmatrix}$

<><><><><><><><><><><><><><><><><><><><><><><><><><><><><><><><><><><><><>

While we can not draw pictures for $\mathbf{R}^k$, $k > 3$, intuitively those for $\mathbf{R}^2$ and $\mathbf{R}^3$ serve as a valid model. The diagram on the right is the generic picture of the projection **p** of vector **b** onto vector **a**. For our purposes we interpret this diagram as follows:

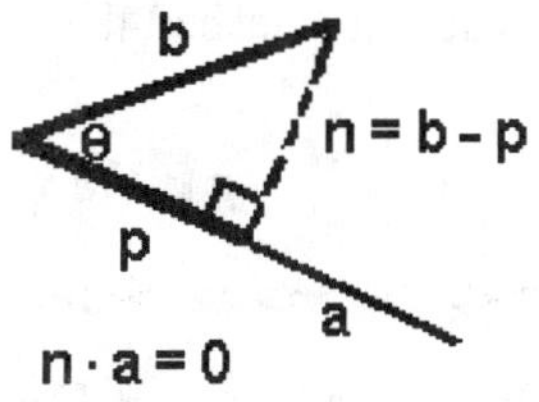

b

θ

$n = b - p$

p

a

$n \cdot a = 0$

The vector b equals p + n, where p is a scalar multiple of a and n is orthogonal to a. We call this the PROJECTION PROPERTY

<><><><><><><><><><><><><><><><><><><><><><><><><><><><><><><><><><><><><>

It is convenient to have a notation for the projection of one vector onto another. We use the following:

$\mathbf{proj_a b}$ = the projection of **b** onto **a**

If **a** and **b** are vectors in $\mathbf{R}^n$, then we may write **b** as the sum of two components: $\mathbf{proj_a b}$ and $\mathbf{n} = \mathbf{b} - \mathbf{proj_a b}$. The projection property gives us

$$\mathbf{b} = \mathbf{proj_a b} + (\mathbf{b} - \mathbf{proj_a b}) = \mathbf{proj_a b} + \mathbf{n}$$

These components can be computed using the dot product:

$$\mathrm{proj_a\, b} = \frac{\mathrm{a \cdot b}}{\mathrm{a \cdot a}} \cdot \mathrm{a} \qquad \text{and} \qquad \left| \mathrm{proj_a\, b} \right| = \frac{|\mathrm{a \cdot b}|}{|\mathrm{a}|}$$

Verify algebraically that $\mathbf{n} = \mathbf{b} - \mathbf{proj_a b}$ is orthogonal to **a**. (Hint: Compute **a**·**n**.)

ANSWER:

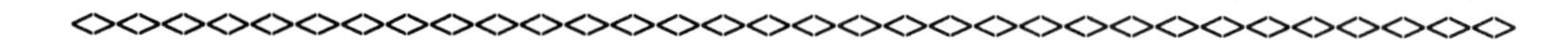

The projection **p** is the component (or piece) of **b** that is in the direction of **a**. Intuitively **p** is the multiple of **a** closest to **b**. Geometrically the dashed line in the figure above is the shortest when it connects the tip of **b** to the tip of **p**. Drawing a line from the tip of **b** to any other point on **a** will result in a dashed line that is longer. This implies that

p is the vector in Span(a) closest to b.

The Distance from a Point to a Plane

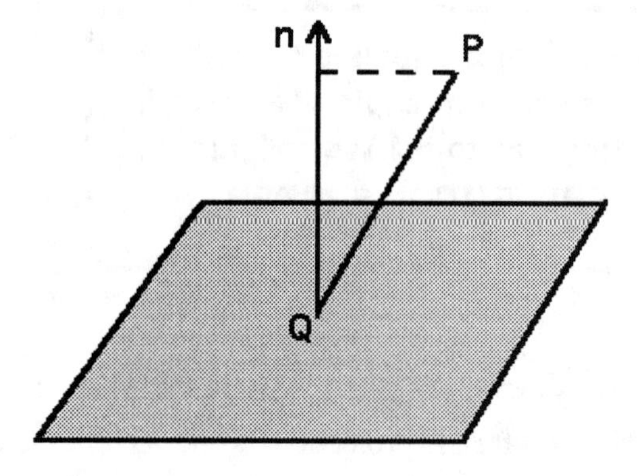

Let **n** be a normal to the plane, **Q** a point in the plane, and **P** any point. Then the distance from **P** to the plane is the length of the projection of the vector **P** - **Q** in the **n** direction.

Write an expression for the distance from point **P** to the plane in terms of **n**, **P**, and **Q**.
ANSWER:

If you have not completed Section 4-2, then do Exercise 9 at the end of this section NOW!

Check your answer to the previous problem by finding the distance from the point **P** = (1,5,-2) to the plane 2x - 2y + z = 5. (Hint: Find a point **Q** in the plane. The correct answer is 5.)
ANSWER:

Use another point **Q** in the plane and see that the answer doesn't change.
ANSWER:

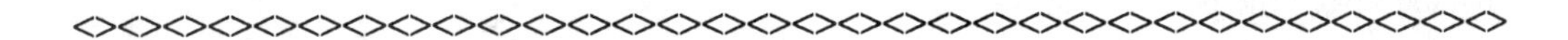

Distance from a Line to a Point

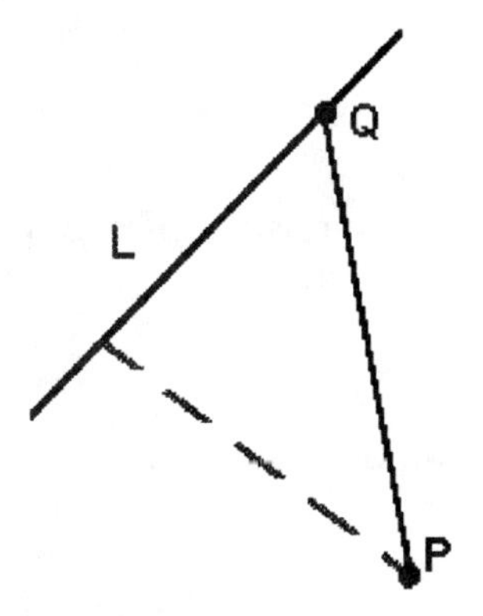

Let **L** be a line in $\mathbf{R}^n$, **Q** a point on the line and **P** any other point.

For a general line, **L**, in $\mathbf{R}^n$ there is no unique perpendicular. In fact, there is an entire (n-1)-dimensional hyperplane perpendicular to the line through a given point. In $\mathbf{R}^3$ this situation is depicted in the following diagram.

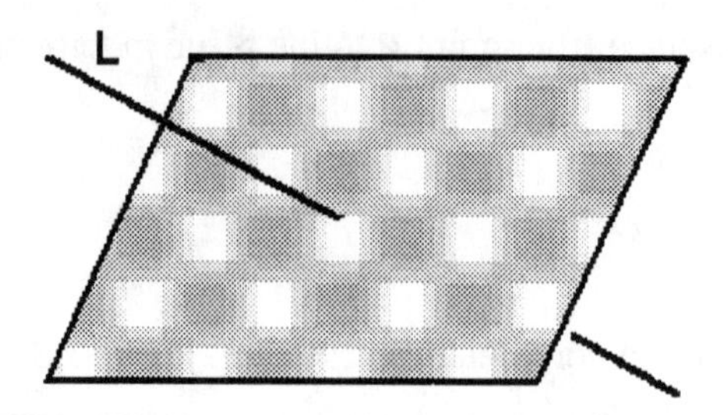

The projection property allows us to write the line from **Q** to **P** as the sum of two components:

$$\mathbf{P - Q = proj_L(P - Q) + ((P - Q) - proj_L(P - Q))}$$

The vector $\mathbf{(P - Q) - proj_L(P - Q)}$ is the component of $(\mathbf{P} - \mathbf{Q})$ that is orthogonal to **L** and $|\,\mathbf{(P - Q) - proj_L(P - Q)}\,|$ is the distance from the point, **P**, to the line, **L**.

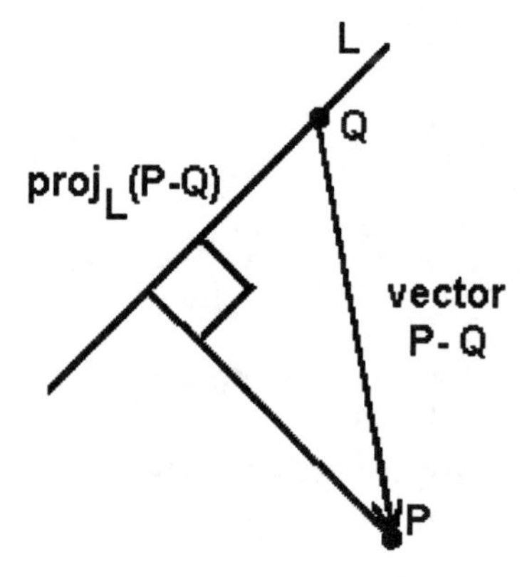

Let **L** be given as $\mathbf{Q} + t\cdot\mathbf{v}$, let **P** be any point. Express the distance from **P** to the line **L** as a function of **P**, **Q**, and **v**.

ANSWER:

Check your answer by computing the distance from the point $\mathbf{P} = (1,0,3,5,-2)$ to the line **L**: $(0,3,7,-3,4) + t\cdot(7,2,1,0,-1)$. (Show your work.)

ANSWER:

The correct answer is

$$\frac{3}{55}\cdot\sqrt{769}\cdot\sqrt{55}=11.218$$

To show that your answer does not depend on the particular point **Q** chosen on line **L**, proceed as follows. Do not choose a value for t, rather use a 'generic point' **Q** = (0,3,7,-3,4) + t·(7,2,1,0,-1). Set up the expression for computing the distance as before, but now use Maple to do the computation. Your answer should be independent of t.
ANSWER:

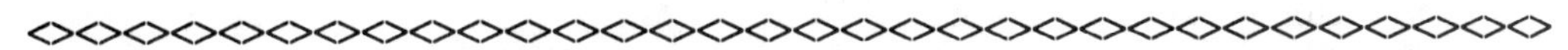

The computation of the distance from a point to a plane and the distance from a point to a line contain the basic ideas upon which all the applications in this chapter are based. Make sure that you understand *projection* and *the projection property* before proceeding.

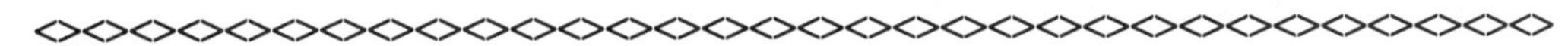

Exercises

1. Find the distance from the point (3,-2) to the line y = 5x + 7 using projections.
ANSWER:

2. Let $\mathbf{u} := \begin{bmatrix} 3 \\ 7 \\ -2 \\ 4 \\ 0 \end{bmatrix}$ and $\mathbf{v} := \begin{bmatrix} 2 \\ 5 \\ 1 \\ -1 \\ 2 \end{bmatrix}$

a) Compute $\mathbf{proj_v u}$.
ANSWER:

b) Use the projection property to express **u** as a sum of a vector in the direction **v** and a vector orthogonal to **v**. Compute the two vectors.
ANSWER:

c) Find the component of **v** in the **u** direction.
ANSWER:

d) Use the projection property to express **v** as a sum of a vector in the direction **u** and a vector orthogonal to **u**. Compute the two vectors.
ANSWER:

3. Compute the distance from the point **P** = (3,5,-2) to the line through (2,2,-5) with direction vector (1,0,5).
ANSWER:

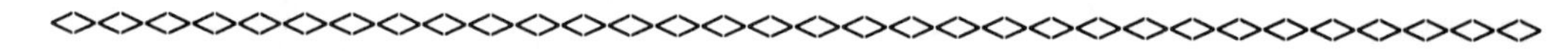

Simple Regression

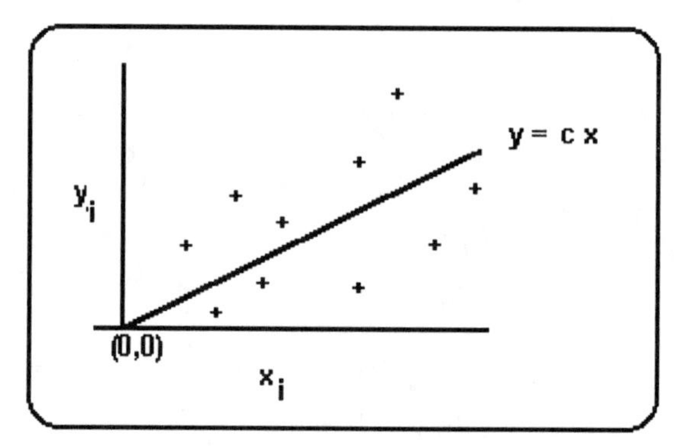

Let (x_i,y_i) i = 1, ..., n be a collection of data points. Find the line through the origin which best 'fits' these points; i.e., find c such that y = cx is closest to the data points.

Discuss what closest might mean. (There are at least five different ways to interpret closest.)
ANSWER: (You can specify algebraic criteria or use *paintbrush* to depict your measures graphically. Briefly discuss your choices.)

<><><><><><><><><><><><><><><><><><><><><><><><><><><><>

The meaning of closest that we choose here is that of minimizing the squares of the distance of the y values from the lines. This is called **least squares.** There is nothing inherently superior about this choice of 'closest' other than the fact that we can use linear algebra to compute the value of c which gives the 'closest' line. Pictorially we minimize the sum of the squares of the vertical distances shown below.

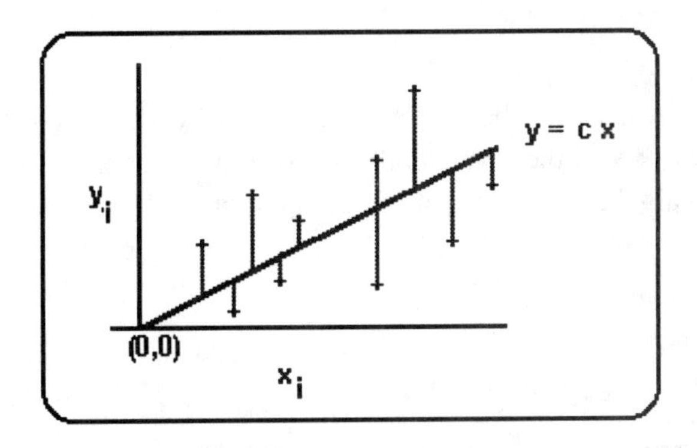

Algebraically we minimize $\sum_i (y_i - cx_i)^2 = (|y - c\cdot x|)^2$

If we write think of the points x_i and y_i as components of vectors, then based on the projection property, we can write $\mathbf{y} = c\cdot\mathbf{x} +$ **difference**. If we choose **difference** to be orthogonal to $c\cdot\mathbf{x}$ this will minimize the length of **difference**. (If **difference** is orthogonal to $c\cdot\mathbf{x}$ it is also orthogonal to **x**. Why?) To do this let c be such that $c\cdot\mathbf{x}$ is the projection of **y** onto the **x** vector. This number is called the ***regression coefficient***.

Look back at what we have done here. We began with n points in $\mathbf{R}^2$ and the problem of finding the best fit line of the form y = cx to these points. To solve this problem, we changed it into an equivalent problem about $\mathbf{R}^n$ that we knew how to solve using projections. Don't look for the vectors **x** and **y** in the above picture since they are in $\mathbf{R}^n$ and not in $\mathbf{R}^2$.

<><><><><><><><><><><><><><><><><><><><><><><><><><><><><><><><>

What is the formula for c? (Hint: Treat equation $y = c\cdot x +$ difference as though y, x, and difference are vectors then take the dot product of both sides with **x** and solve for c.)
ANSWER:

Check your answer to make sure that if $y_i = bx_i$ for all values of i then the regression coefficient must be b.

Compute the regression coefficient for the following data points:

$x := (1\ 2\ 3\ 4\ 5\ 6\ 7\ 8)^T \qquad y := (-2\ -6\ -4\ 12\ 10\ 8\ -14\ 16)^T$

ANSWER:

<><><><><><><><><><><><><><><><><><><><><><><><><><><><><><><><>

Exercises

4. Interchange the role of **x** and **y** in the above examples. Now we want to find the regression coefficient k such that $\mathbf{x} = k \cdot \mathbf{y}$. Either use paintbrush to show what is minimized in the least squares sense or give an algebraic expression for the quantity to be minimized.

ANSWER:

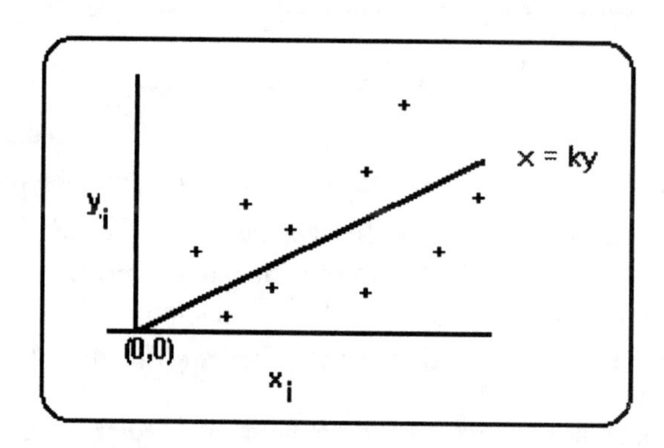

Find the regression coefficient k such that $|\mathbf{x} - k \cdot \mathbf{y}|$ is a minimum? Explain the significance of k in the picture above.

ANSWER:

<><><><><><><><><><><><><><><><><><><><><><><><><><><><><><><><><>

5. There is still another possibility for the best fit line determined by a set of data. This is indicated by the picture below in which we try to minimize the sums of the lengths of the lines perpendicular to $\mathbf{y} = c \cdot \mathbf{x}$.

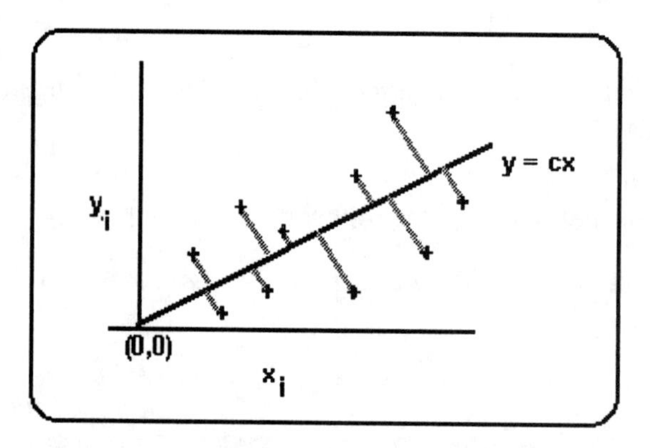

Let the line be $y = cx$ and the points (x_i, y_i) be as given above. Then the distance from a data point (x_i,y_i) to the line $y = cx$ is computed using projections. Proceed as follows:

What is a direction vector for the line $y = cx$?

ANSWER:

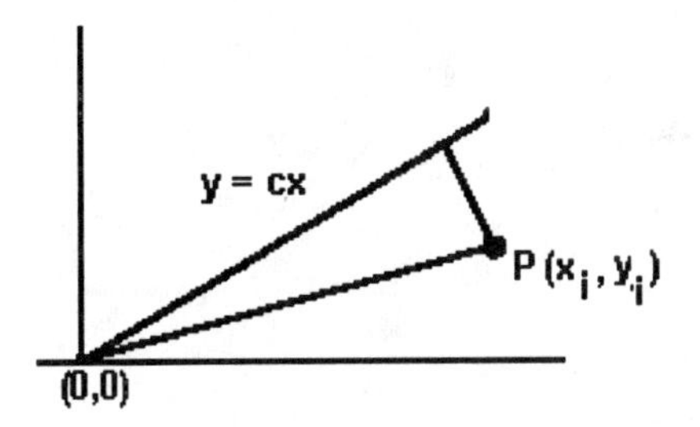

Find a unit vector perpendicular to the line $y = cx$.

ANSWER:

Show that the distance from (x_i,y_i) to the line, $y = cx$ is

$$\left| \frac{c \cdot x_i - y_i}{\sqrt{1 + c^2}} \right|$$

by computing the length of the component of the vector (x_i,y_i) in the direction orthogonal to the line $y = cx$.

ANSWER:

The total distance is therefore equal to

$$D(c) = \sum_i \left| \frac{c \cdot x_i - y_i}{\sqrt{1 + c^2}} \right|$$

Compute the coefficient c for the procedure immediately above for the following data points. That is, find the value of c that minimizes D(c). (Hint: Graph D(c) and estimate the value for c. Refine the graph by narrowing the range for c.)

$$x := (1 \;\; 2 \;\; 3 \;\; 4 \;\; 5 \;\; 6 \;\; 7 \;\; 8)^T$$

$$y := (-2 \;\; -6 \;\; -4 \;\; 12 \;\; 10 \;\; 8 \;\; -14 \;\; 16)^T$$

$i := 1..8 \qquad c := -5, -4.9..5$

$$D(c) := \sum_i \left| \frac{c \cdot x_i - y_i}{\sqrt{1 + c^2}} \right|$$

ANSWER:

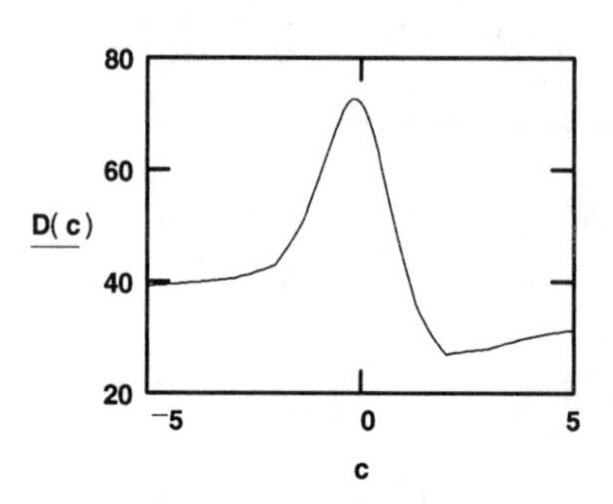

<><><><><><><><><><><><><><><><><><><><><><><><><><><><><>

6. Hooke's Law for springs says that the distance a spring is stretched is proportional to the applied force F; that is, $F = k \cdot d$, where k is the proportionality constant which is called the spring constant in physics. An experiment to determine the spring constant produced the following data. The spring was stretched 2, 5, 7, and 12 inches when forces of 1.5, 4, 6, and 9.4 pounds were applied. Determine the least squares approximation to the spring constant k.

ANSWER:

<><><><><><><><><><><><><><><><><><><><><><><><><><><><><>

7. Once brakes are applied in a moving vehicle, the distance needed to stop is roughly proportional to the velocity squared; $d = kv^2$, where k is the proportionality constant. The following data was acquired from a set of experiments:

$$(v,d) = ((25,440), (40,1125), (55,2200), (65,2950), (80,4485)).$$

Determine the proportionality constant using least squares.

ANSWER:

8. Which of the three choices of *best fit* line given above in the material preceding Exercise 1, in Exercise 1 and in Exercise 2 is best? Discuss.
ANSWER:

<><><><><><><><><><><><><><><><><><><><><><><><><><><><><><><>

If you have done Section 4-2 skip the next problem.

9. Finding a Normal Vector to a Plane in $\mathbf{R}^3$
A plane in $\mathbf{R}^3$ has an equation of the form: $Ax + By + Cz = D$.

For many applications it is important to find a ***normal vector*** to the plane. A normal vector is a vector in $\mathbf{R}^3$ which is *orthogonal* to all lines lying in the plane. Suppose that (x_1, y_1, z_1) and (x_2, y_2, z_2) are two points lying in the plane. Then both points satisfy the equation of the plane; that is,

$$Ax_1 + By_1 + Cz_1 = D$$

$$Ax_2 + By_2 + Cz_2 = D$$

Subtract the two equations to get

$$A\cdot(x_2 - x_1) + B\cdot(y_2 - y_1) + C\cdot(z_2 - z_1) = 0$$

Write this expression as a dot product of two vectors, one of which is a direction vector for a line that lies in the plane. Indicate which vector is the direction vector and explain why.
ANSWER:

Using the preceding development, state a vector in $\mathbf{R}^3$ that is guaranteed to be normal to the plane.
ANSWER:

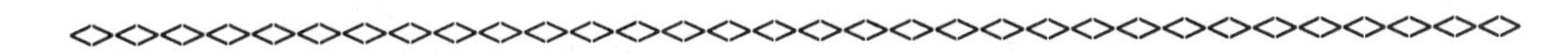

Your answer above should indicate that the coefficients, (A, B, C), of the standard form

$$Ax + By + Cz = D$$

of the equation of the plane written in x,y,z order give a normal vector to the plane.

<><><><><><><><><><><><><><><><><><><><><><><><><><><><><><><>

Determine a vector normal to each of the following planes.

$5\cdot x + 4\cdot y - z = -4$ $\qquad$ $6\cdot x - 3z = 5$

ANSWER:

normal vector: _________ $\qquad$ normal vector: _________

Is the normal vector to a plane unique?

ANSWER:

<><><><><><><><><><><><><><><><><><><><><><><><><><><><><><><>

10. (Individual) Write a summary of the MATHEMATICS that you learned in this section.

Section 5-2

Projections into a Subspace: Orthonormal Basis

Name:______________

Score:____________

Overview

In the previous section you discovered how to project a vector in $\mathbf{R}^n$ into a 1-dimensional subspace. In this section those methods are extended to a procedure for calculating a projection into any subspace. Using this process a method is derived for replacing a basis with an orthonormal basis.

<><><><><><><><><><><><><><><><><><><><><><><><><><><><><><><><><>

Let U be a subspace of $\mathbf{R}^n$. We denote the set of all vectors in $\mathbf{R}^n$ that are orthogonal to each of the vectors in U by $U^\perp$ (pronounced U perp). More precisely, a vector $\mathbf{x}$ is a member of $U^\perp$ if and only if $\mathbf{x}\cdot\mathbf{u} = 0$ for all vectors $\mathbf{u}$ that belong to U.

Show that $U^\perp$ is a subspace. (Show that it is closed under addition and scalar multiplication.)

ANSWER:

Our goal is to show that for any subspace U every vector in $\mathbf{R}^n$ can be written as the sum $\mathbf{x} + \mathbf{y}$ where $\mathbf{x}$ is a member of U and $\mathbf{y}$ is a member of $U^\perp$. In the examples in the previous section U was a one dimensional subspace. For example in $\mathbf{R}^3$, if U is the subspace of all multiples of the vector (A,B,C) then $U^\perp$ is the plane, $Ax + By + Cz = 0$.

What is $(U^\perp)^\perp$?

ANSWER:

If the dimension of U is between 1 and (n-1) then the calculations are more complicated than in the previous section. In particular we need to choose a *'**good basis**'* for U to aid in the calculations.

<><><><><><><><><><><><><><><><><><><><><><><><><><><><><><><><><><>

A set of vectors $\mathbf{u}_1, \mathbf{u}_2, \ldots, \mathbf{u}_n$ is said to be ***orthogonal*** if $\mathbf{u}_i\cdot\mathbf{u}_j = 0$ for $i \neq j$.
The set is called ***orthonormal*** if, in addition, each vector is a unit vector. We express this by

$$\mathbf{u}_i\cdot\mathbf{u}_j = 0, \quad i \neq j$$
$$\mathbf{u}_i\cdot\mathbf{u}_i = 1$$

A matrix **A** is called ***orthogonal*** if its columns form an orthonormal set. (***Beware*** of the terminology; we do *not* call a matrix orthonormal when its columns are an orthonormal set.)

$$\mathbf{W} := \begin{pmatrix} 0.389249472080761 & 0.793856749290387 & 0.454684502132616 & -0.107416542540542 \\ -0.155699788832305 & -0.424470343498125 & 0.889622791045771 & 0.064449925524325 \\ 0.778498944161523 & -0.28352026760371 & -0.039492426438919 & 0.558566021210816 \\ 0.467099366496914 & -0.330503626235181 & -0.016542110697057 & -0.819946274726133 \end{pmatrix}$$

Verify that **W** is an orthogonal matrix. One can do this, of course, by computing, $\mathbf{W}^{<i>} \cdot \mathbf{W}^{<j>}$ for $i \geq j$. This means we need to do 10 calculations. There is a way of checking with one calculation involving matrix multiplication.

ANSWER:

Examine $\mathbf{W}^{-1}$. Is there any obvious relationship between **W** and $\mathbf{W}^{-1}$?

ANSWER:

If **A** is an orthogonal matrix what is $\mathbf{A}^{-1}$?

ANSWER:

If the columns of **A** are orthonormal, must the rows be orthonormal? Explain.

ANSWER:

<><><><><><><><><><><><><><><><><><><><><><><><><><><><><><><>

Let $\mathbf{u}_1, \mathbf{u}_2, ..., \mathbf{u}_m$ be an orthonormal basis for the subspace U of $\mathbf{R}^n$. (We show below that every subspace has an orthonormal basis.) Let **q** be a vector in U. Then, since $\mathbf{u}_1, \mathbf{u}_2, ..., \mathbf{u}_m$ is a basis, there exist coefficients $a_1, a_2, ..., a_m$ so that $\mathbf{q} = a_1 \cdot \mathbf{u}_1 + ... + a_m \cdot \mathbf{u}_m$. To determine the values of the coefficients, $a_1, a_2, ..., a_m$, compute the dot product $\mathbf{u}_i \cdot \mathbf{q}$ for each i. What is the formula for the coefficients?

ANSWER:

Using the formula for the a_j express **q** as a linear combination of the orthonormal basis, $\mathbf{u}_1, \mathbf{u}_2, ..., \mathbf{u}_m$.

ANSWER:

If $\mathbf{u}_1, \mathbf{u}_2, ..., \mathbf{u}_m$ is an orthogonal basis (but not necessarily an orthonormal basis) then $\mathbf{q} = b_1 \cdot \mathbf{u}_1 + ... + b_m \cdot \mathbf{u}_m$. What are the coefficients $\mathbf{b}_i$?

ANSWER:

Let **w** be a vector in $\mathbf{R}^n$, let U be a subspace of $\mathbf{R}^n$, and let $\mathbf{u}_1, ..., \mathbf{u}_m$ be an orthonormal basis for U. We compute the projection of **w** onto each of these basis vectors.

$$\text{proj}_{u_j} w = (w \cdot u_j) \cdot u_j \,, j = 1, ..., m$$

We call these projections onto the basis vectors of subspace U the ***components of w in the directions*** $\mathbf{u}_j$.

Let $\mathbf{w}_U$ be the sum of the components of **w** in the directions $\mathbf{u}_i$ for i = 1, ..., m.

$$w_U = \sum_{j=1}^{m} \text{proj}_{u_j} w = \sum_{j=1}^{m} (w \cdot u_j) \cdot u_j$$

w_U is called the component of w in the subspace U

Show that $(\mathbf{w} - \mathbf{w}_U)$ is an element of $U^{\perp}$. That is, $(\mathbf{w} - \mathbf{w}_U)$ is orthogonal to every vector in U. (Hint: Show $(\mathbf{w} - \mathbf{w}_U)$ is orthogonal to each of the basis vectors $\mathbf{u}_j$ then provide an argument to verify that $(\mathbf{w} - \mathbf{w}_U)$ is orthogonal to every vector in U.)

ANSWER:

Since $\mathbf{w} = \mathbf{w}_U + (\mathbf{w} - \mathbf{w}_U)$ we have ***decomposed*** the vector **w** into

(the projection of w into the subspace U) + (a vector orthogonal to the subspace U).
This is a generalization of the projection property from Section 5-1.

<><><><><><><><><><><><><><><><><><><><><><><><><><><><><><><><><><><><><><><><><>

Example 1

Let U = Span(**u1, u2**) where **u1** and **u2** are the following vectors in $\mathbf{R}^5$,

$$u1 := \begin{bmatrix} 1 \\ 0 \\ -3 \\ 4 \\ 2 \end{bmatrix} \qquad u2 := \begin{bmatrix} 0 \\ 7 \\ 2 \\ -3 \\ 9 \end{bmatrix} \qquad \text{and let} \qquad w := \begin{bmatrix} 3 \\ 5 \\ -7 \\ 1 \\ 3 \end{bmatrix}$$

Show that **u1** and **u2** form an orthogonal basis for U.

ANSWER:

Convert **u1** and **u2** into unit vectors. (Keep the names **u1** and **u2**.)

ANSWER:

Verify that the pair, **u1** and **u2**, is now an orthonormal set.

ANSWER:

Find the components of **w** in the **u1** and **u2** directions.

ANSWER:

Determine $\mathbf{w}_U$, the component of **w** in the subspace U, and $(\mathbf{w} - \mathbf{w}_U)$ the component of **w** in the subspace $U^{\perp}$.

ANSWER:

By construction $(\mathbf{w} - \mathbf{w}_U)$ is orthogonal to every vector in subspace U. You should have found that $(\mathbf{w} - \mathbf{w}_U)$ is the vector which we display on the right with 15 places.

$$\begin{bmatrix} 1.866666666666667 \\ 2.797202797202797 \\ -4.22937062937063 \\ -2.589277389277389 \\ -2.098834498834499 \end{bmatrix} \quad \text{or} \quad \begin{bmatrix} \frac{28}{15} \\ \frac{400}{143} \\ -\frac{3024}{715} \\ -\frac{5554}{2145} \\ -\frac{4502}{2145} \end{bmatrix}$$

Evaluate the following expression symbolically and interpret the results

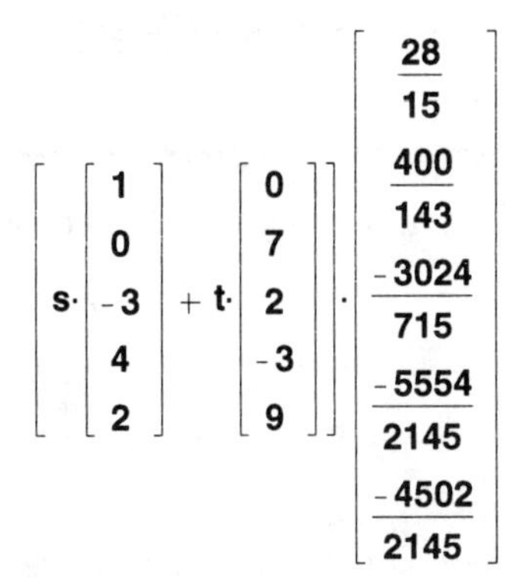

ANSWER:

The component of **w** in U is called ***the projection of* w *into U***. Explain why the projection of **w** into **U** is that vector in U that is closest to **w**.

ANSWER:

<><><><><><><><><><><><><><><><><><><><><><><><><><><><><><>

Example 2

Let U be the subspace spanned by **u1** and **u2** where **u1** and **u2** are the following vectors in $\mathbf{R}^5$,

$$\mathbf{u1} := \begin{bmatrix} 1 \\ 0 \\ -3 \\ 4 \\ 2 \end{bmatrix} \qquad \mathbf{u2} := \begin{bmatrix} 1 \\ 7 \\ -1 \\ 1 \\ 11 \end{bmatrix} \qquad \text{and let} \qquad \mathbf{w} := \begin{bmatrix} 3 \\ 5 \\ -7 \\ 1 \\ 3 \end{bmatrix}$$

Find the components of **w** in U and in $U^{\perp}$. Follow the steps in the previous example. Explain any difficulties encountered.

ANSWER:

<><><><><><><><><><><><><><><><><><><><><><><><><><><><><><>

Explain the difference between the first and second example.

ANSWER:

<><><><><><><><><><><><><><><><><><><><><><><><><><><><><><>

For any subspace of R^n there is an orthonormal basis.

If you can construct an orthogonal basis for U, how can you construct an orthonormal basis?
ANSWER:

<><><><><><><><><><><><><><><><><><><><><><><><><><><><><><><><>

To show that there is an orthonormal basis for a subspace, U, start with any basis for U and construct an orthogonal basis for U. The orthogonal basis is then converted to an orthonormal basis by making each vector a unit vector.

Assume that **u1**, **u2**, ..., **um** are a basis for U. Construct a new basis **v1**, **v2**, ..., **vm** such that

1. **v1**, **v2**, ..., **vm** is an orthogonal set, and

2. Span(**u1**, ..., **uj**) = Span(**v1**, ..., **vj**) for j = 1, ..., m.

Begin by setting **v1** = **u1**. This obviously satisfies both properties 1 and 2.
To get **v2** subtract the component of **u2** in the **v1** direction from **u2**.

$$\mathbf{v2} := \mathbf{u2} - \left(\frac{\mathbf{u2}\cdot\mathbf{v1}}{\mathbf{v1}\cdot\mathbf{v1}}\right)\cdot\mathbf{v1}$$

Then as was shown above **v2** is orthogonal to **v1**.

Span(**v1**, **v2**) = Span(**v1**, **u2**) (since **v2** is a linear combination of **v1** and **u2**)
= Span(**u1**, **u2**) (since **v1** = **u1**)

To get **v3** subtract the component of **u3** in the **v1** direction and the component of **u3** in the **v2** direction from **u3**. The result is orthogonal to both **v1** and **v2**.

$$\mathbf{v3} := \mathbf{u3} - \frac{\mathbf{u3}\cdot\mathbf{v1}}{\mathbf{v1}\cdot\mathbf{v1}}\cdot\mathbf{v1} - \frac{\mathbf{u3}\cdot\mathbf{v2}}{\mathbf{v2}\cdot\mathbf{v2}}\cdot\mathbf{v2}$$

By an argument similar to that given in the previous step, it follows that
Span(**v1**, **v2**, **v3**) = Span(**v1**, **v2**, **u3**) = Span(**u1**, **u2**, **u3**).
Continue in this way setting

$$\mathbf{vj} := \mathbf{uj} - \sum_{i=1}^{j-1} \frac{\mathbf{uj}\cdot\mathbf{vi}}{\mathbf{vi}\cdot\mathbf{vi}}\cdot\mathbf{vi}$$

By construction, the **v**'s are an orthogonal basis for the subspace U. Construct an orthonormal basis by ***normalizing*** each vector. That is, make each vector a unit vector. Thus the set of vectors

$$\frac{v1}{|v1|}, \frac{v2}{|v2|}, \ldots, \frac{vm}{|vm|}$$

is an orthonormal basis for U.

This construction is called the ***Gram-Schmidt Orthogonalization procedure***.

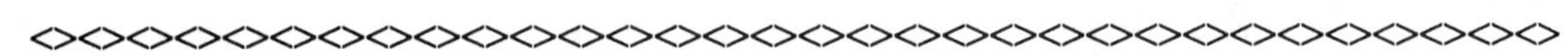

Example 3

Determine an orthonormal basis for the subspace Span(**u1**, **u2**, **u3**) in $\mathbf{R}^4$ where

$$u1 := \begin{pmatrix} 1 \\ 0 \\ 0 \\ 0 \end{pmatrix} \qquad u2 := \begin{pmatrix} 2 \\ 3 \\ 5 \\ -2 \end{pmatrix} \qquad u3 := \begin{pmatrix} 0 \\ 1 \\ 2 \\ 0 \end{pmatrix}$$

Step 1. $v1 := u1$

Step 2. $$v2 := u2 - \frac{u2 \cdot v1}{v1 \cdot v1} \cdot v1$$

Step 3. $$v3 := u3 - \frac{u3 \cdot v1}{v1 \cdot v1} \cdot v1 - \frac{u3 \cdot v2}{v2 \cdot v2} \cdot v2$$

Display the **v**-basis.

$$v1 = \begin{bmatrix} 1 \\ 0 \\ 0 \\ 0 \end{bmatrix} \qquad v2 = \begin{bmatrix} 0 \\ 3 \\ 5 \\ -2 \end{bmatrix} \qquad v3 = \begin{bmatrix} 0 \\ -0.026 \\ 0.289 \\ 0.684 \end{bmatrix}$$

Verify that the **v**'s are orthogonal.

ANSWER:

From the **v**-basis construct an orthonormal basis.

ANSWER:

◇◇◇

Exercises

1. Give an example of an orthonormal basis for $\mathbf{R}^3$.

ANSWER:

2. The vectors **u1**, **u2**, and **u3** given below are a basis for $\mathbf{R}^3$. Use the Gram-Schmidt process to find an orthonormal basis from these vectors.

$$\mathbf{u1} := \begin{pmatrix} 1 \\ 1 \\ 0 \end{pmatrix} \qquad \mathbf{u2} := \begin{pmatrix} 0 \\ 1 \\ -1 \end{pmatrix} \qquad \mathbf{u3} := \begin{pmatrix} -1 \\ 1 \\ 0 \end{pmatrix}$$

ANSWER:

3. Can a subspace have more than one orthonormal basis? Explain.

ANSWER:

4. Find an orthonormal basis for the subspace of solutions to 2w + 5x - 3y - z = 0.

ANSWER:

5. Use the basis in Example 3 above to find the projection of **x** into the subspace Span(**u1**, **u2**, **u3**), where =========> $\mathbf{x} := \begin{pmatrix} 1 \\ 1 \\ 1 \\ 1 \end{pmatrix}$

ANSWER:

6. Refer to the subspace in Exercise 4 above. Find the projection of $\mathbf{x} := \begin{pmatrix} 1 \\ 1 \\ 1 \\ 1 \end{pmatrix}$ onto the subspace.

ANSWER:

7. In Chapter 4 we saw that the subspace of Exercise 4 above was a hyperplane with normal vector **n** given below. Redo Exercise 6 by calculating the component, $\mathbf{x_n}$ of **x** in the direction of the normal **n.**

$$\mathbf{n} := \begin{pmatrix} 2 \\ 5 \\ -3 \\ -1 \end{pmatrix}$$

Then note that $(\mathbf{x} - \mathbf{x_n})$ is the component of **x** in the hyperplane. If you don't get the same answer by both methods, find your error.

ANSWER:

8. Assume that the set of vectors **p1**, **p2**, ..., **pk** are orthogonal vectors in $\mathbf{R}^n$ and none of the vectors is the zero vector. Show that they are linearly independent.

ANSWER:

Explain why we can say every orthonormal set is linearly independent.

ANSWER:

9. (Individual) Write a summary of the MATHEMATICS that you learned from this section.

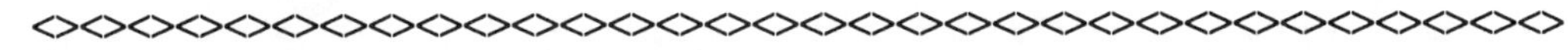

A Matrix Factorization Using the Gram-Schmidt Process

The Gram-Schmidt process constructs an orthonormal set from a linearly independent set. We can use Gram-Schmidt to show that every nonsingular matrix can be written as the product of an orthogonal matrix **Q** and an upper triangular matrix **R**. This is known as the **QR-factorization** and is important in computational linear algebra (i.e., deriving fast algorithms for machine computations involving large matrices).

Suppose we have a nonsingular n by n matrix **A**. The columns of **A**, denoted **A1**, **A2**, ..., **An**, are linearly independent. Using the Gram-Schmidt process we can compute a set of orthonormal vectors **q1**, **q2**, ..., **qn** with the additional property that Span(**A1**, ..., **Aj**) = Span(**q1**, ..., **qj**) for j = 1, ..., n.

Let **Q** be the matrix whose columns are the vectors **q1**, **q2**, ..., **qn**.

For any vector **c**, we have

$$Q \cdot c = c_1 \cdot q1 + c_2 \cdot q2 + \ldots + c_n \cdot qn$$

Since the **qi**'s are a basis, there are scalars $r_{i,j}$ such that

$$Aj = \sum_{i=1}^{n} r_{i,j} \cdot qi \qquad (1)$$

In particular, since Span(**A1**, ..., **Aj**) = Span(**q1**, ..., **qj**) we have $r_{i,j} = 0$ for $i > j$ which implies that **R** is upper triangular. Equation (1) says

$$\mathbf{Col}_j(\mathbf{A}) = \mathbf{Q} \cdot \mathbf{Col}_j(\mathbf{R})$$

hence **A** = **Q**·**R** where **Q** is orthogonal and **R** is upper triangular.

<><><><><><><><><><><><><><><><><><><><><><><><><><><><><><><><><><><><><>

Example

For $\mathbf{A} := \begin{pmatrix} 1 & 2 & 5 & 3 \\ 3 & 0 & 2 & 5 \\ 5 & 0 & 1 & -5 \\ 7 & 1 & 0 & -9 \end{pmatrix}$ determine the QR-factorization.

We first apply the Gram-Schmidt process:

Since $\mathbf{Q}^{<i>} = \mathbf{q}_i$ we compute the matrix **Q** directly.

$$Q^{<1>} := \frac{A^{<1>}}{\left|A^{<1>}\right|}$$

$$Q^{<2>} := \frac{A^{<2>} - (A^{<2>} \cdot Q^{<1>}) \cdot Q^{<1>}}{\left|A^{<2>} - (A^{<2>} \cdot Q^{<1>}) \cdot Q^{<1>}\right|}$$

$$Q^{<3>} := \frac{A^{<3>} - (A^{<3>} \cdot Q^{<1>}) \cdot Q^{<1>} - (A^{<3>} \cdot Q^{<2>}) \cdot Q^{<2>}}{\left|A^{<3>} - (A^{<3>} \cdot Q^{<1>}) \cdot Q^{<1>} - (A^{<3>} \cdot Q^{<2>}) \cdot Q^{<2>}\right|}$$

$$Q^{<4>} := \frac{A^{<4>} - (A^{<4>} \cdot Q^{<1>}) \cdot Q^{<1>} - (A^{<4>} \cdot Q^{<2>}) \cdot Q^{<2>} - (A^{<4>} \cdot Q^{<3>}) \cdot Q^{<3>}}{\left|A^{<4>} - (A^{<4>} \cdot Q^{<1>}) \cdot Q^{<1>} - (A^{<4>} \cdot Q^{<2>}) \cdot Q^{<2>} - (A^{<4>} \cdot Q^{<3>}) \cdot Q^{<3>}\right|}$$

Display the matrix **Q** ==> $Q = \begin{bmatrix} 0.109 & 0.942 & 0.293 & -0.121 \\ 0.327 & -0.16 & 0.662 & 0.655 \\ 0.546 & -0.267 & 0.364 & -0.706 \\ 0.764 & 0.124 & -0.586 & 0.241 \end{bmatrix}$

Verify that **Q** is orthogonal;
compute $\mathbf{Q}^T \cdot \mathbf{Q}$. ==>

$$Q^T \cdot Q = \begin{bmatrix} 1 & 0 & 0 & 0 \\ 0 & 1 & 0 & 0 \\ 0 & 0 & 1 & 0 \\ 0 & 0 & 0 & 1 \end{bmatrix}$$

We can solve the preceding equations for the original vectors **A1, A2, A3, A4** in terms of the orthonormal vectors $\mathbf{Q}^{<1>}$, $\mathbf{Q}^{<2>}$, $\mathbf{Q}^{<3>}$, $\mathbf{Q}^{<4>}$. (Earlier in this section we found that with an orthonormal basis the coefficients needed to express a vector **v** in terms of such a basis are just the dot products of the **v** with the basis vectors.)

Thus $$A^{<j>} = \sum_i \left(A^{<j>} \cdot Q^{<i>} \right) \cdot Q^{<i>}$$ for $i := 1..4$

$j := 1..4$

and $$R_{i,j} := A^{<j>} \cdot Q^{<i>}$$

$$R = \begin{bmatrix} 9.165 & 0.982 & 1.746 & -7.638 \\ 0 & 2.009 & 4.124 & 2.24 \\ 0 & 0 & 3.153 & 7.642 \\ 0 & 0 & 0 & 4.272 \end{bmatrix}$$

and

$$Q \cdot R = \begin{bmatrix} 1 & 2 & 5 & 3 \\ 3 & 0 & 2 & 5 \\ 5 & 0 & 1 & -5 \\ 7 & 1 & 0 & -9 \end{bmatrix}$$ which is **A**. $$A = \begin{bmatrix} 1 & 2 & 5 & 3 \\ 3 & 0 & 2 & 5 \\ 5 & 0 & 1 & -5 \\ 7 & 1 & 0 & -9 \end{bmatrix}$$

<><><><><><><><><><><><><><><><><><><><><><><><><><><><><><><>

If **A** is not square but has linearly independent columns then we can still perform a QR-factorization. Describe matrices **Q** and **R** in this case.
ANSWER:

<><><><><><><><><><><><><><><><><><><><><><><><><><><><><><><>

Exercises

1. Determine the QR-factorization of the matrix **B** given below.

$$\mathbf{B}^{<1>} := \begin{pmatrix} 1 \\ 1 \\ 0 \end{pmatrix} \qquad \mathbf{B}^{<2>} := \begin{pmatrix} 1 \\ 0 \\ 1 \end{pmatrix} \qquad \mathbf{B}^{<3>} := \begin{pmatrix} 1 \\ 1 \\ -1 \end{pmatrix}$$

ANSWER:

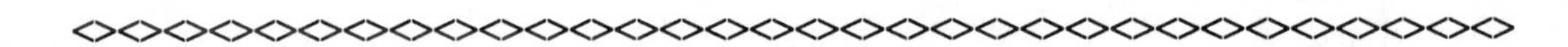

2. If **A** is nonsingular, then **A** has a QR-factorization with **Q** nonsingular. Using this factorization, explain how to solve the linear system $\mathbf{A} \cdot \mathbf{x} = \mathbf{b}$. (This process is an alternative to Gaussian elimination.)

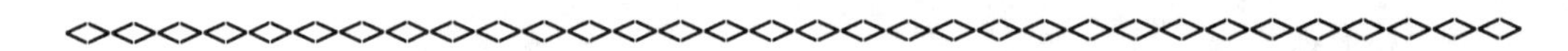

3. If in the Gram-Schmidt process $\mathbf{A}^{<3>}$ is in Span($\mathbf{A}^{<1>}$, $\mathbf{A}^{<2>}$) explain how the process described in Exercise 2 will fail. (Assume that $\mathbf{A}^{<1>}$ and $\mathbf{A}^{<2>}$ are linearly independent.)

Section 5-3 Least Square Solution to A·x = b

Name:________________

Score:_____________

Overview

If the system of equations $\mathbf{A}\cdot\mathbf{x} = \mathbf{b}$ does not have a solution we would like to find a vector, $\mathbf{p}$, that is closest to a solution (i.e., $\mathbf{A}\cdot\mathbf{p}$ is closest to $\mathbf{b}$). Closest means that $|\mathbf{A}\cdot\mathbf{p} - \mathbf{b}|$ is a minimum. The pseudo inverse of **A** provides a tool to find such a vector. Applications to regression and curve fitting are given.

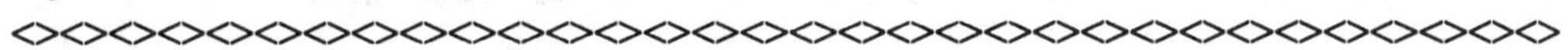

Recall the Oil Refinery Model

A company runs **three refineries**. Each refinery produces **three petroleum-based products**: heating oil, diesel oil and gasoline. We represent this model as a matrix as follows:

■	Refinery1	Refinery2	Refinery3
HeatingOil	4	2	2
DieselOil	2	5	2
Gasoline	1	2.5	5

This matrix is interpreted as follows:

From one barrel of input Refinery 1 produces 4 units of Heating Oil, 2 units of Diesel Oil and 1 unit of Gasoline. The columns corresponding to the other refineries are interpreted similarly.

Let $\mathbf{x}_i$ be the number of barrels of petroleum used by the ith refinery.

Then the number of units of Gasoline produced is $\mathbf{x}_1 + 2.5\mathbf{x}_2 + 5\mathbf{x}_3$. Write an expression for number of units of Diesel Oil produced.

ANSWER:

How many barrels of petroleum should be processed by each refinery if there is demand for 600 units of Heating Oil, 800 units of Diesel Oil and 1000 units of Gasoline? To answer this question we consider the following system of equations.

$$4\cdot x_1 + 2\cdot x_2 + 2\cdot x_3 = 600$$

$$2\cdot x_1 + 5\cdot x_2 + 2\cdot x_3 = 800$$

$$1\cdot x_1 + 2.5\cdot x_2 + 5\cdot x_3 = 1000$$

This system can be expressed as the vector equation.

$$x_1 \cdot M^{\langle 1\rangle} + x_2 \cdot M^{\langle 2\rangle} + x_3 \cdot M^{\langle 3\rangle} = \begin{pmatrix} 600 \\ 800 \\ 1000 \end{pmatrix} \qquad \text{or} \qquad M \cdot x = b$$

where $M := \begin{pmatrix} 4 & 2 & 2 \\ 2 & 5 & 2 \\ 1 & 2.5 & 5 \end{pmatrix}$ and $b := \begin{pmatrix} 600 \\ 800 \\ 1000 \end{pmatrix}$

Explain why there is a unique solution to this refinery problem and find it.

ANSWER:

Now suppose that there is a malfunction at the second refinery and we can not produce any oil products there. Determine the linear system corresponding to this situation and explain why there is no solution to this problem.

ANSWER:

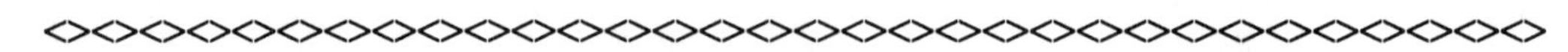

General problem: Find the 'best' answer.

What does 'best' mean?

In our context 'best' means the vector that produces a result ***closest*** to the desired result.

What does 'closest' mean?

There are many different meanings for closest. We will explore two of these in terms of the refinery problem.

Let's return to the refinery problem.

With the closing of the second refinery we can remove it from consideration in our problem. This results in three equations in two unknowns. (x is the number of barrels of petroleum used by Refinery 1 and y the number used by Refinery 3.)

$$R \cdot \begin{pmatrix} x \\ y \end{pmatrix} = b \qquad \text{where} \qquad R := \begin{pmatrix} 4 & 2 \\ 2 & 2 \\ 1 & 5 \end{pmatrix} \qquad \text{and} \qquad b := \begin{pmatrix} 600 \\ 800 \\ 1000 \end{pmatrix}$$

Our goal is to choose x and y to minimize $\mathbf{R}\cdot\begin{pmatrix} x \\ y \end{pmatrix} - \mathbf{b}$

In this example we minimize

$$\begin{pmatrix} 4 & 2 \\ 2 & 2 \\ 1 & 5 \end{pmatrix}\cdot\begin{pmatrix} x \\ y \end{pmatrix} - \begin{pmatrix} 600 \\ 800 \\ 1000 \end{pmatrix} \quad \text{which equals} \quad \begin{pmatrix} 4\cdot x + 2\cdot y - 600 \\ 2\cdot x + 2\cdot y - 800 \\ x + 5\cdot y - 1000 \end{pmatrix}$$

One solution is to minimize the sum of the absolute values of the components. (This is the absolute value distance.)

$$g(x,y) := |4\cdot x + 2\cdot y - 600| + |2\cdot x + 2\cdot y - 800| + |x + 5\cdot y - 1000|$$

Another solution is to use the **standard** distance function, the square root of the sum of the squares. The solution using this distance function is called the **least square solution**.

$$f(x,y) := \left|\begin{pmatrix} 4\cdot x + 2\cdot y - 600 \\ 2\cdot x + 2\cdot y - 800 \\ x + 5\cdot y - 1000 \end{pmatrix}\right|$$

<== Expanding this symbolically we get the expression below.

$$f(x,y) := \sqrt{(|4\cdot x + 2\cdot y - 600|)^2 + (|2\cdot x + 2\cdot y - 800|)^2 + (|x + 5\cdot y - 1000|)^2}$$

One approach to determining the values of x and y that minimize the distance defined by g(x,y) or that defined by f(x,y) is to use a search procedure. We describe this next.

A SEARCH: Find the pair of integers that minimizes these two quantities by searching. The easiest way to search is to look at various values in the lattice of integer points and try to find a minimum. We will set up a prototype search that you can modify.

$i := 1..5 \quad j := 1..5$ **<== Setting indices**

$x := 50 \quad y := 170$ **<== Setting lower left corner of grid.**

$x_i := x + 5\cdot i \quad y_j := y + 5\cdot j$ **<== Generating the grid.**

$M_{i,j} := g(x_i, y_j)$ **<== Evaluating function g at grid point and storing results in matrix M.**

$$M = \begin{bmatrix} 440 & 395 & 350 & 315 & 340 \\ 405 & 360 & 335 & 330 & 355 \\ 390 & 365 & 340 & 345 & 370 \\ 395 & 370 & 345 & 360 & 385 \\ 400 & 375 & 350 & 375 & 400 \end{bmatrix}$$

<== Display values at the grid points.

At what grid points did the smallest value of g occur?

ANSWER:

Choose three different sets of starting values for x and y and record the values of x and y where g was a minimum.

ANSWERS:

Repeat the search to find the values of (x,y) that minimize f(x,y). Record your results below.

ANSWER:

Search procedures such as the ones you have just completed are not very efficient ways to solve problems. You were able to perform the search because we were dealing with a two dimensional array of numbers. If we had a six dimensional array (corresponding to a function in six variables) it would have been much more difficult. As a result mathematicians try to develop minimization procedures other than searches. Both of the above definitions of closest are reasonable and there is no inherent reason to choose one over the other - even though the answers are quite different. However the Euclidean length or **least-square** solution is more easily computed using the methods of linear algebra and this is the one we discuss.

Restatement of problem

Given a system of linear equations $\mathbf{A}\cdot\mathbf{x} = \mathbf{b}$, find a vector $\mathbf{p}$ such that $|\mathbf{A}\cdot\mathbf{p} - \mathbf{b}|$ is a minimum. If the system is consistent then any solution $\mathbf{q}$ would make $\mathbf{A}\cdot\mathbf{q} - \mathbf{b} = \mathbf{0}$ and $|\mathbf{A}\cdot\mathbf{p} - \mathbf{b}|$ would be a minimum for $\mathbf{p} = \mathbf{q}$.

For the purpose of this discussion we will always assume that the system has more equations than variables. This is often called **over constrained** or **over determined**.

This problem is very similar to those considered in the previous section. In that section, we found the vector that was '**closest**' to a subspace by taking the projection of the given vector into the subspace. Here we wish to find vector $\mathbf{p}$ such that $|\mathbf{A}\cdot\mathbf{p} - \mathbf{b}|$ is as small as possible. If $\mathbf{A}$ is an m by n matrix, the set of all $\mathbf{A}\cdot\mathbf{p}$ is simply the column space of $\mathbf{A}$ (i.e., the span of the columns) since

$$A\cdot p = p_1\cdot A^{<1>} + p_2\cdot A^{<2>} + \dots + p_n\cdot A^{<n>}$$

The column space of $\mathbf{A}$, denoted Col($\mathbf{A}$), is a subspace of $\mathbf{R}^m$.

To minimize $|\mathbf{A}\cdot\mathbf{p} - \mathbf{b}|$ we choose **p** such that $\mathbf{A}\cdot\mathbf{p}$ is the projection of **b** into Col(**A**). We could solve this problem by finding an orthogonal basis for Col(**A**) and calculating the projection of **b** directly. It turns out there is an easier way to solve this problem. To motivate this alternative procedure we first reexamine the refinery problem.

In the refinery problem we want the 'best' solution to $\mathbf{R}\cdot\mathbf{x} = \mathbf{b}$ where

$$R := \begin{pmatrix} 4 & 2 \\ 2 & 2 \\ 1 & 5 \end{pmatrix} \qquad b := \begin{pmatrix} 600 \\ 800 \\ 1000 \end{pmatrix}$$

The best solution is the component of **b** that lies in Col(**A**), the subspace of linear combinations of the columns of **R**. We solve for the input, the scalars of the linear combination, that yields this output.

More precisely, our goal is to find the 'solution' (x,y) such that

$$\boxed{\mathrm{proj}_{\mathrm{Col(R)}}b = R\cdot\begin{pmatrix} x \\ y \end{pmatrix}}$$

To determine x and y we use the projection property:

$$\mathbf{b} - \mathbf{proj}_{\mathrm{Col}(\mathbf{R})}\mathbf{b} \text{ must be in } \mathrm{Col}(\mathbf{R})^{\perp}$$

Thus it follows that $b - R\cdot\begin{pmatrix} x \\ y \end{pmatrix}$ must be orthogonal to each column of **R**.

$$R^{\langle 1\rangle}\cdot\left[b - R\cdot\begin{pmatrix} x \\ y \end{pmatrix}\right] = 0 \quad \text{and} \quad R^{\langle 2\rangle}\cdot\left[b - R\cdot\begin{pmatrix} x \\ y \end{pmatrix}\right] = 0$$

Since the dot product gives the same numerical value as a row times a column we can rewrite the preceding expressions

$$R^{\langle 1\rangle^{T}}\cdot\left[b - R\cdot\begin{pmatrix} x \\ y \end{pmatrix}\right] = 0 \quad \text{and} \quad R^{\langle 2\rangle^{T}}\cdot\left[b - R\cdot\begin{pmatrix} x \\ y \end{pmatrix}\right] = 0$$

Next we combine these expressions into a single equation as a matrix-vector product.

$$R^{T}\cdot\left[b - R\cdot\begin{pmatrix} x \\ y \end{pmatrix}\right] = 0$$

Solve this directly for vector $\begin{pmatrix} x \\ y \end{pmatrix}$ (Hint: Expand and use an inverse.)

Explain each step since your solution will yield a general method.

ANSWER:

Solve for x and y using the refinery information stored in matrix **R** and vector **b**.

ANSWER:

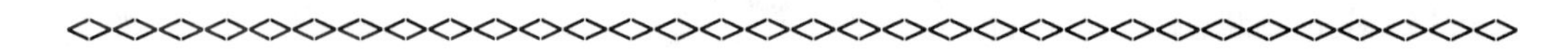

General Solution

If $\mathbf{A} \cdot \mathbf{p} - \mathbf{b}$ is orthogonal to the columns of **A** then explain why

$$\mathbf{A}^{\mathsf{T}} \cdot (\mathbf{A} \cdot \mathbf{p} - \mathbf{b}) = 0$$

ANSWER:

Expanding the previous expression, we obtain: $\mathbf{A}^{\mathsf{T}} \cdot \mathbf{A} \cdot \mathbf{p} = \mathbf{A}^{\mathsf{T}} \cdot \mathbf{b}$

If $\mathbf{A}^{\mathsf{T}} \cdot \mathbf{A}$ is invertible we can solve for **p** $\mathbf{p} = \left(\mathbf{A}^{\mathsf{T}} \cdot \mathbf{A}\right)^{-1} \cdot \mathbf{A}^{\mathsf{T}} \cdot \mathbf{b}$

A does not have to be invertible or even square for $\mathbf{A}^{\mathsf{T}} \cdot \mathbf{A}$ to be invertible.

Example: Display $\mathbf{B}^{\mathsf{T}} \cdot \mathbf{B}$ and its determinant below.

$$\mathbf{B} := \begin{pmatrix} 1 & 2.3 \\ 0 & 0 \\ -3 & 1 \\ .57 & 11 \end{pmatrix} \qquad \mathbf{B}^{\mathsf{T}} \cdot \mathbf{B} = \begin{pmatrix} 10.325 & 5.57 \\ 5.57 & 127.29 \end{pmatrix} \qquad \left|\mathbf{B}^{\mathsf{T}} \cdot \mathbf{B}\right| = 1283.232$$

The (i,j)th entry of $\mathbf{A}^{\mathsf{T}} \cdot \mathbf{A}$ is the dot product of the ith column of **A** and the jth column of **A**. In particular this shows that $\mathbf{A}^{\mathsf{T}} \cdot \mathbf{A}$ is symmetric.

$$A := \begin{pmatrix} 2 & 0 \\ 1 & 1 \\ 0 & -1 \\ 3 & 2 \end{pmatrix} \qquad C := A^T \cdot A \qquad C$$

<== **Display C.**

<><><><><><><><><><><><><><><><><><><><><><><><><><><><><><><><><><><><><><><><>

If $A^T \cdot A$ is invertible, then the pseudo inverse or generalized inverse of A is $A^+ = (A^T \cdot A)^{-1} \cdot A^T$.

Show mathematically and not by example that if **A** is invertible the pseudo inverse is the standard inverse.

ANSWER:

More generally, prove that if $\mathbf{A}^+$ exists, it is always a 'left inverse' for **A**. That is, $\mathbf{A}^+ \cdot \mathbf{A} = \mathbf{I}$.

ANSWER:

Return to the refinery problem. Compute $\mathbf{R}^+$ and then $\mathbf{R}^+ \cdot \mathbf{b}$. Verify that the pseudo inverse yields the answer you derived earlier for the refinery problem.

ANSWER:

If $A^T \cdot A$ is invertible, then the least squares solution to the system of equations $Ax = b$ is

$$A^+ \cdot b$$

where A^+ is the pseudo inverse of A.

When is $\mathbf{A}^T \cdot \mathbf{A}$ invertible?

In Section 3-3 we showed that $\mathbf{A}^T \cdot \mathbf{A}$ is nonsingular if and only if the columns of **A** are linearly independent. Thus $\mathbf{A}^+$ exists whenever the columns of **A** are linearly independent.

<><><><><><><><><><><><><><><><><><><><><><><><><><><><><><><><><><><><><><><><>

A major application of least squares approximations is the determination of coefficients of equations that can be used to approximate a set of data. The equations generated using the least squares approach are called **REGRESSION MODELS** of the data.

Suppose that we made m observations on each of the quantities

$$x_1, x_2, \ldots, x_n, y$$

The observations on the x-quantities are arranged in an m by n matrix **X** where column 1 corresponds to the observed values of x_1, column 2 corresponds to the observed values of x_2, and so on. We further put the corresponding observations of variable y into a vector **y**. We would like to find the 'best fit' for the observations when the linear model has the form

$$y = q_1 \cdot x_1 + q_2 \cdot x_2 + \ldots + q_n \cdot x_n + r$$

That is, we want to find coefficients q1, q2, ..., qn, r so that vector

$$c = \begin{pmatrix} q_1 & q_2 & \cdot & \cdot & \cdot & q_n & r \end{pmatrix}^T$$

is the ***least squares approximation*** to the linear system $[\mathbf{X} \mid \mathbf{1}] \cdot \mathbf{z} = \mathbf{y}$.

The rows of the coefficient matrix $[\mathbf{X} \mid \mathbf{1}]$ correspond to observations of the x's and the 1 to the coefficient of the constant term r in the model above.

For example suppose we make 7 observations of 3 quantities x_1, x_2, x_3 which are given in the columns of matrix **X** and 7 observations of another quantity y which appear in vector **y**.

$$X := \begin{bmatrix} .2 & .18 & 1.1 \\ .4 & 2 & 1.3 \\ 0 & .01 & 2 \\ -.2 & 3 & 3 \\ 3 & 0 & 1 \\ -2 & 1 & .2 \\ 1 & .2 & 0 \end{bmatrix} \qquad y := \begin{bmatrix} 11 \\ 15 \\ 10 \\ 8 \\ 5 \\ 11 \\ 9 \end{bmatrix}$$

If the model for the data is to have the form

$$y = q_1 \cdot x_1 + q_2 \cdot x_2 + q_3 \cdot x_3 + r$$

then we would like to find values for the q's such that

$$q_1 \cdot .2 + q_2 \cdot .18 + q_3 \cdot 1.1 = 11$$

$$q_1 \cdot .4 + q_2 \cdot 2 + q_3 \cdot 1.3 = 15$$

etc

The coefficient matrix is

$$\mathbf{A} := \mathbf{augment}\left(\mathbf{X}, \begin{bmatrix} 1 \\ 1 \\ 1 \\ 1 \\ 1 \\ 1 \\ 1 \end{bmatrix}\right)$$

and the right side is vector **y**.

Find the least squares approximation to this system and hence the coefficients of best fit for this linear model.

ANSWER:

Geometrically describe the model you just computed.

ANSWER:

Another application of least squares is to build polynomial models to a set of ordered pairs. This is called **POLYNOMIAL FITTING**.

Suppose we graph a set of ordered pairs (x_i, y_i). If a straight line is a good model for the data in our set we determine coefficients a and b so that

$$ax_i + b = y_i$$

It is highly likely that the system of equations in the unknowns a and b will be inconsistent. Hence we look for the least squares approximation to find values of the coefficients a and b that give 'the best' least squares line to fit (or approximate) the data. The resulting linear system is

$$\begin{bmatrix} x_1 & 1 \\ x_2 & 1 \\ . & . \\ . & . \\ . & . \\ x_n & 1 \end{bmatrix} \cdot \begin{pmatrix} a \\ b \end{pmatrix} = \begin{bmatrix} y_1 \\ y_2 \\ . \\ . \\ . \\ y_n \end{bmatrix}$$

Once we find the least squares approximation, we use a and b for the equation of the line $y = ax + b$, which we call the least squares line.

a) Find the least squares line for the following data.

$$x := (2\ \ 1\ \ -2\ \ 3\ \ 5\ \ 7\ \ 1)^T \qquad y := (2\ \ 0\ \ 4\ \ 8\ \ 11\ \ 6\ \ 3)^T$$

ANSWER:

b) To graph the least squares line and the data in part (a) we proceed as follows. You must supply the values for a and b as computed above.

$$a := \blacksquare \qquad b := \blacksquare \qquad i := 1..7 \qquad z(t) := a \cdot t + b$$

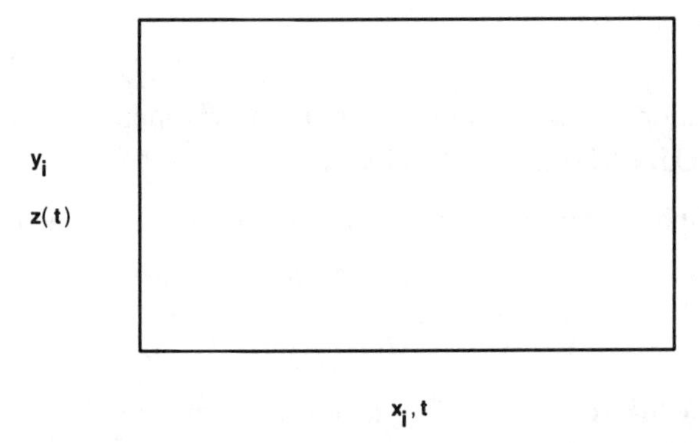

c) If instead of using a line to build a model for the data in the ordered pairs we use a quadratic $y = ax^2 + bx + c$, then the coefficient matrix has the form

$$\begin{bmatrix} (x_1)^2 & x_1 & 1 \\ (x_2)^2 & x_2 & 1 \\ . & . & . \\ . & . & . \\ . & . & . \\ (x_n)^2 & x_n & 1 \end{bmatrix}$$

The unknowns are a, b and c and the right side is the vector **y**.

Compute the determinant of the matrix $\mathbf{M}(t) = \mathbf{L} - t \cdot \mathbf{I}$ symbolically and call the answer f(t).

$$\mathbf{M}(t) := \begin{pmatrix} -t & .5 & .8 & .2 \\ .8 & -t & 0 & 0 \\ 0 & .9 & -t & 0 \\ 0 & 0 & .8 & -t \end{pmatrix}$$

ANSWER:

Use the root command to determine the roots of f(t). Set c equal to the root that corresponds to 1 + the growth rate of the mammal population studied above.

ANSWER:

Now solve $(\mathbf{L} - c \cdot \mathbf{I}) \cdot \mathbf{x} = \mathbf{0}$ to find a stable population distribution $\mathbf{x}$. That is, find a nonzero solution to the homogeneous system $(\mathbf{L} - c \cdot \mathbf{I}) \cdot \mathbf{x} = \mathbf{0}$ by using rref.

rref(M(c)) =

We will, of course, not find a unique solution. How do we compute the distribution based upon the solutions that we find? Do it and explain.

ANSWER:

<><><><><><><><><><><><><><><><><><><><><><><><><><><><><><>

CHAPTER 6

EIGENVALUES AND THEIR APPLICATIONS

$$\mathbf{eigenvals}\left(\begin{bmatrix} 0 & .4 & .7 & .2 \\ .8 & 0 & 0 & 0 \\ 0 & .85 & 0 & 0 \\ 0 & 0 & .75 & 0 \end{bmatrix}\right) = \begin{bmatrix} 0.96184 \\ -0.35712 - 0.54841i \\ -0.35712 + 0.54841i \\ -0.2476 \end{bmatrix}$$

$$\mathbf{v} := \mathbf{eigenvec}\left(\begin{bmatrix} 0 & .4 & .7 & .2 \\ .8 & 0 & 0 & 0 \\ 0 & .85 & 0 & 0 \\ 0 & 0 & .75 & 0 \end{bmatrix}, 0.962\right)$$

$$\mathbf{w} := \frac{\mathbf{v}}{\sum \mathbf{v}} \qquad \mathbf{w} = \begin{bmatrix} 0.31848 \\ 0.26489 \\ 0.23409 \\ 0.18253 \end{bmatrix}$$

Section 6-1 The Leslie Population Model

Name:_______________

Score:____________

Overview

The study of population growth is increasingly important. Mathematical models have been proposed for particular species. Some of these models involve functions that are continuous over time such as

$$P(t) = P_0 e^{kt}$$

where P_0 is the initial population and k is a measure of the population change. This and similar models can be studied using differential equations. Such models consider the population as a whole. In this section we are more interested in changes within different age groups of a population. We use matrix theory to develop the Leslie population model to study the size and distribution of age groups of a population over time.

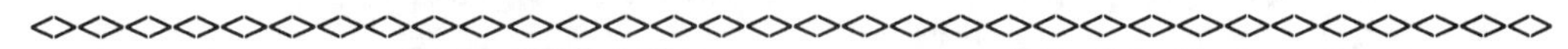

It is important for a wide variety of public policy questions to know ***the number of people in particular age groups at some future time***. Issues such as social security, pension benefits, and school class size revolve around forecasts concerning age-class populations. We develop a model for age-class groups.

Demography is the science dealing with the study of statistics of human populations.

From a demographic viewpoint it suffices to study the size and distribution of females. (We do not limit ourselves to human populations.)

<><><><><><><><><><><><><><><><><><><><><><><><><><><><><><>

Example 1

We study a small marsh mammal whose life span is assumed to be 18 - 24 months. We consider age groups of 6 months and suppose the population is given by the following table.

Age (months)	0 - 6	6-12	12-18	18-24
Number of females	8	12	10	6

To predict the populations in each age group over time we need information about the birth rate and death rate for each of the age groups. Gathering this information may require long studies and, of course, varies with the species. We assume that this information has been collected and that the birth rate reflects the average number of females born to each female in the population within the age group. Information on death rates is also collected, and we use it in the following way to predict new populations.

$$\text{Survival Rate} = 1 - \text{Death Rate}$$

(The birth rate is the *probability* that a female in the group has a female child and the survival rate is the *probability* that a female in the group survives to the next age group.)

The following table contains the birth and survival rates for our marsh mammal.

Age in Months

Age (months)	0 - 6	6-12	12-18	18-24
Number of females	8	12	10	6
Birth Rate	0	.5	.8	.2
Survival Rate	.8	.9	.8	0

We see that no female survives after 24 months.
The number of females surviving to the next age group is

For example: (.8)8 = 6.4 survive from age group 0 - 6 to age group 6 - 12
(.9)12 = 10.8 survive from age group 6 - 12 to age group 12 - 18

The fractional pieces of females surviving to an age group must be retained since the population may be given in some unit like thousands or millions. In addition the birth and survival rates are probabilistic quantities and were found by averaging data over a long period of time. We do not expect these rates to be exact at any one time, but we do expect them to accurately reflect changes over a long period of time. To obtain a population estimate at a particular time we would round off the values after we apply the particular unit associated with the data, but we retain the fractional parts as we progress through successive stages of our prediction.

We described the survivors in the age groups above, but we still need to know how many females enter the age group 0 - 6. This is the sum of all the births in each age group. Using the birth rates as a vector and the populations as a vector we compute their dot product.

$$8(0) + 12(.5) + 10(.8) + 6(.2) = 15.2$$

Fill in the following table which gives the populations after 1 six month period:

Age (months)	0 - 6	6 - 12	12 - 18	18 - 24
Number of females				

What is the new (total) population? ___________

What is the (total) growth in the population? ___________

Fill in the following table which gives the populations after 2 six month periods (that is, after a total of 12 months):

Age (months)	0-6	6-12	12 -18	18 - 24
Number of females				

What is the new (total) population? ___________

What is the (total) growth in the population? ___________

<><><><><><><><><><><><><><><><><><><><><><><><><><><><><><><><><>

The calculations performed in the Example 1 can be summarized in matrix form. A matrix containing the birth and survival rates is constructed below. This matrix is called the **Leslie matrix** and is usually denoted by **L**. The populations of the age groups are arranged in a vector called the **age distribution vector**. From the previous example the initial populations are given by the vector

$$X0 := (8 \;\; 12 \;\; 10 \;\; 6)^T$$

What is the age distribution vector **X1** after one 6-month period?

ANSWER:

Let matrix **L** be defined as follows:

Assume that a population is divided into n age groups.

Notation: b_k is the **birth rate** for the kth group

s_k is the **survival rate** for the kth group

The Leslie matrix has the form

$$L = \begin{bmatrix} b_1 & b_2 & b_3 & . & . & . & b_n \\ s_1 & 0 & 0 & 0 & 0 & 0 & 0 \\ 0 & s_2 & 0 & 0 & 0 & 0 & 0 \\ 0 & 0 & s_3 & 0 & 0 & 0 & 0 \\ 0 & 0 & 0 & . & 0 & 0 & 0 \\ 0 & 0 & 0 & 0 & . & 0 & 0 \\ 0 & 0 & 0 & 0 & 0 & s_{n-1} & 0 \end{bmatrix}$$

For the marsh mammal described in Example 1, the Leslie matrix is

$$\mathbf{L} := \begin{pmatrix} 0 & .5 & .8 & .2 \\ .8 & 0 & 0 & 0 \\ 0 & .9 & 0 & 0 \\ 0 & 0 & .8 & 0 \end{pmatrix}$$

Using matrix multiplication the age distribution vectors **X1, X2, X3, ...**

are computed as

$$\mathbf{X1} := \mathbf{L} \cdot \mathbf{X0}$$
$$\mathbf{X2} := \mathbf{L} \cdot \mathbf{X1}$$
$$\mathbf{X3} := \mathbf{L} \cdot \mathbf{X2}$$

....

In our work with Markov chains we learned that this type of iterative process can be rewritten using matrix powers as follows:

$$\mathbf{xj} = \mathbf{L}^{\mathbf{j}} \cdot \mathbf{X0}$$

You computed **X1** and **X2** above without the use of the Leslie matrix. Use the Leslie matrix to verify that your result is correct. Then compute **X5**, **X10** and **X20**.

ANSWER:

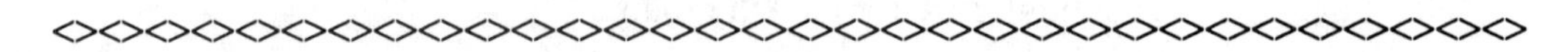

To study the long term behavior of populations it is convenient to use a matrix as a data structure to hold the information on the populations of the age classes. To do this in Mathcad we use the following technique.

Let $\mathbf{X}^{<j>}$ be the vector of populations in the jth time period (i.e., $\mathbf{X}^{<j>}_k$ is the population in age group k during the jth time period). We always assume that j is a positive integer since we measure the population yearly, in six month periods, monthly, etc. In this structure $\mathbf{X}^{<1>}$ is the initial population vector. How is $\mathbf{X}^{<10>}$ computed?

ANSWER:

$\mathbf{X}^{<10>}$ is the population after how many time periods?

ANSWER:

Example 2

Investigate the long term behavior of the marsh mammal from Example 1. In particular answer the following two questions. (We indicate how to proceed with the computations below.)

1. Let $\mathbf{TOTALPOP}_j$ be the total population in time period j.

Consider the ratio as j gets large.

$$\frac{TOTALPOP_{j+1}}{TOTALPOP_j} \rightarrow c$$

Our goal is to investigate if there is a constant c that is the limit of the ratio. If the limit does exist, the value c-1 is called the ***growth rate of the population***.

Mathcad will compute the sum of the components in a vector when we select the icon on the right from the matrix palette.

2. Does the population distribution tend to stabilize over time? The population distribution at time t is **X(t) / TOTALPOP(t)**. This gives the percentage of the total population in each age group at time t.

Our computational procedure:

First compute the population vectors over 20 six month periods and store this information column-wise into matrix **X**.

$$L := \begin{pmatrix} 0 & .5 & .8 & .2 \\ .8 & 0 & 0 & 0 \\ 0 & .9 & 0 & 0 \\ 0 & 0 & .8 & 0 \end{pmatrix}$$

$j := 1 .. 21$ **<== Setting number of time periods.**

$X^{<1>} := (8 \;\; 12 \;\; 10 \;\; 6)^T$ **<== Initial population vector.**

$X^{<j+1>} := L \cdot X^{<j>}$ **<== The iteration procedure.**

X =

	1	2	3	4	5	6	7
1	8	15.2	13.44	12.416	15.053	14.459	14.721
2	12	6.4	12.16	10.752	9.933	12.042	11.567
3	10	10.8	5.76	10.944	9.677	8.94	10.838
4	6	8	8.64	4.608	8.755	7.741	7.152

To determine a growth rate for the entire population first compute the total population in each time period by summing the columns of matrix **X**.

$\text{TOTALPOP}_j := \sum X^{\langle j \rangle}$ **<== Display this vector below.**

$\text{TOTALPOP}^T =$

	1	2	3	4	5	6	7	8
1	36	40.4	40	38.72	43.418	43.182	44.278	46.742

You should see that the total initial population is 36; after 1 six month period the total population is 40.4, and so on.

To determine the growth rate of the marsh mammal, we want to determine the limit, as t gets large, of the ratio

$$\frac{\text{TOTALPOP}_{j+1}}{\text{TOTALPOP}_j}$$

If the limit exists and equals c, the growth rate is c - 1. To compute this in Mathcad we use the following:

$$j := 1 .. 20$$

$$c_j := \frac{\text{TOTALPOP}_{j+1}}{\text{TOTALPOP}_j}$$

$c^T =$

	1	2	3	4	5	6	7	8
1	1.122	0.99	0.968	1.121	0.995	1.025	1.056	1.018

<== Display the vector of c-values.

Based upon your inspection of the vector **C**, make a conjecture (to two decimal places) for the value of c (c should be the limit of $\mathbf{C}_j$ as j gets large).
ANSWER:

What is the approximate growth rate of the mammal population? (Express your answer as a percent.)
ANSWER:

To answer the second question posed above, proceed as follows. The ***distribution*** of the population at a given time is obtained by computing the percent of the total population at that time in each of the age groups. Let $\mathbf{dist}^{<j>}$ be the vector which represents the population distribution in period j. We compute this as

$$\mathrm{dist}^{<j>} := \frac{X^{<j>}}{\mathrm{TOTALPOP}_j}$$

$$\mathrm{dist}^{<1>} = \begin{bmatrix} 0.222 \\ 0.333 \\ 0.278 \\ 0.167 \end{bmatrix}$$

<== Display this vector. The data in this vector gives the initial distribution of the population: (approximately) 22.2% in ages 0 - 6 months, 33.3% in ages 6 - 12 months, 27.8 % in ages 12 - 18 months, and 16.7% in ages 18 - 24 months.

We say that the distribution of the population ***stabilizes*** if for large values of j (which means after a long period of time)

$$\left|\mathrm{dist}^{<j+1>} - \mathrm{dist}^{<j>}\right| \to 0$$

Explain why if this condition holds we say that **the population distribution stabilizes**.
ANSWER:

By examining the graphs of the coordinates of the vector $\mathrm{dist}^{<j>}$ as j increases we get a visual picture of what it means for the population to stabilize.

$j := 1 .. 20 \quad k := 1 .. 4$

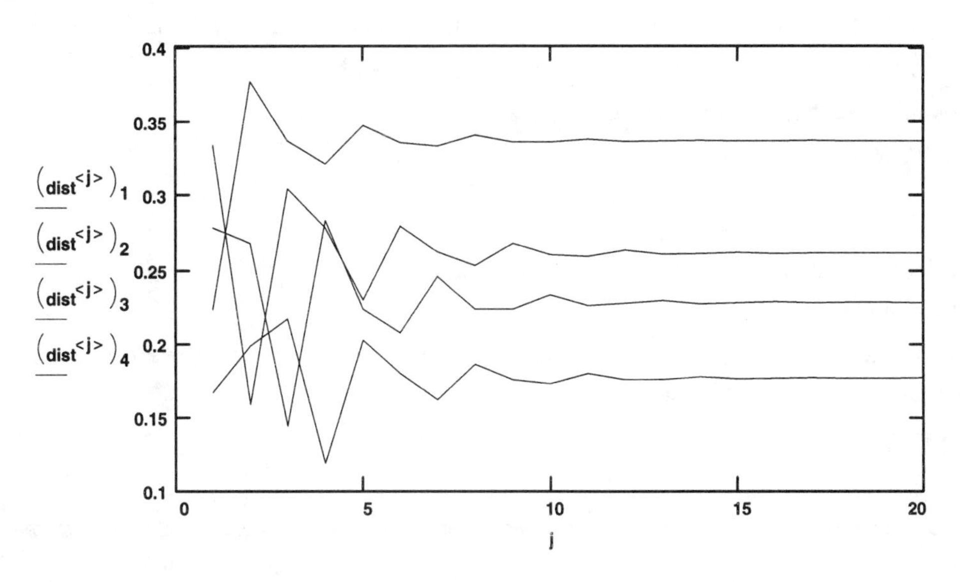

We can also examine what it means to stabilize by scrolling through the values of **dist** in the following table.

dist =

	1	2	3	4	5	6	7	8
1	0.222	0.376	0.336	0.321	0.347	0.335	0.332	0.34
2	0.333	0.158	0.304	0.278	0.229	0.279	0.261	0.252
3	0.278	0.267	0.144	0.283	0.223	0.207	0.245	0.223
4	0.167	0.198	0.216	0.119	0.202	0.179	0.162	0.185

Next compute the distance between successive distributions.

$j := 1 .. 19$

$$\mathrm{diff}_j := \left| \mathrm{dist}^{\langle j+1 \rangle} - \mathrm{dist}^{\langle j \rangle} \right|$$

Display this vector. ===> $\mathrm{diff}^T =$

	1	2	3	4	5	6
1	0.2354	0.1958	0.1719	0.1161	0.0583	0.0454

If we had done 100 iterations instead of just 20, the values in **diff** would become much smaller.
Does the distribution of the marsh mammal population stabilize?

ANSWER:

What is the (approximate) stable distribution of the mammal population? (Express your answer in percentages for each age group.)

ANSWER:

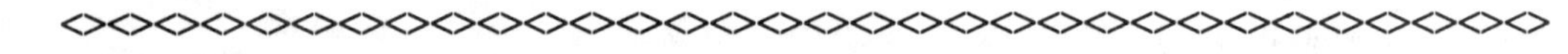

Algebraic solution of the above problem

Assume that the population tends to a stable distribution over time. Call this **Stabledist**.

The vector **Stabledist** must satisfy: **L**·**Stabledist** = c·**Stabledist**.

Explain why **L**·**Stabledist** must equal c·**Stabledist.**

ANSWER:

Stabledist must be a solution to matrix equation **L**·**x** = c·**x** where c is 1 + the growth rate, as discussed above.

Does **L**·**x** = c·**x** have a nontrivial solution? This is the same as asking if there is a value for c such that (**L** - c·**I**) is a singular matrix. Why?

ANSWER:

<><><><><><><><><><><><><><><><><><><><><><><><><><><><><><><><><><><><><><>

Let $\mathbf{I} := \mathbf{identity}(4)$ For the Leslie matrix of the marsh mammal used in Example 1 we display (**L** - c·**I**) below.

$$\begin{pmatrix} 0 & .5 & .8 & .2 \\ .8 & 0 & 0 & 0 \\ 0 & .9 & 0 & 0 \\ 0 & 0 & .8 & 0 \end{pmatrix} - \begin{pmatrix} c & 0 & 0 & 0 \\ 0 & c & 0 & 0 \\ 0 & 0 & c & 0 \\ 0 & 0 & 0 & c \end{pmatrix} = \begin{pmatrix} -c & .5 & .8 & .2 \\ .8 & -c & 0 & 0 \\ 0 & .9 & -c & 0 \\ 0 & 0 & .8 & -c \end{pmatrix}$$

We want to find c such that the above matrix is singular. We first outline the procedure and then indicate how to perform each step.

1. Compute the determinant to get a polynomial in c.
2. Find the roots of the polynomial.
3. Determine which of the roots to use for c so that c = 1 + growth rate.
4. Use the value of c from Step 3 to find the stable distribution.

==

Finding roots of functions in Mathcad

Roots of an expression

root(f(x),x) Returns the value of x that makes the function f equal to zero.

Arguments:

- **f** is a scalar valued function of any number of variables.
- **x** is a scalar variable found in **f**.

The **root** function requires a guess value to start the root finding process. For functions with many roots, the root returned depends on the guess value.

Mathcad computes the root to an accuracy based upon the variable **TOL**. For our examples we would like high accuracy so always set

$$\mathbf{TOL} \equiv 10^{-15}$$

The symbol ≡ means that this definition holds for the entire document.

You can either do this directly as we have done or open the **Math** menu at the top and then click on **Built-in variables**. Change the value of **TOL**.

Example

Let $f(x) := x^3 + 2 \cdot x^2 - 15 \cdot x + 3$

$x := 10$ <===== ***initial guess needed by the root command***

$\text{root}(f(x), x) = 2.867$

If we compute the value of the function at the value of root displayed above, we will not get zero. This is because we have rounded off the answer.

$f(2.867) = 2.264 \cdot 10^{-4}$

This will cause problems when we compute the rref. To overcome this problem we should always set the root equal to a variable and use the variable instead of its value.

$x := 10$

$c := \text{root}(f(x), x)$

$c = 2.867$

$f(c) = 0$

Change the initial guess above to -10 and then to 0 to see how the answer varies. You should see that f has three real roots. One way to determine the number of roots and their values is to graph the function:

$t := -8, -7.9 .. 8$

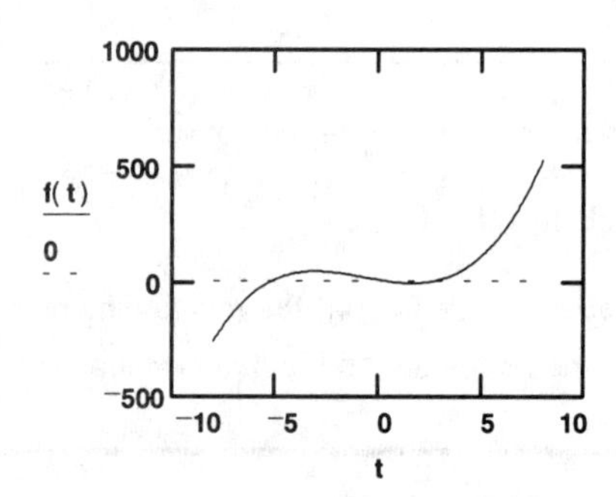

By clicking on the graph and choosing **Crosshair...** from the **X-Y Plot** menu you should see that there is one root around -5 and two additional roots between 0 and 5. To examine the curve between 0 and 5 choose **Zoom** from the **X-Y Plot** menu and drag the rectangle over the portion of the graph you wish to zoom in on.

==

Compute the determinant of the matrix **M**(t) = **L** - t·**I** symbolically and call the answer f(t).

$$M(t) := \begin{pmatrix} -t & .5 & .8 & .2 \\ .8 & -t & 0 & 0 \\ 0 & .9 & -t & 0 \\ 0 & 0 & .8 & -t \end{pmatrix}$$

ANSWER:

Use the root command to determine the roots of f(t). Set c equal to the root that corresponds to 1 + the growth rate of the mammal population studied above? Explain.

ANSWER:

Now solve (**L** - c·**I**)·**x** = **0** to find a stable population distribution **x**. That is, find a nonzero solution to the homogeneous system (**L** - c·**I**)·**x** = **0** by using rref.

rref(M(c)) =

We will, of course, not find a unique solution. How do we compute the distribution based upon the solutions that we find? Do it and explain.

ANSWER:

<><><><><><><><><><><><><><><><><><><><><><><><><><><><><><><><><>

Exercises

1. Let the Leslie matrix for the marsh mammal be the following matrix, determine the growth rate, and the long term population distribution. Show your work.

$$L := \begin{pmatrix} 0 & .5 & .75 & .2 \\ .8 & 0 & 0 & 0 \\ 0 & .7 & 0 & 0 \\ 0 & 0 & .7 & 0 \end{pmatrix}$$

ANSWER:

2. (Individual) The calculations that we have just done are in many ways similar to the calculations that we did earlier with Markov Chains. Discuss the similarities and differences.

Section 6-2 Eigenvalues and Eigenvectors

Name:________________

Score:_____________

Overview

In Section 6-1 the Leslie population model was used to compute a stable distribution and the growth rate of a population. This was similar to our earlier Markov Chain calculations where the goal was to calculate a long term stable weather distribution. We have solved both of these problems by iteration. However, both can be solved algebraically by using eigenvalues and eigenvectors. This section introduces the concepts of eigenvalues and eigenvectors and develops an algebraic procedure for computing them.

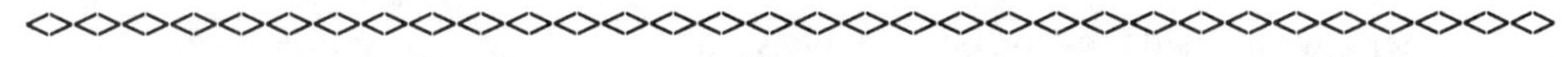

Let **A** be an n by n matrix. If there exists a number λ and a nonzero vector **v** such that $\mathbf{A}\cdot\mathbf{v} = \lambda\cdot\mathbf{v}$ we say that λ is an **eigenvalue** of **A** and **v** is an **eigenvector** associated with λ.

Show that if **v** is an eigenvector, so is $k\cdot\mathbf{v}$ for any nonzero real number k.

ANSWER:

<><><><><><><><><><><><><><><><><><><><><><><><><><><><><><><><>

$\mathbf{A}\cdot\mathbf{v} = \lambda\cdot\mathbf{v}$ is the same as $\mathbf{A}\cdot\mathbf{v} - \lambda\cdot\mathbf{v} = \mathbf{0}$. (The right side is the zero column vector.)

Let **I** be the n by n identity matrix then we know that $\mathbf{v} = \mathbf{I}\cdot\mathbf{v}$, so substituting in the equation above we get the equivalent equation:

$$\mathbf{A}\cdot\mathbf{v} - \lambda\cdot\mathbf{I}\cdot\mathbf{v} = \mathbf{0}$$

after factoring this becomes

$$(\mathbf{A} - \lambda\cdot\mathbf{I})\cdot\mathbf{v} = \mathbf{0}$$

This is a homogeneous system and therefore has a nontrivial solution only if the coefficient matrix is singular. The coefficient matrix

$$\mathbf{A} - \lambda\cdot\mathbf{I}$$

is singular if and only if its determinant

$$|\mathbf{A} - \lambda\cdot\mathbf{I}| = 0$$

Thus we have the following ***two-step process*** to solve the eigenproblem:

i) Find values of λ so that $\quad |\mathbf{A} - \lambda\cdot\mathbf{I}| = 0$

These values of λ are called the ***eigenvalues***.

Warning: The eigenvalues may be complex numbers.

ii) For each eigenvalue λ, find a basis for the solution space of

$$(\mathbf{A} - \lambda \cdot \mathbf{I}) \cdot \mathbf{v} = \mathbf{0}$$

Any nonzero vector, **v**, in the solution space for this equation has the property that $\mathbf{A} \cdot \mathbf{v} = \lambda \cdot \mathbf{v}$. As a result any nonzero linear combination of the basis vectors for this subspace will be an eigenvector corresponding to the eigenvalue λ.

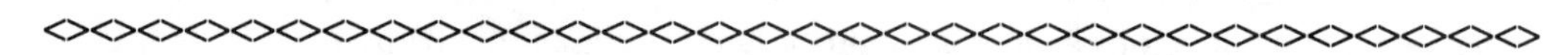

In Chapter 2 we learned how to compute a basis for the solution space of a homogeneous system. So the only remaining question from the procedure outlined above is **how to compute the eigenvalues.**

Compute the following determinants by 'hand' using *the diagonal trick.*

$$\left|\begin{pmatrix} 2-\lambda & 7 \\ 5 & 1-\lambda \end{pmatrix}\right| \qquad \left|\begin{pmatrix} 6-\lambda & 0 & 4 \\ 2 & -\lambda & 5 \\ 1 & 3 & 1-\lambda \end{pmatrix}\right|$$

ANSWERS:

If you computed correctly you should see that the first determinant is a polynomial of degree 2 in λ and the second is a polynomial of degree 3 in λ.

Use Maple to compute the following determinants. In each case, describe the resulting function.

$$\left|\begin{bmatrix} 11 & 0 & -13 & 11 & 1 \\ 7 & 32 & 12 & 5 & 6 \\ -9 & 1 & 0 & 8 & 9 \\ 0 & 7 & 5 & 17 & -23 \\ 41 & 9 & -9 & 0 & 4 \end{bmatrix} - \lambda \cdot \mathrm{identity}(5)\right|$$

ANSWER:

$$\left|\begin{bmatrix} 3 & 3 & 15 & 4 & 1 & 0 & 3 \\ 6 & 7 & -21 & 27 & -8 & 1 & 0 \\ 3 & 21 & 7 & 13 & 21 & 7 & 7 \\ 7 & -3 & 4 & 0 & 5 & 3 & 5 \\ 6 & 5 & 3 & -7 & 21 & 9 & 0 \\ -4 & 0 & 5 & -7 & -36 & -8 & 5 \\ -10 & 1 & 0 & 2 & 7 & 2 & 3 \end{bmatrix} - \lambda\cdot \mathbf{identity}(7)\right|$$

ANSWER:

Make a guess. What type of function is |**A** - λ·**identity**(n)| when **A** is an n by n matrix? (See also Exercise 5.)

ANSWER:

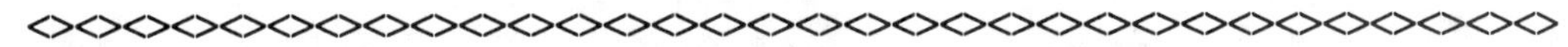

The function, |**A** - λ·**I**| is called the ***characteristic polynomial*** of the matrix **A**.

The process of determining λ such that (**A** - λ·**I**) is singular is the same as finding the roots of the nth degree polynomial |**A** - λ·**I**|.

How many roots does such an nth degree polynomial have?

ANSWER:

Must all the roots be real numbers? Explain.

ANSWER:

<><><><><><><><><><><><><><><><><><><><><><><><><><><><><><><><><><><><><>

Experiment 1

Compute the eigenvalues and the corresponding eigenvectors of the matrix

$$\mathbf{A} := \begin{pmatrix} 0 & -2 & 1 \\ 1 & 3 & -1 \\ 0 & 0 & 1 \end{pmatrix}$$

Computing the eigenvalues:

Find |**A** - λ·**I** | symbolically

$$\left|\begin{pmatrix} 0 & -2 & 1 \\ 1 & 3 & -1 \\ 0 & 0 & 1 \end{pmatrix} - \lambda\cdot \mathbf{identity}(3)\right|$$

$$-5\cdot\lambda + 4\cdot\lambda^2 - \lambda^3 + 2$$

<== Note that since the equation is a cubic polynomial there will be 3 eigenvalues but some may be repeated.

The characteristic equation is

$$-5\cdot\lambda + 4\cdot\lambda^2 - \lambda^3 + 2=0$$

To find the roots of this polynomial we can factor it, either by hand or using Maple, or we can use Mathcad's **root** command. (See Section 6-1 for a description of this command.) We first set the tolerance used to compute the roots.

$TOL \equiv 10^{-15}$

Compute the roots of $\qquad f(\lambda) := -5\cdot\lambda + 4\cdot\lambda^2 - \lambda^3 + 2$

You should have found that f(λ) has roots at 1 and 2.
To verify that these are the only roots we can examine the graph of f(λ).

$\lambda := -2, -1.9 .. 4$

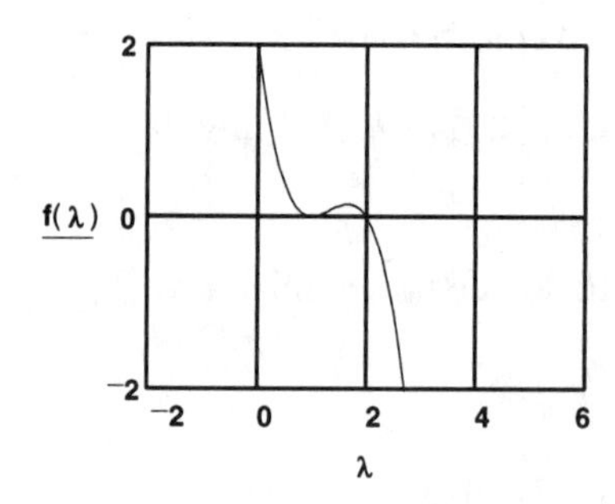

The roots of the characteristic polynomial are the points where the graph crosses or is tangent to the horizontal axis. When λ equals 2 the curve crosses the axis and we can see that 2 is a **simple** root of the equation. When λ equals 1 the curve is tangent to the axis but does not cross it. This indicates that 1 is a **multiple** root. Since the characteristic polynomial is of degree 3 we can conclude that 1 is a root of multiplicity two. **Thus the set of eigenvalues of the matrix A is 1, 2.**

Fact: If λ is a multiple root of the polynomial p(x) then λ is also a root of the derivative, p'(x). This is clear in the graph above but it can be shown to be true in general.

Another way of reaching the same conclusion would have been to factor the polynomial by hand or using Maple. To do this, box the expression below and click on Factor Expression in the **Symbolic** menu.

ANSWER: $\qquad -5\cdot\lambda + 4\cdot\lambda^2 - \lambda^3 + 2$

Obtaining the corresponding eigenvectors:

Take the individual eigenvalues one at a time and find a basis for the solution space of the related homogeneous system.

Case $\lambda = 1$

The homogeneous system is $(\mathbf{A} - \lambda \cdot \mathbf{I}) \cdot \mathbf{x} = \mathbf{0}$ which here is $(\mathbf{A} - 1 \cdot \mathbf{I}) \cdot \mathbf{x} = \mathbf{0}$.
We proceed by finding the rref of matrix $(\mathbf{A} - 1 \cdot \mathbf{I})$.

$$\mathbf{Q} := \mathbf{A} - 1 \cdot \mathbf{identity}(3)$$

$$\mathrm{rref}(\mathbf{Q}) = \begin{pmatrix} 1 & 2 & -1 \\ 0 & 0 & 0 \\ 0 & 0 & 0 \end{pmatrix}$$ **<== Press F9.**

Finding the eigenvectors corresponding to the eigenvalue 1 in this case is the same as finding a basis for the solution space of the homogeneous system $(\mathbf{A} - 1 \cdot \mathbf{I}) \cdot \mathbf{x} = \mathbf{0}$ which here is equivalent to the single equation

$$x_1 + 2 \cdot x_2 - x_3 = 0$$

Compute a basis for this solution space.

ANSWER:

Verify that the basis vectors are eigenvectors by multiplying each of them by the matrix **A**. What should the result be?

ANSWER:

Case $\lambda = 2$

Repeat the above process to find a basis for the solution space of homogeneous system $(\mathbf{A} - 2 \cdot \mathbf{I}) \cdot \mathbf{x} = \mathbf{0}$.

ANSWER:

Form a matrix **P** whose columns are the three eigenvectors determined above. Compute the product $\mathbf{P}^{-1}\cdot\mathbf{A}\cdot\mathbf{P}$ below.
ANSWER:

Describe the resulting matrix in terms of the eigenvalues of matrix **A**.
ANSWER:

Interchange the order of the columns of matrix **P** and repeat the computation. Describe the resulting matrix again.
ANSWER:

There is a pattern of association between the columns **P** and the diagonal entries of $\mathbf{P}^{-1}\cdot\mathbf{A}\cdot\mathbf{P}$. Describe the pattern.
ANSWER:

In Experiment 1 the dimension of the solution space to $(\mathbf{A} - 1\cdot\mathbf{I})\cdot\mathbf{x} = \mathbf{0}$ was 2, the same as the multiplicity of the root 1. This need not always be the case. The fact that $(\mathbf{A} - \lambda\cdot\mathbf{I})$ is singular only ensures that there is *some* nonzero solution. (The dimension of the solution space is at least one.) Compute the dimension of the subspace of all eigenvectors corresponding to the eigenvalue 1 in each of the following. (Hint: Use rref.)

$$\begin{pmatrix} 1 & 1 & 1 \\ 0 & 1 & 1 \\ 0 & 0 & 1 \end{pmatrix} \qquad \begin{pmatrix} 1 & 0 & 1 \\ 0 & 1 & 1 \\ 0 & 0 & 1 \end{pmatrix} \qquad \begin{pmatrix} 1 & 0 & 0 \\ 0 & 1 & 0 \\ 0 & 0 & 1 \end{pmatrix}$$

ANSWER:

<><><><><><><><><><><><><><><><><><><><><><><><><><><><><><><><><><><><>

Experiment 2 shows that a matrix with all real entries can have complex eigenvalues and complex eigenvectors.

Experiment 2

Let $\mathbf{A} := \begin{pmatrix} 0 & 1 \\ -1 & 0 \end{pmatrix}$

Find the characteristic polynomial and factor it by hand to show that the eigenvalues are $\lambda = i$ and $\lambda = -i$, where i is the square root of -1.

<== Show your work!

Determine by 'hand' an eigenvector corresponding to each of the eigenvalues.

<== Show your work!

<><><><><><><><><><><><><><><><><><><><><><><><><><><><><><>

Matrix **A** in Example 2 is the matrix of a rotation through $-\pi/2$. More generally,

consider the matrix of a rotation through t radians $\begin{pmatrix} \cos(t) & -\sin(t) \\ \sin(t) & \cos(t) \end{pmatrix}$

Show that the characteristic equation is

$$-2\cdot\cos(t)\cdot\lambda + \lambda^2 + 1$$

ANSWER:

Using the quadratic formula we see that the roots of this equation are

$$\frac{2\cdot\cos(t) + \sqrt{4\cdot\cos(t)^2 - 4}}{2} \quad \text{and} \quad \frac{2\cdot\cos(t) - \sqrt{4\cdot\cos(t)^2 - 4}}{2}$$

The roots will therefore be complex unless $(\cos(t))^2 \geq 1$. For what values of t is this true? Remember that an eigenvector is a vector that gets mapped into a scalar multiple of itself. For what values of the angle t, does the rotation through t map some vector into a multiple of itself?

ANSWER:

<><><><><><><><><><><><><><><><><><><><><><><><><><><><><><>

Exercises

1. The objective of this exercise is to introduce you to the Mathcad commands that will compute eigenvalues and eigenvectors 'automatically'. That is, these commands skip all the details of the characteristic equation, finding its roots, and finding bases of associated homogeneous systems.

Click on **Help** in the menu bar at the top of the screen. Then click on the following items o successive screens that appear. (To backup a screen, click on **Back**.)

Index

Search

eigenvalues

Go To

Once you have read and thought about the **eigenvals** and **eigenvec** commands click on the **File** menu and chose **Exit**.

(The word *normalize* means that an eigenvector is found and is divided by its length so tha it becomes a unit vector.)

What restrictions are placed on the matrix **M** in the **eigenvals** and **eigenvec** commands?

<== List them here.

Compute the eigenvalues and eigenvectors of the matrix **A** using the Mathcad functions.

$$\mathbf{A} := \begin{pmatrix} 0 & -2 & 1 \\ 1 & 3 & -1 \\ 0 & 0 & 1 \end{pmatrix}$$

Check the results with those you obtained in Experiment 1. In particular let Mathcad find a eigenvector corresponding to each eigenvalue and compare the results with those in the experiment. Remember that any nonzero scalar multiple of an eigenvector is another eigenvector. Explain how the results here are different from above.

ANSWER:

==

Roundoff error can impose difficulties in computing eigenvectors. Unless a high degree of accuracy is retained in computing the eigenvalues, the determinant $|\mathbf{A} - \lambda\cdot\mathbf{I}|$ will not be zero and rref($|\mathbf{A} - \lambda\cdot\mathbf{I}|$) will be the identity. In particular we note that copying and pasting results only maintains the accuracy shown on the screen. We illustrate these problems with an example.

$$A := \begin{pmatrix} 1 & 2 & 0 \\ -3 & 7 & -2 \\ 4 & 5 & 1 \end{pmatrix}$$

$$\text{eigenvals}(A) = \begin{pmatrix} 0.242 \\ 4.379 - 3.115i \\ 4.379 + 3.115i \end{pmatrix}$$

$$|A - 0.242\cdot\text{identity}(3)| = 0.011$$

$$\text{rref}(A - 0.242\cdot\text{identity}(3)) = \begin{pmatrix} 1 & 0 & 0 \\ 0 & 1 & 0 \\ 0 & 0 & 1 \end{pmatrix}$$

To overcome these problems, use a variable with greater accuracy for λ, as follows:

$$\lambda := \text{eigenvals}(A)_1$$

$$|A - \lambda\cdot\text{identity}(3)| = 1.167\cdot 10^{-14}$$

$$AA := \text{rref}(A - \lambda\cdot\text{identity}(3))$$

$$AA = \begin{pmatrix} 1 & 0 & 0.36 \\ 0 & 1 & -0.136 \\ 0 & 0 & 0 \end{pmatrix}$$

The eigenvector is $$w := \begin{bmatrix} -\left[\left(AA^{<3>}\right)_1\right] \\ -\left[\left(AA^{<3>}\right)_2\right] \\ 1 \end{bmatrix}$$

$$\lambda\cdot w = \begin{pmatrix} -0.0872004755 \\ 0.0330312524 \\ 0.2424065997 \end{pmatrix} \qquad A\cdot w = \begin{pmatrix} -0.0872004755 \\ 0.0330312524 \\ 0.2424065997 \end{pmatrix}$$

==

2. A Leslie population model has the matrix

$$\mathbf{L} := \begin{bmatrix} 0 & .6 & .75 & .6 & 0 \\ .65 & 0 & 0 & 0 & 0 \\ 0 & .8 & 0 & 0 & 0 \\ 0 & 0 & .9 & 0 & 0 \\ 0 & 0 & 0 & .7 & 0 \end{bmatrix}$$

a) Find the growth rate using the eigen commands in Mathcad. Carefully explain your procedure. (See Section 6-1.)
ANSWER:

b) Find the long term percentage distribution using the eigen commands in Mathcad. Carefully explain your procedure.
ANSWER:

3. The matrix $\mathbf{A} := \begin{pmatrix} .2 & .5 & .4 \\ .7 & .5 & .2 \\ .1 & 0 & .4 \end{pmatrix}$ is the transition matrix for a certain Markov

chain model. Find its steady state distribution using the eigen commands in Mathcad. Explain your procedure.
ANSWER:

<><><><><><><><><><><><><><><><><><><><><><><><><><><><><><><><><><><><><><>

4. In some of the previous exercises we saw how eigenvalues and eigenvectors have been used to calculate long term behavior. The theoretical properties of eigenvalues and eigenvectors have also been studied extensively. Here we state several properties and ask that you supply arguments for inferences that can be drawn from the statements. You must clearly explain how you make the inferences; supply reasons based on previous facts.

a) ***det(M) = the product of its eigenvalues*** (If an eigenvalue appears as a repeated root of the characteristic equation it is included in the product with the same multiplicity.)

This follows from the combination of two statements

(1) For any polynomial the constant term is the product of the roots.

(2) The constant term of the characteristic equation is the determinant. This is immediate since the constant term is the value of the characteristic polynomial when $\lambda = 0$ and $|\mathbf{M} - 0 \cdot \mathbf{I}| = |\mathbf{M}|$.

Suppose that a matrix **M** is singular, what can you say about its eigenvalues?
What fact about the eigenvalues of **M** would lead you to conclude that **M** is nonsingular?
ANSWERS:

b) Suppose that **M** is a diagonal matrix. What are its eigenvalues and eigenvectors?
ANSWER:

c) Let 8 be an eigenvalue of matrix **A**. Let S be the set consisting of the zero vector and ***all*** eigenvectors of **A** corresponding to 8. (i.e., if **z** is in S then $\mathbf{A} \cdot \mathbf{z} = 8 \cdot \mathbf{z}$) Prove that S is a subspace.
ANSWER:

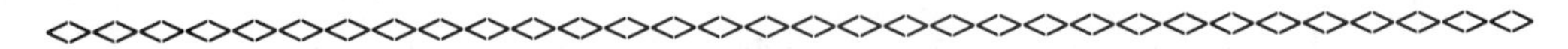

5. Near the beginning of this section you were ask to formulate a conjecture about the type of function obtained for the characteristic polynomial of an n by n matrix. The appropriate response should have been a polynomial of degree n. Explain why this is true by referring to the discussion in Section 3-1 of the determinant as a sum of products of the entries of a matrix.
ANSWER:

<><><><><><><><><><><><><><><><><><><><><><><><><><><><><><><><><><><>

6. The following exploration considers the relationship between the eigenvalues and eigenvectors of a matrix and its transpose. We use the results to analyze the behavior of Markov chains.

a) $\begin{pmatrix} 1 & 0 \\ 0 & 0 \end{pmatrix}$ b) $\begin{pmatrix} 1 & 1 \\ 1 & 0 \end{pmatrix}$ c) $\begin{pmatrix} 0 & -1 & 0 \\ 1 & 0 & 0 \\ 0 & 0 & 3 \end{pmatrix}$ d) $\begin{bmatrix} 5 & \frac{1}{2} & -\frac{1}{2} \\ -3 & \frac{1}{2} & \frac{3}{2} \\ 1 & -\frac{3}{2} & \frac{7}{2} \end{bmatrix}$

i) Compute the characteristic polynomial for each matrix.

ANSWER:

ii) Compute the characteristic polynomial of the transpose of each matrix.

ANSWERS:

iii) Conjecture a relationship between the characteristic polynomial of a matrix and the characteristic polynomial of its transpose.

ANSWER:

iv) The characteristic polynomial of the matrix $\mathbf{A}$ is $|\mathbf{A} - \lambda \cdot \mathbf{I}|$ and the characteristic polynomial of $\mathbf{A}^T$ is $|\mathbf{A}^T - \lambda \cdot \mathbf{I}|$. Use the properties of the determinant to verify the relationship you conjectured above.

ANSWER:

v) Without making computations, conjecture a relationship between the eigenvalues of a matrix and the eigenvalues of its transpose. Give an argument that verifies your conjecture.

ANSWER:

vi) For an eigenvalue λ, a corresponding eigenvector of **A** is in the solution space of $(\mathbf{A} - \lambda \cdot \mathbf{I}) \cdot \mathbf{x} = \mathbf{0}$, while an eigenvector of $\mathbf{A}^T$ is in the solution space of $(\mathbf{A}^T - \lambda \cdot \mathbf{I}) \cdot \mathbf{x} = \mathbf{0}$. Does this mean that **A** and $\mathbf{A}^T$ have the same eigenvectors? Experiment with the matrices c) and d) given above before answering.

ANSWER:

vii) Let **M** be the matrix of a Markov Chain. Explain why the vector $(1,1,1,...,1)^T$ is an eigenvector for $\mathbf{M}^T$.

ANSWER:

viii) What is the corresponding eigenvalue?

ANSWER:

ix) Can you conclude that there is always a stable vector, **s**, for **M**? (i.e., a vector **s** such that $\mathbf{M} \cdot \mathbf{s} = \mathbf{s}$)

ANSWER:

x) What is the meaning of this vector?

ANSWER:

<><><><><><><><><><><><><><><><><><><><><><><><><><><><><><><><><><><><><><><><><>

7. Let **P** be an invertible matrix and let **A** be any square matrix of the same size. Explain mathematically why $\mathbf{P} \cdot \mathbf{A} \cdot \mathbf{P}^{-1}$ and **A** have the same characteristic polynomial and hence the same eigenvalues. (Do not use a specific example.)

ANSWER:

8. Let **P** be an invertible matrix and let **A** be any square matrix of the same size. If **v** is an eigenvector of **A** with eigenvalue λ show that $\mathbf{P}\cdot\mathbf{v}$ is an eigenvector of $\mathbf{P}\cdot\mathbf{A}\cdot\mathbf{P}^{-1}$ with eigenvalue λ.

ANSWER:

***Note*:** Two n by n matrices **A** and **B** are called **similar** if there exists a nonsingular matrix **P** such that $\mathbf{B} = \mathbf{P}^{-1}\cdot\mathbf{A}\cdot\mathbf{P}$.

Complete the following:

Similar matrices have the same ____________________.

9. (Individual) Write a summary of the MATHEMATICS that you learned in this section.

Section 6-3 Eigenvector Bases

Name:________________

Score:_____________

Overview

Let **M** be an n by n matrix. We are interested in determining when the eigenvectors of **M** form a basis for $\mathbf{R}^n$. We examine some conditions that ensure that this is the case. In those cases, we study the properties of **M** by using the eigenvector basis.

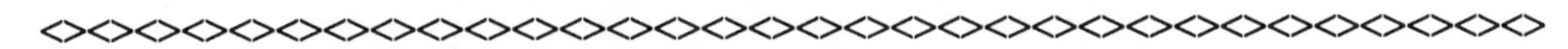

In Section 6-2 we showed that if **v** is an eigenvector for a matrix **A** then $k \cdot \mathbf{v}$ ($k \neq 0$) is also an eigenvector for **A**. So, in general the eigenvectors of **A** need not be linearly independent. One of our goals in this section is to show that eigenvectors that correspond to ***different*** eigenvalues are linearly independent. We begin by considering two such eigenvectors.

Let λ_1 and λ_2 be ***distinct*** eigenvalues of an n by n matrix **A** and let **v1** and **v2** be the corresponding eigenvectors, then **v1** and **v2** are linearly independent.

To show that **v1** and **v2** are linearly independent we must show that if $b1 \cdot \mathbf{v1} + b2 \cdot \mathbf{v2} = \mathbf{0}$ then $b1 = 0$ and $b2 = 0$.

Assume $\quad \mathbf{b1 \cdot v1 + b2 \cdot v2 = 0} \qquad (1)$

Explain why $\quad \mathbf{b1 \cdot \lambda_1 \cdot v1 + b2 \cdot \lambda_2 \cdot v2 = 0} \qquad (2)$

Hint: Compute $\mathbf{A} \cdot (b1 \cdot \mathbf{v1} + b2 \cdot \mathbf{v2})$.

ANSWER:

Putting (1) and (2) together we have a system of vector equations

$$\mathbf{b1 \cdot v1 + b2 \cdot v2 = 0}$$
$$\mathbf{b1 \cdot \lambda_1 \cdot v1 + b2 \cdot \lambda_2 \cdot v2 = 0}$$

where $\quad \mathbf{v1} = \begin{bmatrix} v1_1 \\ v1_2 \\ \cdots \\ v1_n \end{bmatrix} \quad$ and $\quad \mathbf{v2} = \begin{bmatrix} v2_1 \\ v2_2 \\ \cdots \\ v2_n \end{bmatrix}$

We can rewrite this system as a matrix equation as follows.

Equation $b1 \cdot v1 + b2 \cdot v2 = 0$ is

$$b1 \cdot \begin{bmatrix} v1_1 \\ v1_2 \\ \ldots \\ v1_n \end{bmatrix} + b2 \cdot \begin{bmatrix} v2_1 \\ v2_2 \\ \ldots \\ v2_n \end{bmatrix} = \begin{pmatrix} 0 \\ 0 \\ \ldots \\ 0 \end{pmatrix} \quad \text{which is the same as} \quad \begin{bmatrix} v1_1 & v2_1 \\ v1_2 & v2_2 \\ \ldots & \ldots \\ v1_n & v2_n \end{bmatrix} \cdot \begin{pmatrix} b1 \\ b2 \end{pmatrix} = \begin{pmatrix} 0 \\ 0 \\ \ldots \\ 0 \end{pmatrix} \qquad (3)$$

and $b1 \cdot \lambda_1 \cdot v1 + b2 \cdot \lambda_2 \cdot v2 = 0$ is the same as

$$\begin{bmatrix} v1_1 & v2_1 \\ v1_2 & v2_2 \\ \ldots & \ldots \\ v1_n & v2_n \end{bmatrix} \cdot \begin{pmatrix} \lambda_1 \cdot b1 \\ \lambda_2 \cdot b2 \end{pmatrix} = \begin{pmatrix} 0 \\ 0 \\ \ldots \\ 0 \end{pmatrix} \qquad (4)$$

(3) and (4) can be combined into

$$\begin{bmatrix} v1_1 & v2_1 \\ v1_2 & v2_2 \\ \ldots & \ldots \\ v1_n & v2_n \end{bmatrix} \cdot \begin{pmatrix} b1 & \lambda_1 \cdot b1 \\ b2 & \lambda_2 \cdot b2 \end{pmatrix} = \begin{pmatrix} 0 & 0 \\ 0 & 0 \\ \ldots & \ldots \\ 0 & 0 \end{pmatrix} \qquad (5)$$

The matrix $\begin{pmatrix} b1 & \lambda_1 \cdot b1 \\ b2 & \lambda_2 \cdot b2 \end{pmatrix} = \left[\begin{pmatrix} b1 & 0 \\ 0 & b2 \end{pmatrix} \cdot \begin{pmatrix} 1 & \lambda_1 \\ 1 & \lambda_2 \end{pmatrix} \right]$

which transforms (5) into

$$\begin{bmatrix} v1_1 & v2_1 \\ v1_2 & v2_2 \\ \ldots & \ldots \\ v1_n & v2_n \end{bmatrix} \cdot \left[\begin{pmatrix} b1 & 0 \\ 0 & b2 \end{pmatrix} \cdot \begin{pmatrix} 1 & \lambda_1 \\ 1 & \lambda_2 \end{pmatrix} \right] = \begin{pmatrix} 0 & 0 \\ 0 & 0 \\ \ldots & \ldots \\ 0 & 0 \end{pmatrix} \qquad (6)$$

Since $\lambda_1 \neq \lambda_2$, the matrix $\begin{pmatrix} 1 & \lambda_1 \\ 1 & \lambda_2 \end{pmatrix}$ is nonsingular.
Explain why this is true.

ANSWER:

Multiplying both sides of Equation (6) on the right by $\begin{pmatrix} 1 & \lambda_1 \\ 1 & \lambda_2 \end{pmatrix}^{-1}$ gives

$$\begin{bmatrix} v1_1 & v2_1 \\ v1_2 & v2_2 \\ \dots & \dots \\ v1_n & v2_n \end{bmatrix} \cdot \begin{pmatrix} b1 & 0 \\ 0 & b2 \end{pmatrix} = \begin{pmatrix} 0 & 0 \\ 0 & 0 \\ \dots & \dots \\ 0 & 0 \end{pmatrix} \qquad (7)$$

This is the same as

$$b1 \cdot \begin{bmatrix} v1_1 \\ v1_2 \\ \dots \\ v1_n \end{bmatrix} = \begin{pmatrix} 0 \\ 0 \\ \dots \\ 0 \end{pmatrix} \quad \text{and} \quad b2 \cdot \begin{bmatrix} v2_1 \\ v2_2 \\ \dots \\ v2_n \end{bmatrix} = \begin{pmatrix} 0 \\ 0 \\ \dots \\ 0 \end{pmatrix}$$

But this can only be the case if b1 = 0 and b2 = 0. Explain why.

ANSWER:

This completes the proof that the eigenvectors corresponding to the distinct eigenvalues λ_1 and λ_2 are linearly independent.

<><><><><><><><><><><><><><><><><><><><><><><><><><><><><><><><><><><><><>

The above result is a special case of the following more general fact which is proved in an exercise below.

> Let **M** be an n by n matrix. Let **v1**, **v2**, ..., **vk** be eigenvectors of **M** corresponding to eigenvalues, λ1, λ2, ..., λk. If λ1, λ2, ..., λk are distinct, **v1**, **v2**, ..., **vk** are linearly independent.

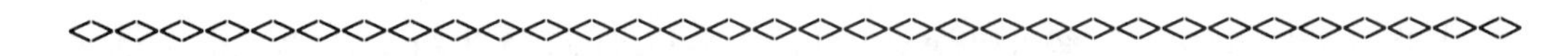

If the matrix **M** has k distinct real eigenvalues, then we may (1) find a real eigenvector corresponding to each eigenvalue and (2) these eigenvectors are linearly independent.

The eigenvalues of $\mathbf{A} := \begin{pmatrix} -262 & 114 & 30 & -20 \\ -150 & 86 & 30 & -20 \\ 168 & -84 & 56 & 56 \\ 2202 & -894 & -138 & 232 \end{pmatrix}$ are $\begin{pmatrix} -112 \\ 140 \\ 28 \\ 56 \end{pmatrix}$

Compute an eigenvector corresponding to each eigenvalue and verify that these eigenvectors are linearly independent.

ANSWER:

The eigenvalues of $\mathbf{B} := \begin{pmatrix} 3 & -2 & -2 \\ -2 & 3 & -2 \\ 0 & 0 & 5 \end{pmatrix}$ are $\begin{pmatrix} 1 \\ 5 \\ 5 \end{pmatrix}$

Are there 3 linearly independent eigenvectors of **B**?

The command **eigenvec(B**,5) will only produce one eigenvector. If there are others you must find them directly.

ANSWER:

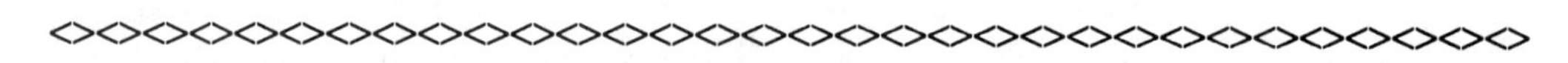

The eigenvalues of $\mathbf{C} := \begin{bmatrix} \frac{5}{2} & \frac{-3}{2} & \frac{-5}{2} \\ -2 & 3 & -2 \\ \frac{1}{2} & \frac{-1}{2} & \frac{11}{2} \end{bmatrix}$ are $\begin{pmatrix} 1 \\ 5 \\ 5 \end{pmatrix}$

Are there three linearly independent eigenvectors of **C**?

ANSWER:

<><><><><><><><><><><><><><><><><><><><><><><><><><><><><><><>

Explain why the following statement is true.

> If the n by n matrix **A** has n distinct real eigenvalues then **A** has n linearly independent eigenvectors. These eigenvectors then form a basis for $\mathbf{R}^n$.

ANSWER:

Note: A matrix can have repeated eigenvalues and still have n linearly independent eigenvectors (e.g., identity(n)).

Assume that the n by n matrix **M** has n linearly independent eigenvectors . Let **V** be the n by n matrix whose ith column is the ith eigenvector of **M**.

Explain why **V** is nonsingular.

ANSWER:

Let **D** be the diagonal matrix whose (diagonal) entries are the eigenvalues of **M**. What is the product $\mathbf{V} \cdot \mathbf{D}$?

ANSWER:

Explain why $\mathbf{M} \cdot \mathbf{V} = \mathbf{V} \cdot \mathbf{D}$. (Hint: Consider the columns.)

ANSWER:

Since $\mathbf{M} \cdot \mathbf{V} = \mathbf{V} \cdot \mathbf{D}$ and **V** is nonsingular, it follows that $\mathbf{M} = \mathbf{V} \cdot \mathbf{D} \cdot \mathbf{V}^{-1}$ and

$\mathbf{M}^2 = \mathbf{V} \cdot \mathbf{D} \cdot \mathbf{V}^{-1} \cdot \mathbf{V} \cdot \mathbf{D} \cdot \mathbf{V}^{-1} = \mathbf{V} \cdot \mathbf{D}^2 \cdot \mathbf{V}^{-1}$

What is the corresponding expression for $\mathbf{M}^3$? For $\mathbf{M}^{20}$?

ANSWER:

More generally what is the expression for $\mathbf{M}^k$?

ANSWER:

If **D** is a diagonal matrix, describe the entries of $\mathbf{D}^k$?

ANSWER:

If **M** has n linearly independent eigenvectors, **vj**, with corresponding eigenvalues λ_j, explain how to compute **M** to any power ***without repeated multiplication of* M**.
ANSWER:

Determine $\mathbf{V}^{-1}\cdot\mathbf{M}\cdot\mathbf{V}$.
ANSWER:

An n by n matrix **M** is said to be ***diagonalizable*** if there is a nonsingular matrix **P** such that $\mathbf{P}^{-1}\cdot\mathbf{M}\cdot\mathbf{P} = \mathbf{D}$, where **D** is a diagonal matrix. Another way to say this is that **M** is similar to a diagonal matrix.

Explain why **M** is diagonalizable if it has n distinct real eigenvalues.
ANSWER:

<><><><><><><><><><><><><><><><><><><><><><><><><><><><><><><><><><><><><><>

In Chapter 2 we studied the bases of a space. We saw that $\mathbf{R}^n$ had dimension n, which meant that a basis for $\mathbf{R}^n$ contained n linearly independent vectors that were also a spanning set. In fact, it is enough to find n linearly independent vectors in $\mathbf{R}^n$ to obtain a basis. It follows that:

If the n by n matrix **M** has n linearly independent real eigenvectors, they form a basis for $\mathbf{R}^n$.

<><><><><><><><><><><><><><><><><><><><><><><><><><><><><><><><><><><><><><>

For a general n by n matrix **A**, it is not possible to know by inspection if the eigenvalues are real and distinct and hence form a basis for $\mathbf{R}^n$. However, there is an important special case.

> If **A** is a **symmetric matrix** then all the eigenvalues are real and the associated eigenvectors form an **orthogonal** basis for $\mathbf{R}^n$.

The eigenvalues of a symmetric matrix need not be distinct, but there always exist eigenvectors that form an orthogonal basis for $\mathbf{R}^n$. We discuss the details behind this statement in the exercises at the end of the section.

We can create a matrix whose columns are eigenvectors as follows: Let **A** be the original matrix. Our new matrix will be **EIG_A**

$$j := 1 .. \mathrm{rows}(A)$$

$$EIG\_A^{\langle j \rangle} := \mathrm{eigenvec}\left(A, \mathrm{eigenvals}(A)_j\right)$$

Be careful! The function **eigenvec** will return only one eigenvector per eigenvalue even if there are two or more linearly independent eigenvectors corresponding to a particular eigenvalue. As a result the matrix **EIG_A** will be singular if **A** has a repeated eigenvalue even though **A** may have n linearly independent eigenvectors. (Experiment with the identity matrix and eigenvalue 1.)

For each of the following matrices use the eigen commands in Mathcad to compute the eigenvalues and the matrix of eigenvectors (see Mathcad hint above).

$$A := \begin{pmatrix} 2 & 0 & -3 & 1 \\ 0 & 7 & 5 & 0 \\ -3 & 5 & 2 & -1 \\ 1 & 0 & -1 & 0 \end{pmatrix} \quad \text{and} \quad B := \begin{bmatrix} 1 & -2 & 7 & 3 & 1 \\ -2 & 4 & -4 & 0 & 2 \\ 7 & -4 & -5 & 5 & -3 \\ 3 & 0 & 5 & 2 & 0 \\ 1 & 2 & -3 & 0 & 7 \end{bmatrix}$$

ANSWER:

Use the matrices of eigenvectors that you have calculated to show that the eigenvectors of each of the above matrices form an orthogonal set.

ANSWER:

The **eigenvec** command will only calculate one of the two eigenvectors of **C** corresponding to the eigenvalue -3. Calculate **directly** two linearly independent eigenvectors corresponding to the eigenvalue -3.

$$C := \begin{pmatrix} 0 & 3 & 3 \\ 3 & 0 & 3 \\ 3 & 3 & 0 \end{pmatrix}$$

ANSWER:

Compute two orthogonal eigenvectors of **C** corresponding to the eigenvalue -3.

ANSWER:

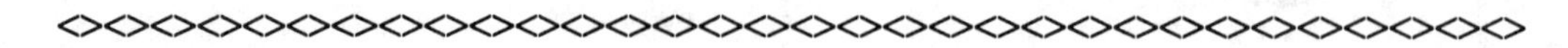

Exercises

1. Given vectors, **v1**, **v2**, ..., **vn** and real scalars $\lambda_1, \lambda_2, ..., \lambda_n$ there is a matrix, **M**, such that $\lambda_1, \lambda_2, ..., \lambda_n$ are the eigenvalues of **M** and **v1**, **v2**, ..., **vn** are the corresponding eigenvectors.

Let $\quad \mathbf{v1} := \begin{pmatrix} 2 \\ 3 \\ 1 \end{pmatrix} \quad \mathbf{v2} := \begin{pmatrix} -1 \\ 4 \\ 0 \end{pmatrix} \quad \mathbf{v3} := \begin{pmatrix} 7 \\ -5 \\ 2 \end{pmatrix}$

$$\lambda_1 := 2 \qquad \lambda_2 := -5 \qquad \lambda_3 := -1$$

Find **M** such that **v1, v2** and **v3** are eigenvectors corresponding to eigenvalues, λ_1, λ_2, and λ_3.

ANSWER:

Explain how, in general, to find such a matrix whenever the eigenvalues are real and the eigenvectors are linearly independent.

ANSWER:

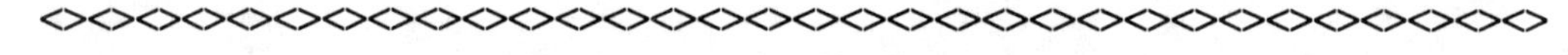

2. In this section we showed that eigenvectors corresponding to a pair of distinct eigenvalues were linearly independent. We now generalize this statement.

Let **A** be an n by n matrix with n ***real*** and ***distinct*** eigenvalues, $\lambda_1, \lambda_2, ..., \lambda_n$.

Let **vj** be an eigenvector corresponding to λ_j. We would like to conclude that

v1, v2, ..., **vn** are linearly independent and hence a basis for $\mathbf{R}^n$.

We give the proof when n = 4. In so doing one can isolate the essential feature of the proof for greater values of n.

Suppose that we have a linear combination of the v's with coefficients, b1, ..., b4.

$$\mathbf{b1 \cdot v1 + b2 \cdot v2 + b3 \cdot v3 + b4 \cdot v4 = 0} \qquad (1')$$

We want to show that (1') is true only when the coefficients, b1, b2, b3, and b4 are all zero. Evaluate each of the following:

$$\mathbf{A \cdot (b1 \cdot v1 + b2 \cdot v2 + b3 \cdot v3 + b4 \cdot v4)} \qquad (2')$$

$$\mathbf{A^{2} \cdot (b1 \cdot v1 + b2 \cdot v2 + b3 \cdot v3 + b4 \cdot v4)} \qquad (3')$$

$$\mathbf{A^{3} \cdot (b1 \cdot v1 + b2 \cdot v2 + b3 \cdot v3 + b4 \cdot v4)} \qquad (4')$$

ANSWER:

Since $\mathbf{b1 \cdot v1 + b2 \cdot v2 + b3 \cdot v3 + b4 \cdot v4 = 0}$

Expressions (2'), (3'), and (4') must also equal zero.

We may combine these in the same way that we did earlier in this section to obtain the following expression.

$$\begin{bmatrix} v1_1 & v2_1 & v3_1 & v4_1 \\ v1_2 & v2_2 & v3_2 & v4_2 \\ v1_3 & v2_3 & v3_3 & v4_3 \\ v1_4 & v2_4 & v3_4 & v4_4 \end{bmatrix} \cdot \begin{pmatrix} b1 & 0 & 0 & 0 \\ 0 & b2 & 0 & 0 \\ 0 & 0 & b3 & 0 \\ 0 & 0 & 0 & b4 \end{pmatrix} \cdot \begin{bmatrix} 1 & \lambda_1 & (\lambda_1)^2 & (\lambda_1)^3 \\ 1 & \lambda_2 & (\lambda_2)^2 & (\lambda_2)^3 \\ 1 & \lambda_3 & (\lambda_3)^2 & (\lambda_3)^3 \\ 1 & \lambda_4 & (\lambda_4)^2 & (\lambda_4)^3 \end{bmatrix} = \begin{pmatrix} 0 & 0 & 0 & 0 \\ 0 & 0 & 0 & 0 \\ 0 & 0 & 0 & 0 \\ 0 & 0 & 0 & 0 \end{pmatrix}$$

If the matrix of λ's is nonsingular, we may multiply both sides on the right by its inverse and conclude that

$$\begin{bmatrix} v1_1 & v2_1 & v3_1 & v4_1 \\ v1_2 & v2_2 & v3_2 & v4_2 \\ v1_3 & v2_3 & v3_3 & v4_3 \\ v1_4 & v2_4 & v3_4 & v4_4 \end{bmatrix} \cdot \begin{pmatrix} b1 & 0 & 0 & 0 \\ 0 & b2 & 0 & 0 \\ 0 & 0 & b3 & 0 \\ 0 & 0 & 0 & b4 \end{pmatrix} = \begin{pmatrix} 0 & 0 & 0 & 0 \\ 0 & 0 & 0 & 0 \\ 0 & 0 & 0 & 0 \\ 0 & 0 & 0 & 0 \end{pmatrix}$$

But this can only hold if b1 = b2 = b3 = b4 = 0. Explain why.

ANSWER:

The matrix of λ's is nonsingular if its determinant is nonzero. Compute the determinant symbolically and use Maple to factor it. Why can you conclude that it is nonsingular?

ANSWER:

It is clear that for other values of n linear independence is equivalent to showing a corresponding matrix of λ's is nonsingular. Maple is unable to evaluate and factor any such matrices when n is larger than 4. However, the corresponding fact is true.

The matrix of λ's is called the Vandermonde matrix. It is clear that if $\lambda_i = \lambda_j$ for any distinct values of i and j there will be two equal rows in the matrix and the determinant will be zero. This implies that $(\lambda_i - \lambda_j)$ is a factor of the determinant. One can show that the determinant equals

$$-\left[\prod_{(i<j)} (\lambda_i - \lambda_j)\right]$$

From which it follows that the matrix is nonsingular if the λ's are distinct.

<><><><><><><><><><><><><><><><><><><><><><><><><><><><><><><>

3. In this exercise we investigate whether or not a matrix **A** has a square root; that is, can we find a matrix **R** such that $\mathbf{R}^2 = \mathbf{A}$.

Find all the square roots of

$$\mathbf{D} := \begin{pmatrix} 4 & 0 & 0 & 0 \\ 0 & 1 & 0 & 0 \\ 0 & 0 & 9 & 0 \\ 0 & 0 & 0 & 64 \end{pmatrix}$$

How many such matrices are there?

ANSWER:

Suppose that **A** is an n by n matrix all of whose eigenvalues are real, distinct and non-negative. Explain how to find the square roots of **A**. Hint: Use the fact that A is diagonalizable.

ANSWER:

Compute the square roots of $\quad \mathbf{A} := \begin{pmatrix} 25 & 0 & -24 & 0 \\ 11 & 4 & -46 & 5 \\ 0 & 0 & 1 & 0 \\ 32 & 0 & -88 & 9 \end{pmatrix}$

ANSWER:

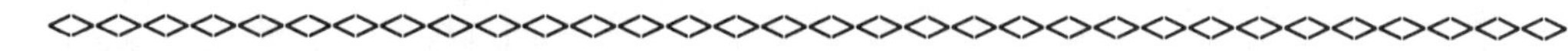

Complex vectors and matrices

To prove that the eigenvalues of a symmetric matrix are real we need first to establish some properties of vectors and matrices whose entries are complex numbers.

Throughout this text we have limited our discussion to real vectors and matrices and in particular to $\mathbf{R}^n$. Almost everything we have done (we shall see an exception soon) could have been done using complex numbers instead of real numbers. Define $\mathcal{C}^n$ to be the space of n-tuples of ***complex*** numbers. In this space, scalar multiplication means multiplication by complex numbers. We add vectors and do scalar multiplication in the same way that we do in a real space. Subspaces are defined similarly as are linear independence, spanning sets, and a basis.

For the sake of completeness we review briefly a few facts about complex numbers. A complex number is a number of the form a + ib. (To enter i in Mathcad type 1i with no space or operation between.) a is called the real part and ib the imaginary part. The conjugate of a complex number a + ib is defined to be a - ib.

> Conjugates are written in Mathcad using the x-bar button on the calculator palette (palette in Mathcad 5.0).
> or by using Maple.
>
> $$\overline{(3+5\cdot i)} = 3 - 5i$$
>
> The real and imaginary parts of a complex number may be computed in Mathcad using the functions Re(z) and Im(z).
>
> $\mathbf{Re}(3+5\cdot i) = 3 \qquad\qquad \mathbf{Im}(3+5\cdot i) = 5$

The magnitude or absolute value of a complex number c is defined by

$$(|\mathbf{c}|)^2 = \bar{\mathbf{c}}\cdot\mathbf{c}$$

In Section 1-7 we defined a linear function as follows: A function, f, is linear if $f(a\cdot\mathbf{u} + b\cdot\mathbf{v}) = a\cdot f(\mathbf{u}) + b\cdot f(\mathbf{v})$ for all vectors **u** and **v** and all scalars a and b. The definition of a linear function between complex vector spaces is defined in the same way except that the scalars are now complex numbers.

Let $f(a + ib) = a - ib$. This is called the conjugate mapping of $\mathcal{C}$ to $\mathcal{C}$. Explain why f is **not** linear.

ANSWER:

Explain why $\bar{c}\cdot c \geq 0$ and why $|c| = 0$ if and only if $c = 0$.

ANSWER:

Let **M** be a matrix whose entries are complex. Define the ***conjugate*** of **M** to be the matrix whose entries are the conjugates of the entries of **M.**

$$\overline{\begin{pmatrix} i & 1 - i & 4 \\ 2 & 3 + 2\cdot i & 5\cdot i \end{pmatrix}} = \begin{pmatrix} -1\cdot i & 1 + i & 4 \\ 2 & 3 - 2i & -5\cdot i \end{pmatrix}$$

Use Maple to verify that the conjugate of the product is the same as the product of the conjugates. To do this evaluate each of the following expressions symbolically.

ANSWER:

$$\left(\overline{(a + b\cdot i)}\right)\cdot\left(\overline{(c + d\cdot i)}\right) \qquad \overline{((a + b\cdot i)\cdot(c + d\cdot i))}$$

Using the fact that $i^2 = -1$ compute the following product by hand. Use Mathcad to check your work

$$\begin{pmatrix} i & 3 & 1 - i \\ 2 - 3\cdot i & 5 + i & 3 \\ 0 & 2 & 5 \end{pmatrix}\cdot\begin{pmatrix} i \\ 0 \\ 1 \end{pmatrix}$$

ANSWER:

Compute the inverse of the following matrix first using Mathcad and then using Maple.

$$\begin{pmatrix} i & 3 & 1-i \\ 2-3\cdot i & 5+i & 3 \\ 0 & 2 & 5 \end{pmatrix}$$

ANSWER:

Use Maple to compute the dot product of the following two vectors. (Mathcad gives the 'wrong' answer.)

$$u := \begin{pmatrix} i \\ 1-i \\ 3 \end{pmatrix} \qquad v := \begin{pmatrix} 1+3\cdot i \\ 1+i \\ 0 \end{pmatrix}$$

Compute both **u·v** and **v·u** using Maple.

$$\begin{pmatrix} i \\ 1-i \\ 3 \end{pmatrix} \cdot \begin{pmatrix} 1+3\cdot i \\ 1+i \\ 0 \end{pmatrix} \qquad \begin{pmatrix} 1+3\cdot i \\ 1+i \\ 0 \end{pmatrix} \cdot \begin{pmatrix} i \\ 1-i \\ 3 \end{pmatrix}$$

Guess how the dot product is defined for complex vectors?

Try $$\begin{pmatrix} a1+b1\cdot i \\ a2+b2\cdot i \\ a3+b3\cdot i \end{pmatrix} \cdot \begin{pmatrix} 1 \\ 1 \\ 1 \end{pmatrix} \qquad \begin{pmatrix} 1 \\ 1 \\ 1 \end{pmatrix} \cdot \begin{pmatrix} a1+b1\cdot i \\ a2+b2\cdot i \\ a3+b3\cdot i \end{pmatrix}$$

ANSWER:

You might wonder why the definition of the dot product has changed. The answer is so that the properties given in Chapter 1 can be preserved. In particular, the following property must hold.

Let **v** be a vector in $\mathbb{C}^n$ we define the length of **v**, |**v**|, by

$$(|\mathbf{v}|)^2 = \mathbf{v} \cdot \mathbf{v}$$

It then follows that $\mathbf{v} = \mathbf{0}$ if and only if $|\mathbf{v}| = 0$.

Calculate $\begin{pmatrix} i \\ 1 \end{pmatrix} \cdot \begin{pmatrix} i \\ 1 \end{pmatrix}$ using the 'old' definition of dot product and then using **Maple**.

ANSWER:

By now you should have discovered that $\mathbf{u} \cdot \mathbf{v} = \mathbf{u}^T \cdot \overline{\mathbf{v}}$

<><><><><><><><><><><><><><><><><><><><><><><><><><><><><><><><><><>

Warning:

The dot product could have just as easily been defined to be

$$\overline{(\mathbf{u}^T)} \cdot \mathbf{v}$$

Some linear algebra texts define the dot product as Maple does and some define it in the other way. If you have to use the complex inner product you should be sure you know which definition is being used.

<><><><><><><><><><><><><><><><><><><><><><><><><><><><><><><><><><>

If **u** and **v** are real vectors, does this definition of the dot product agree with our previous definition?
ANSWER:

Let c be a complex number and let **u** and **v** be vectors in $\mathbb{C}^n$. Complete the following relationships

$(c \cdot \mathbf{u}) \cdot \mathbf{v} = \_\_\_\_ \cdot (\mathbf{u} \cdot \mathbf{v})$ $\qquad$ $\mathbf{u} \cdot (c \cdot \mathbf{v}) = \_\_\_\_ \cdot (\mathbf{u} \cdot \mathbf{v})$

We need one more piece of information: Let **A** be a real n by n matrix and let **u** and **v** be vectors in $\mathcal{C}^n$. Explain why

$$(\mathbf{A}\cdot\mathbf{u})\cdot\mathbf{v}=\mathbf{u}\cdot\left(\mathbf{A}^{\mathrm{T}}\cdot\mathbf{v}\right) \qquad (1'')$$

Be careful to distinguish between the dot product and the matrix-vector product in the previous expression. They are both denoted by · .

ANSWER:

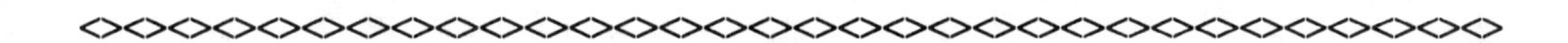

For an n by n symmetric real matrix **A** we will show that

(i) all its eigenvalues are real.

(ii) all the eigenvectors are real.

(iii) the eigenvectors can be selected to form an orthogonal basis for $\mathbf{R}^n$.

(i) To see that all the eigenvalues are real we use (1") above together with the facts that **A** is real and $\mathbf{A}^{\mathrm{T}} = \mathbf{A}$.

Suppose that **u** is an eigenvector of **A** corresponding to the eigenvalue λ, then $\mathbf{A}\cdot\mathbf{u} = \lambda\cdot\mathbf{u}$, and

$$(\mathbf{A}\cdot\mathbf{u})\cdot\mathbf{u}=\mathbf{u}\cdot\left(\mathbf{A}^{\mathrm{T}}\cdot\mathbf{u}\right)=\mathbf{u}\cdot(\mathbf{A}\cdot\mathbf{u})$$

The left hand side gives $(\mathbf{A}\cdot\mathbf{u})\cdot\mathbf{u}=(\lambda\cdot\mathbf{u})\cdot\mathbf{u}=\lambda\cdot(\mathbf{u}\cdot\mathbf{u})=\lambda\cdot(|\mathbf{u}|)^2$

The right hand side gives $\mathbf{u}\cdot(\mathbf{A}\cdot\mathbf{u})=\mathbf{u}\cdot(\lambda\cdot\mathbf{u})=\overline{\lambda}\cdot(\mathbf{u}\cdot\mathbf{u})=\overline{\lambda}\cdot(|\mathbf{u}|)^2$

Since **u** is not the zero vector, it follows that $\lambda=\overline{\lambda}$

But this is the same as the fact that λ is real.

(ii) Next we show that we can always choose the eigenvector **y** of a real symmetric matrix to be real. Suppose **A** is real and symmetric and $\mathbf{A}\cdot\mathbf{y} = \lambda\cdot\mathbf{y}$. Then **y** is in the solution space of $(\mathbf{A} - \lambda\cdot\mathbf{I})\cdot\mathbf{x} = \mathbf{0}$. Since **A** and λ are both real we can find a basis for the solution space in the same way that we did for other real matrices. This implies that we can find real eigenvectors since they form a basis for the solution space.

(iii) To see that all the eigenvectors are orthogonal we use (1) above together with the facts that $\mathbf{A}$ is real and $\mathbf{A}^T = \mathbf{A}$.

We first assume that the eigenvectors correspond to different eigenvalues.

Suppose that $\mathbf{A}\cdot\mathbf{u} = \lambda_1\cdot\mathbf{u}$ and $\mathbf{A}\cdot\mathbf{v} = \lambda_2\cdot\mathbf{v}$ where $\lambda_1 \neq \lambda_2$ then

$$(\mathbf{A}\cdot\mathbf{u})\cdot\mathbf{v} = \mathbf{u}\cdot\left(\mathbf{A}^T\cdot\mathbf{v}\right) = \mathbf{u}\cdot(\mathbf{A}\cdot\mathbf{v})$$

The left hand side gives $\lambda_1\cdot(\mathbf{u}\cdot\mathbf{v})$.

The right hand side gives $\lambda_2\cdot(\mathbf{u}\cdot\mathbf{v})$.

Since $\lambda_1 \neq \lambda_2$ it follows that $\mathbf{u}\cdot\mathbf{v} = 0$.

We must now show that if the eigenvalue λ has multiplicity m as a root of the characteristic equation of $\mathbf{A}$, then there exist m orthogonal eigenvectors corresponding to λ.

If λ is an eigenvalue of multiplicity m of a real symmetric matrix $\mathbf{A}$, then the dimension of the solution space $(\mathbf{A} - \lambda\cdot\mathbf{I})\cdot\mathbf{x} = \mathbf{0}$ is m. (The proof of this fact requires more ideas than we have developed in this course.)

This implies that any basis for the solution space $(\mathbf{A} - \lambda\cdot\mathbf{I})\cdot\mathbf{x} = \mathbf{0}$ has m elements in it. We can apply the Gram-Schmidt procedure to these vectors to get an orthogonal basis for the solution space.

Exercise. (Individual) Write a summary of the MATHEMATICS that you learned in this section.

Section 6-4 Dominant Eigenvalues: Geometric Series & Principal Component Analysis

Name:________________

Score:____________

Overview

Let $\mathbf{M}$ be a square matrix. In this section we examine $\mathbf{M}^k$ as k gets large. We have already done this for Markov chains and for the Leslie population model. If $\mathbf{M}$ has one eigenvalue that is larger in magnitude than all the other eigenvalues it is called the ***dominant eigenvalue***. The dominant eigenvalue and its eigenvector determine $\mathbf{M}^k$ for large k. The dominant eigenvalue is also used in principal component analysis which is an important tool in statistics.

<><><><><><><><><><><><><><><><><><><><><><><><><><><><><><><><><><><><>

In the weather model from Chapters 0 and 1 with transition matrix $\mathbf{A}$ and initial weather distribution $\mathbf{x_0}$, the long term behavior was determined by examining the limit of the sequence

$$\mathbf{x_0},\ \mathbf{A}\cdot\mathbf{x_0},\ \mathbf{A}^2\cdot\mathbf{x_0},\ \ldots,\ \mathbf{A}^k\cdot\mathbf{x_0},\ \ldots$$

In the Leslie population model in Section 6-1 with Leslie matrix $\mathbf{L}$ and initial population distribution $\mathbf{d_0}$, the limit of the sequence

$$\mathbf{d_0},\ \mathbf{L}\cdot\mathbf{d_0},\ \mathbf{L}^2\cdot\mathbf{d_0},\ \ldots,\ \mathbf{L}^k\cdot\mathbf{d_0},\ \ldots$$

was used to obtain a stable distribution.

In Section 6-2 we saw that the limiting behavior for both of these models could be studied in terms of eigenvalues and eigenvectors. For the Markov model we sought a stable weather state $\mathbf{s}$ such that $\mathbf{A}\cdot\mathbf{s} = \mathbf{s}$; i.e., an eigenvector of the transition matrix $\mathbf{A}$ corresponding to the eigenvalue 1. Similarly, in the Leslie model we wanted a stable population distribution $\mathbf{p}$ so that $\mathbf{L}\cdot\mathbf{p} = c\cdot\mathbf{p}$, where c is 1 + the growth rate of the population. The vector $\mathbf{p}$ is an eigenvector of $\mathbf{L}$ corresponding to the eigenvalue c.

In this section we investigate the limit of the sequence of powers, $\mathbf{M}$, $\mathbf{M}^2$, ..., $\mathbf{M}^k$, ..., of a square matrix $\mathbf{M}$, without reference to a particular model. The goal is to relate the behavior of the sequence to the set of eigenvalues associated with matrix $\mathbf{M}$. Computationally we use a procedure known as the ***scaled power method***.

The Scaled Power Method

Let $\mathbf{M}$ be an n by n matrix and let $\mathbf{x}$ be a nonzero vector in $\mathbf{R}^n$. Consider the sequence of vectors:

$$\mathbf{x},\quad \mathbf{M}\cdot\mathbf{x},\quad \mathbf{M}^2\cdot\mathbf{x} = \mathbf{M}\cdot(\mathbf{M}\cdot\mathbf{x}),\quad \ldots,\quad \mathbf{M}^k\cdot\mathbf{x} = \mathbf{M}\cdot(\mathbf{M}^{k-1}\cdot\mathbf{x}),\ \ldots$$

Each member of this sequence is obtained from the previous one by multiplication by the matrix $\mathbf{M}$. The sequence is just the images of vector $\mathbf{x}$ by powers of M, hence the name ***power method.***

To keep the size of the elements in successive images under control we make the vector a unit vector before the next multiplication by matrix **M**. (Hence the use of the term scaled.)

Set $$c^{<k>} := M\cdot\left(\frac{M^{k-1}\cdot x}{\left|M^{k-1}\cdot x\right|}\right) \qquad (1)$$

In the Markov model we were interested in finding a limiting value for $\mathbf{c}^{<k+1>}$. In the Leslie Model we wanted to find $\mathbf{c}^{<k+1>}$ and a constant, r, such that $|\mathbf{c}^{<k+1>} - r\cdot\mathbf{c}^{<k>}| \to \mathbf{0}$ as $k \to \infty$. In either case, the sequence of λ_k defined below converges (to 1 in the first example and to r in the second).

$$\lambda_k := \frac{\left|M\cdot c^{<k>}\right|}{\left|c^{<k>}\right|} \qquad (2)$$

The scaled power method is an algorithm that is used to compute the dominant eigenvalue of a matrix. It is based upon the sequences given in (1) and (2).

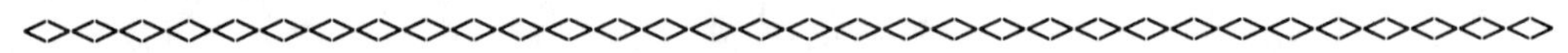

Explorations

$k := 1..80$ $\qquad$ $j := 1..5$

$$M := \begin{pmatrix} 1 & 3 & 1 \\ 2 & -1 & 3 \\ 1 & 2 & 0 \end{pmatrix} \qquad x := \begin{pmatrix} 2 \\ 5 \\ -1 \end{pmatrix}$$

$$c^{<k>} := M\cdot\left(\frac{M^{k-1}\cdot x}{\left|M^{k-1}\cdot x\right|}\right) \qquad \lambda_k := \frac{\left|M\cdot c^{<k>}\right|}{\left|c^{<k>}\right|}$$

c =

	1	2	3	4	5
1	2.92119	0.78446	3.23498	1.21847	3.29623
2	-0.7303	3.53009	-0.21567	3.37848	0.33265
3	2.19089	0.39223	2.15666	0.72	2.17732

λ^T =

	1	2	3	4	5	6
1	3.63741	3.89394	3.66295	3.9644	3.75161	4.00868

$$\text{eigenvals}(M) = \begin{pmatrix} 4 \\ -0.586 \\ -3.414 \end{pmatrix}$$

Experiment by changing the entries in the matrix, **M**, and/or the vector, **x**.

Did the columns of **c** and the entries of λ always converge?.

What would happen if **x** were an eigenvector of **M**?

What is the relation between the limit of λ and the eigenvalues of **M**; the limit of **c** and the eigenvectors?

Report on the result of your experiment.

ANSWER:

<><><><><><><><><><><><><><><><><><><><><><><><><><><><><><><><><><><><><>

Exercises

1. For $\mathbf{M} := \begin{bmatrix} \frac{-13}{4} & \frac{-3}{4} & \frac{-3}{4} \\ -3 & \frac{10}{4} & 3 \\ \frac{1}{4} & 0 & \frac{2}{4} \end{bmatrix}$ investigate the limit of $\mathbf{M}^k$, using the scaled power method.

$k := 1..50$ $\qquad$ $cc^{<1>} := \begin{pmatrix} 1 \\ 1 \\ 1 \end{pmatrix}$

$cc^{<k+1>} := \dfrac{M \cdot cc^{<k>}}{\left| M \cdot cc^{<k>} \right|}$ $\qquad$ $\alpha_k := \dfrac{\left| M \cdot cc^{<k>} \right|}{\left| cc^{<k>} \right|}$

$\alpha^T =$

	1	2	3	4	5	6
1	3.12916	4.83682	2.31798	4.69418	2.57342	4.4791

$cc =$

	1	2	3	4	5	6
1	1	-0.8764	0.4959	-0.96606	0.62078	-0.99853
2	1	0.46127	0.86783	0.25405	0.78261	-0.01757
3	1	0.13838	-0.03099	0.0468	-0.04647	0.05128

Summarize your findings about the sequence of vectors, **cc**, and the sequence of scalars, α.

ANSWER:

<><><><><><><><><><><><><><><><><><><><><><><><><><><><><><><><><><><><><>

2. Repeat the procedure in Exercise 1 for the following matrix and initial vector.

$$\mathbf{M} := \begin{pmatrix} .877 & -.48 \\ .48 & .877 \end{pmatrix} \qquad \mathbf{x} := \begin{pmatrix} 1 \\ 1 \end{pmatrix}$$

Summarize your findings about the sequence of vectors and the sequence of scalars.

ANSWER:

Use the function **eigenvals** to determine the eigenvalues for the matrices in Exercises 1 and 2. Record your results below. If the sequence λ_k converged what is the relationship between the eigenvalues and the limit of the λ_k?

ANSWER:

If the sequence λ_k did not converge, examine the eigenvalues and their magnitude. There is a common property in those cases in which the λ_k do not converge. What is this property?

ANSWER:

<><><><><><><><><><><><><><><><><><><><><><><><><><><><><><><><><><>

To understand the behavior of the scaled power method, ***assume that the eigenvectors for an n by n matrix* A *are a basis for* $\mathbf{R}^n$**. This is always the case, for example, when the eigenvalues are real and distinct.

Since the eigenvectors of **A** are a basis in $\mathbf{R}^n$**,** for any vector **x** there exist scalars, c1, c2, ..., cn, so that

$$\mathbf{x} = c1\cdot\mathbf{w}_1 + c2\cdot\mathbf{w}_2 + \ldots + cn\cdot\mathbf{w}_n \qquad (3)$$

where $\mathbf{w}_1, \mathbf{w}_2, \ldots, \mathbf{w}_n$ are linearly independent eigenvectors of **A**.

Use (3) to write an expression for $\mathbf{A}\cdot\mathbf{x}$, $\mathbf{A}^2\cdot\mathbf{x}$, **and** $\mathbf{A}^k\cdot\mathbf{x}$ in terms of the eigenvectors of **A**.

ANSWER:

$$\mathbf{A}\cdot\left(c1\cdot w_1 + c2\cdot w_2 + \ldots + cn\cdot w_n\right) = \blacksquare$$

$$\mathbf{A}^2\cdot\left(c1\cdot w_1 + c2\cdot w_2 + \ldots + cn\cdot w_n\right) = \blacksquare$$

$$\mathbf{A}^k\cdot\left(c1\cdot w_1 + c2\cdot w_2 + \ldots + cn\cdot w_n\right) = \blacksquare$$

Suppose there is a largest eigenvalue (in absolute value) and it is real. Divide the expression for $\mathbf{A}^k \cdot \mathbf{x}$ by the (k-1)st power of this eigenvalue and explain what happens as k gets large.

ANSWER:

With appropriate modifications the computations you made can be generalized to the case where the largest eigenvalue is real even if some of the other eigenvalues are complex. In the case in which the eigenvalue of largest absolute value is real we call it the ***dominant eigenvalue***.

Was there a dominant eigenvalue in the matrix of Exercise 1?

ANSWER:

Was there a dominant eigenvalue in the matrix of Exercise 2?

ANSWER:

If you have done your work correctly you should be able to conclude that

> If **A** has a dominant eigenvalue, λ, then for almost any vector **x**, $\mathbf{M}^n \cdot \mathbf{x} \rightarrow \lambda^n \mathbf{v}$ where **v** is an eigenvector corresponding to λ.
>
> In addition in the scaled power method
>
> $$\mathbf{c}^{<k>} \rightarrow \mathbf{v} \quad \text{and} \quad |\lambda_k| \rightarrow |\text{dominant eigenvalue}|$$

<><><><><><><><><><><><><><><><><><><><><><><><><><><><><><><><>

In Section 1-10 we studied the number of paths of length n in a graph and showed that the number increased exponentially as a function of n.

More precisely if T(**M**,n) is the number of paths of length n in the graph whose adjacency matrix is **M,** then we saw that T(**M**,n) was asymptotic to $C \cdot a^n$ where C and a are positive constants. It follows that as n gets large

$$\frac{T(M,n)}{a^n} \rightarrow C \qquad \text{and} \qquad \frac{T(M,n+1)}{T(M,n)} \rightarrow a$$

For any 5 by 5 matrix, **M**, let $\quad T(M,n) := \begin{bmatrix} 1 \\ 1 \\ 1 \\ 1 \\ 1 \end{bmatrix} \cdot \left[M^n \cdot \begin{bmatrix} 1 \\ 1 \\ 1 \\ 1 \\ 1 \end{bmatrix} \right]$

For each of the following adjacency matrices use the relationship

$\frac{T(M,n+1)}{T(M,n)} \rightarrow a$ to determine a such that $\frac{T(M,n)}{a^n} \rightarrow C$

Record the values of a and C for each adjacency matrix.

$$M1 := \begin{bmatrix} 0 & 0 & 1 & 1 & 1 \\ 1 & 0 & 1 & 0 & 0 \\ 0 & 1 & 0 & 1 & 1 \\ 1 & 0 & 0 & 0 & 1 \\ 1 & 0 & 1 & 0 & 0 \end{bmatrix} \qquad M2 := \begin{bmatrix} 0 & 0 & 1 & 1 & 1 \\ 0 & 0 & 0 & 0 & 1 \\ 1 & 1 & 0 & 0 & 0 \\ 1 & 1 & 1 & 0 & 1 \\ 0 & 0 & 1 & 0 & 0 \end{bmatrix}$$

$$M3 := \begin{bmatrix} 0 & 1 & 1 & 1 & 1 \\ 1 & 0 & 1 & 1 & 1 \\ 1 & 1 & 1 & 1 & 1 \\ 1 & 1 & 0 & 0 & 1 \\ 1 & 1 & 1 & 1 & 0 \end{bmatrix} \qquad M4 := \begin{bmatrix} 0 & 0 & 1 & 0 & 1 \\ 1 & 0 & 0 & 0 & 0 \\ 0 & 1 & 0 & 1 & 0 \\ 1 & 0 & 0 & 0 & 1 \\ 0 & 0 & 1 & 0 & 0 \end{bmatrix}$$

ANSWER:

M1: a = C =
M2: a = C =
M3: a = C =
M4: a = C =

Examine the eigenvalues for **M1, M2, M3** and **M4**. What is the relation between the eigenvalues and the number a? Explain why the relation holds.
ANSWER:

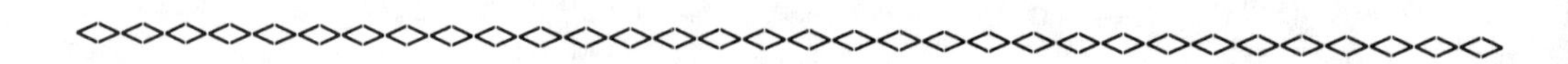

Answer the following questions by examining the eigenvalues of **A**.

Let $\mathbf{A} := \begin{pmatrix} .1 & -.4 & 0 & .8 \\ .5 & .1 & .3 & 0 \\ 0 & .25 & .35 & 0 \\ .2 & -.1 & .25 & .15 \end{pmatrix}$

Does **A** have a dominant eigenvalue? If so, find it.

ANSWER:

For the matrix **A**, what can you conclude about the behavior of the sequence of vectors generated by the scaled power method as k gets large?

ANSWER:

Does $\mathbf{A}^k$ have a limit as k gets large?

ANSWER:

Form a conjecture below that connects the dominant eigenvalue and the limit of $\mathbf{A}^k$ as k gets large.

If **A** is an n by n matrix such that its dominant eigenvalue has magnitude less than 1 then ____ .

<><><><><><><><><><><><><><><><><><><><><><><><><><><><><><><><><><><><><><><>

The Geometric Series of a matrix

For any square matrix **A** we define the geometric series of **A**, **G(A)** by

$$G(A) = I + A + A^2 + A^3 + \ldots = \sum_{j=0}^{\infty} A^j$$

Of course **G(A)** only makes sense if the infinite series converges.

It remains to determine the sum of the infinite geometric series when it does converge.

In the case of real numbers the sum of the series $1 + r + r^2 + r^3 + \ldots$ is $1/(1 - r)$ when $|r| < 1$.

We prove this as follows: Let S_n be the sum of the first n terms then

$$S_n = 1 + r + r^2 + r^3 + \dots + r^{n-1}$$

$$r \cdot S_n = 0 + r + r^2 + r^3 + \dots + r^{n-1} + r^n$$

Subtracting we have

$$(1 - r) \cdot S_n = 1 - r^n$$

Then for $r \neq 1$ $\quad S_n = \frac{1 - r^n}{1 - r}$

It follows that $\lim_{n \to \infty} S_n$ exists provided $\lim_{n \to \infty} r^n$ exists and $r \neq 1$

The limit of r^n, as n goes to infinity, exists if $|r| < 1$ and then $\lim_{n \to \infty} S_n = \frac{1}{1 - r}$

Exactly the same argument works if we replace r by a matrix **A** that has the property that $\mathbf{A}^n$ goes to zero as n goes to infinity. In this case we see that

$(\mathbf{I} - \mathbf{A}) \cdot \mathbf{G}(\mathbf{A}) = \mathbf{I}$ which implies that $\mathbf{G}(\mathbf{A}) = (\mathbf{I} - \mathbf{A})^{-1}$

Assume that G(A) exists. Verify that $(\mathbf{I} - \mathbf{A}) \cdot (\mathbf{I} + \mathbf{A} + \mathbf{A}^2 + \mathbf{A}^3 + \dots) = \mathbf{I}$

ANSWER:

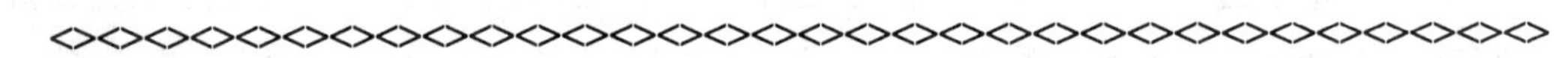

Exercises

3. Let **D** be a diagonal matrix. For what choices of diagonal entries will $\mathbf{D}^n \to \mathbf{0}$?

ANSWER:

4. Let **M** be an invertible n by n matrix and **D** an n by n diagonal matrix. For what choices of diagonal entries in **D** will $(\mathbf{M} \cdot \mathbf{D} \cdot \mathbf{M}^{-1})^n \to \mathbf{0}$? Hint: Calculate $(\mathbf{M} \cdot \mathbf{D} \cdot \mathbf{M}^{-1})^n$ and use the previous exercise.

ANSWER:

5. Let **A** be an n by n matrix such that λ is an eigenvalue with associated eigenvector **v**. Assume that $|\lambda| \geq 1$. What is $|\mathbf{A}^n \cdot \mathbf{v}|$? Can $\mathbf{A}^n \to \mathbf{0}$? Can **G(A)** converge?

ANSWER:

6. For which of the following matrices does the associated geometric series converge? Give a reason for your answer in each case.

$$A := \begin{pmatrix} .3 & .7 & .1 \\ .5 & 0 & .5 \\ .1 & .15 & 0 \end{pmatrix} \qquad B := \begin{pmatrix} .3 & .7 & .5 \\ .5 & 0 & .7 \\ .2 & .4 & 0 \end{pmatrix} \qquad C := \begin{pmatrix} .3 & .7 & .1 \\ .5 & 0 & .5 \\ .2 & .3 & .4 \end{pmatrix}$$

ANSWER:

7. For each matrix in Exercise 6 compute the dominant eigenvalue. Conjecture a relationship between the dominant eigenvalue and convergence of the associated geometric series.

ANSWER:

<><><><><><><><><><><><><><><><><><><><><><><><><><><><><><><><>

We shall not prove the following fact although the above exercises should make it plausible.

If all the eigenvalues of A have magnitude less than 1, then the geometric series G(A) converges.

<><><><><><><><><><><><><><><><><><><><><><><><><><><><><><><><>

It is not always easy to see that the eigenvalues of a matrix have magnitude less than one just by inspecting the matrix. There is, however, an easily verified condition that ensures that the largest eigenvalue in magnitude is less than 1.

The ***sum norm*** of a vector **v** (written $|\mathbf{v}|_S$) equals the sum of the absolute values of the elements of **v**.

In Mathcad notation this is $\overrightarrow{(\Sigma|\mathbf{v}|)}$

(The arrow over this expression means that the absolute value function is applied to each element of the vector rather than to the vector itself.)

$$v := \begin{pmatrix} -2 \\ 3 \\ -7 \\ 5 \end{pmatrix} \qquad \overrightarrow{\left(\sum |v|\right)} = 17 \qquad |v| = 9.327$$

<== **Press F9.**

As the above example shows the sum norm is always greater or equal to the length of the vector. Symbolically this relation is expressed as follows:

$$|v_1| + |v_2| + \dots + |v_n| \geq \sqrt{(v_1)^2 + (v_2)^2 + \dots + (v_n)^2}$$

This can be easily verified by squaring both sides. Explain how this shows that the left hand side is greater than the right hand side.

ANSWER:

Using the sum norm of the vector we define the ***sum norm of a matrix***, **A** (written $\|\mathbf{A}\|_S$) as the maximum of the sum norms of the columns of **A**. That is,

$$(\|A\|)_s = \max\left[\left(\left|A^{<j>}\right|\right)_s\right]$$

where the max is taken over all the columns of **A**.

Let

$$A := \begin{pmatrix} 2 & 2 & -3 & 0 \\ -5 & 6 & 5 & 7 \\ 1 & 0 & 2 & 0 \\ -3 & 1 & 1 & -7 \end{pmatrix} \qquad B := \begin{pmatrix} 2 & 1 & 9 \\ -5 & 4 & 0 \\ 6 & -5 & 1 \end{pmatrix}$$

Calculate the sum norm of **A** and the sum norm of **B** using the following Mathcad operations:

$k := 1..4 \qquad j := 1..3$

$\overrightarrow{\left(\sum \left|A^{<k>}\right|\right)} \qquad \overrightarrow{\left(\sum \left|B^{<j>}\right|\right)}$

<== **Type = and press F9.**

ANSWER:

$$\begin{array}{|c|}\hline 11 \\ \hline 9 \\ \hline 11 \\ \hline 14 \\ \hline \end{array} \qquad \begin{array}{|c|}\hline 13 \\ \hline 10 \\ \hline 10 \\ \hline \end{array}$$

<><><><><><><><><><><><><><><><><><><><><><><><><><><><><><><><><><><><><>

We are now prepared to show that:

|dominant eigenvalue of A| $\leq$ sum norm of A

i) For any vector **v**, $|\mathbf{A}\cdot\mathbf{v}| \leq \|\mathbf{A}\|_s \cdot |\mathbf{v}|$.

Since $|\mathbf{A}\cdot k\mathbf{v}| = |k|\cdot|\mathbf{A}\cdot\mathbf{v}|$ and $|k\mathbf{v}| = |k|\cdot|\mathbf{v}|$ we may assume that $|\mathbf{v}| = 1$.

$$|\mathbf{A}\cdot\mathbf{v}| = \left|v_1\cdot A^{<1>} + v_2\cdot A^{<2>} + \dots + v_n\cdot A^{<n>}\right| \leq \left|v_1\cdot A^{<1>}\right| + \dots + \left|v_n\cdot A^{<n>}\right|$$

$$\left|v_1\cdot A^{<1>}\right| + \dots + \left|v_n\cdot A^{<n>}\right| = \left|v_1\right|\cdot\left|A^{<1>}\right| + \dots + \left|v_n\right|\cdot\left|A^{<n>}\right|$$

$$\left|v_1\right|\cdot\left|A^{<1>}\right| + \dots + \left|v_n\right|\cdot\left|A^{<n>}\right| \leq \left|v_1\right|\cdot\left(\left|A^{<1>}\right|\right)_s + \dots + \left|v_n\right|\cdot\left(\left|A^{<n>}\right|\right)_s$$

Let $|\mathbf{A}^{<j>}|_s$ be a maximum for $j = k$. The sum above can be maximized by choosing **v** to be the unit vector with a 1 in the kth position and 0's elsewhere. Since $|\mathbf{A}^{<k>}|_s = \|\mathbf{A}\|_s$ it follows that $|\mathbf{A}\cdot\mathbf{v}| \leq \|\mathbf{A}\|_s\cdot|\mathbf{v}|$

ii) Use i) to show that |dominant eigenvalue| $\leq \|\mathbf{A}\|_s$.

ANSWER:

The following is a particular application of the above result.

If the sum norm of A < 1 then the dominant eigenvalue of A has magnitude less than 1.

Since the eigenvalues of $\mathbf{A}^T$ are the same as the eigenvalues of **A** it follows that if the sum norm of the rows of **A** (= columns of $\mathbf{A}^T$) are all less than 1 then the dominant eigenvalue is also less than 1 in magnitude.

We combine these results in the following:

> The matrix **A** is said to have the **convergence property** if either
> a) the sum norm of the rows of **A** or
> b) the sum norm of the columns of **A** is less than 1.

Example

$$\mathbf{A} := \begin{pmatrix} .3 & .6 & .5 \\ -.5 & -.2 & -.4 \\ .1 & .15 & 0 \end{pmatrix} \qquad j := 1..3$$

$\overrightarrow{\left(\sum \left|\mathbf{A}^{\langle j \rangle}\right|\right)}$ $\qquad \overrightarrow{\left|\mathbf{eigenvals}(\mathbf{A})\right|}$ **<== Type = and press F9.**

Note that in this example the sum norm of the first two rows is greater than one.

◇◇

If the convergence property holds, the geometric series G(A) converges .

This is a sufficient but not necessary condition; i.e., G(A) may converge when the convergence property is not satisfied. In particular if we replaced the 2nd element in the third row below by 0.25 the convergence property would be lost but the geometric series will still converge.

$$\mathbf{A} := \begin{pmatrix} .3 & .6 & .5 \\ -.5 & .25 & -.4 \\ .1 & .15 & 0 \end{pmatrix} \qquad n := 0..150$$

$$\sum_{n} \mathbf{A}^{n}$$ **<== Type = and press F9.**

◇◇

Principal Components in Statistical Analysis

A financial analyst has collected data on 20 financial measurements $x_1, x_2, ..., x_{20}$ for each of 150 corporations. The analyst computes a measure of the variability $\mathbf{V}_j$ of each variable x_j called the **variance** of x_j. A large variance means that the variable x_j changes considerably from company to company. The analyst also computes a measure of the joint variability $\mathbf{Cov}_{ij}$ of each pair of variables x_i, x_j called the **covariance**. The covariance is proportional to the correlation coefficient (see Section 1-5) and is a measure of the relationship between x_i and x_j.

The analyst would like to create a scale that is a linear combinations of the characteristics that will measure how similar two companies are. Among all possible linear combinations of the variables, the one that does this best is called the **first principal component**.

Let **C** be the matrix where $c_{i,j} = \mathbf{Cov}_{i,j}$ and $c_{i,i} = \mathbf{V}_i$. ***The coordinates of the first principal component are the coordinates of the eigenvector of length one corresponding to the dominant eigenvalue*** (i.e., if the eigenvector, **w**, corresponds to the dominant eigenvalue and has length one, then the first principal component = $\mathbf{w} \cdot \mathbf{x}$ where **x** is the vector whose components are the variables $x_1, \ldots, x_{20}$).

The fraction of the variance explained by the first principal component is:

$$\frac{|\text{dominant\_eigenvalue}|}{\left(\sum_{\text{all\_eigenvalues}} |\text{eigenvalue}|\right)}$$

Suppose the data consists of k variables with n sample points in each and is stored in an n by k matrix, **X**. Then the variance $C_{i,i}$ and the covariance $C_{i,j}$ are defined by

$$C_{i,i} := \frac{1}{n-1} \cdot \left(X^{<i>} - \text{mean}(X^{<i>})\right) \cdot \left(X^{<i>} - \text{mean}(X^{<i>})\right)$$

$$C_{i,j} := \frac{1}{n-1} \cdot \left(X^{<i>} - \text{mean}(X^{<i>})\right) \cdot \left(X^{<j>} - \text{mean}(X^{<j>})\right)$$

Calculate the first principal component for the data contained in the matrix **X**.

$$X := \begin{bmatrix} 0 & -7 & -3 & 0 \\ 3 & 0 & 7 & 3 \\ 7 & 5 & 2 & 4 \\ -5 & 2 & 1 & -5 \\ 2 & 0 & 5 & 5 \\ 1 & 0 & -4 & 4 \\ 1 & 1 & 1 & 3 \\ 9 & 3 & 1 & 2 \\ 5 & 6 & 0 & 1 \\ 0 & -5 & 0 & 0 \end{bmatrix}$$

$i := 1..4$ $j := 1..4$ $n := \text{rows}(X)$

$$C_{i,j} := \frac{1}{n-1} \cdot \left(X^{<i>} - \text{mean}(X^{<i>})\right) \cdot \left(X^{<j>} - \text{mean}(X^{<j>})\right)$$

ANSWER:

This analysis applies as well to other situations in which one would like to encapsulate many variables into a single 'principal' variable.

<><><><><><><><><><><><><><><><><><><><><><><><><><><><><><><><><><><><><><>

The preceding topics used the dominant eigenvalue to determine properties of a matrix and the components of the corresponding eigenvector to provide a scale in first principal component analysis. The following example shows how to view the eigenvector corresponding to the dominant eigenvalue graphically. For a 2 by 2 matrix **A**, if there is a dominant (real) eigenvalue then the successive images of the unit circle under the matrix **A** tend towards a straight line segment whose direction is the direction of the eigenvector corresponding to the dominant eigenvalue.

We construct a sample of vectors from the unit circle by taking a vector every 6 degrees.

$$N := 60 \qquad i := 1 .. N+1$$

Let $\quad A := \begin{pmatrix} .9 & .2 \\ .1 & .6 \end{pmatrix} \qquad u^{\langle i \rangle} := \begin{pmatrix} \cos\left(2\cdot\pi\cdot\frac{i}{N}\right) \\ \sin\left(2\cdot\pi\cdot\frac{i}{N}\right) \end{pmatrix}$

<== Matrix corresponding to the end points of the vectors in the unit circle.

Let **Q** be defined as follows (**Q** will be used to plot the two eigenvectors if the eigenvalues are real and distinct.)

$$j := 1 .. 2 \qquad Q^{\langle 2\cdot j \rangle} := \text{eigenvec}\left(A, \text{eigenvals}(A)_j\right) \qquad Q^{\langle 2\cdot j-1 \rangle} := \begin{pmatrix} 0 \\ 0 \end{pmatrix}$$

$$\lambda := \max\left(\overrightarrow{|\text{eigenvals}(A)|}\right) \qquad k := 1 .. 4$$

$$v(j) := \frac{A^j \cdot u}{\lambda^j}$$

<==== *This normalizes the successive ellipses so that the principal axis is visible.*

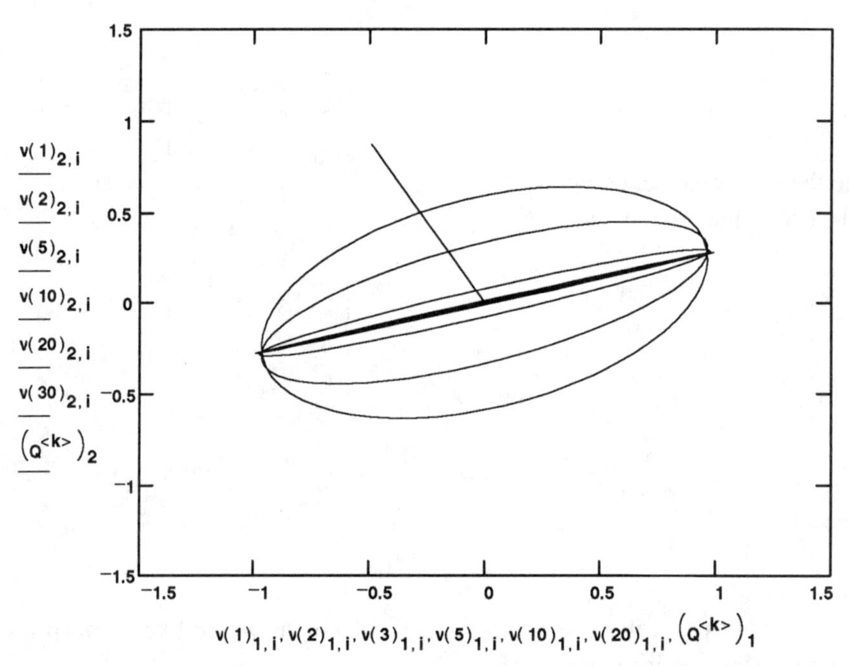

The images of the unit circle under A, A^2, A^5, A^{10}, A^{20} and A^{30}.

Let $\quad \mathbf{x} := \begin{pmatrix} 5 \\ -4 \end{pmatrix}$

The eigenvectors of A are: $\quad \mathbf{u} := \mathbf{eigenvec}(\mathbf{A}, \mathbf{eigenvals}(\mathbf{A})_1)$ and
$\mathbf{w} := \mathbf{eigenvec}(\mathbf{A}, \mathbf{eigenvals}(\mathbf{A})_2)$

$$\mathbf{u} = \begin{pmatrix} 0.963 \\ 0.27 \end{pmatrix} \qquad \mathbf{w} = \begin{pmatrix} -0.49 \\ 0.872 \end{pmatrix}$$

Compute the coordinates of **x** relative to **u** and **w** (i.e., $\mathbf{x} = a \cdot \mathbf{u} + b \cdot \mathbf{v}$).

ANSWER:

What are the coordinates of $\mathbf{A}^n \cdot \mathbf{x}$ relative to **u** and **w**?

ANSWER:

What is the relationship between the picture above and the previous calculation?

ANSWER:

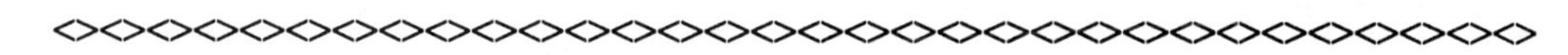

Exercises

8. In the example above there was a dominant eigenvalue (i.e., one eigenvalue whose absolute value was greater than all the other eigenvalues). In the two examples below there is not a dominant eigenvalue. In the first, the two eigenvalues are complex and of the same magnitude. In the second the two eigenvalues are real and of the same magnitude. Replace **A** in the preceding graphical look at the eigenvector by each of the following matrices. Describe the behavior in each case.

$$\begin{pmatrix} .7 & .3 \\ -.3 & .8 \end{pmatrix} \qquad \begin{pmatrix} .5 & 0 \\ .2 & -.5 \end{pmatrix}$$

ANSWER:

<><><><><><><><><><><><><><><><><><><><><><><><><><><><><><><><><><><><><><><><><>

9. The following graph represents a set of trade routes between four cities P_1, P_2, P_3, and P_4 where each edge can be traversed in either direction.

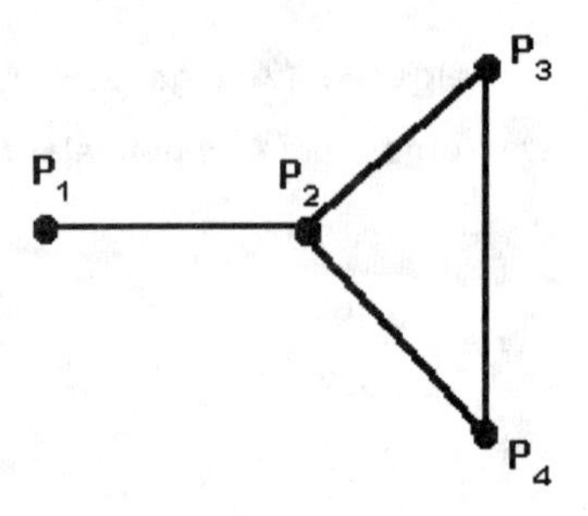

a) Construct the adjacency matrix **A** for this graph.

ANSWER:

b) By inspecting the graph, rank each city in the trade route network by its accessibility to the others. The highest rank to the most accessible. Use 5 for the highest and assign cities with equal accessibility the same value.

City P_1 __________

City P_2 __________

City P_3 __________

City P_4 __________

c) Find the eigenvalues of the adjacency matrix **A** and record them below.

ANSWER:

d) Compute an eigenvector **v** corresponding to the dominant eigenvalue. If |**v**| is not 1, set **u** = **v**/|**v**| otherwise set **u** = **v**. Record **u** below.

ANSWER:

e) Geographers have shown that the entries in vector **u** are a measure of the accessibility of cities 1, 2, 3, and 4 respectively. Does this measure agree with your ranking from part b? Explain why the entries provide such a rating.

ANSWER:

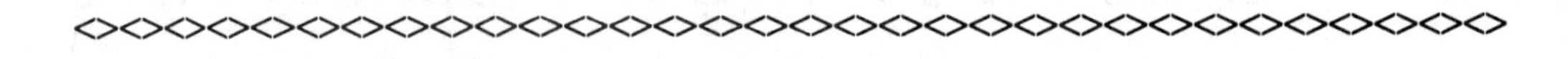

10. Repeat the procedure described in Exercise 9 for the graph shown below.

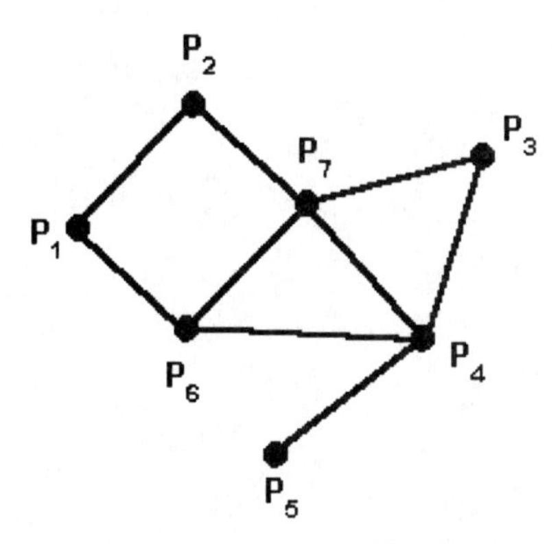

ANSWER:

<><><><><><><><><><><><><><><><><><><><><><><><><><><><><><>

11. (Individual) Write a summary of the MATHEMATICS that you learned in this section.

Section 6-5 Applications of the Geometric Series: Leontief Models and Markov Chains

Name:_______________

Score:____________

Overview

In the previous section we explored the geometric series of a matrix and discovered a criterion that ensured convergence. In this section the geometric series is used to study the Leontief model and Markov chains.

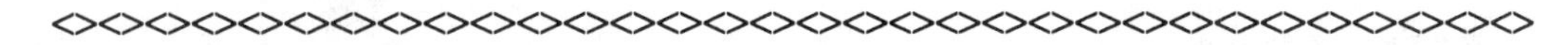

Leontief Model

The Leontief model was introduced in Section 0-2 and studied earlier in Sections 1-9 and 2-4. Our goal here is to show that the Leontief Input Constraint (see below) implies that an equilibrium exists. To review the Leontief model open the file leontief.mcd.

> **Leontief Input Constraint:**
> The sum of the coefficients in each column of **D** is < 1.

D is the matrix of industrial demands and **c** is consumer demand. An equilibrium solution is a supply vector, **x**, such that $\mathbf{x} = \mathbf{D} \cdot \mathbf{x} + \mathbf{c}$.

We can rewrite: $\mathbf{x} = \mathbf{D} \cdot \mathbf{x} + \mathbf{c}$ as $(\mathbf{I} - \mathbf{D}) \cdot \mathbf{x} = \mathbf{c}$ where **I** is the Identity matrix.
If $(\mathbf{I} - \mathbf{D})$ is invertible then there is an equilibrium solution $\mathbf{x} = (\mathbf{I} - \mathbf{D})^{-1} \cdot \mathbf{c}$.

The Leontief Input Constraint is equivalent to the fact that the sum norm of **D** is less than one. As we saw in the previous section this implies that all the eigenvalues of **D** are less than one in magnitude which in turn implies that the geometric series **G(D)** converges. But if **G(D)** converges it must converge to $(\mathbf{I} - \mathbf{D})^{-1}$ thus $(\mathbf{I} - \mathbf{D})$ is invertible and the Input Constraint implies that an equilibrium exists.

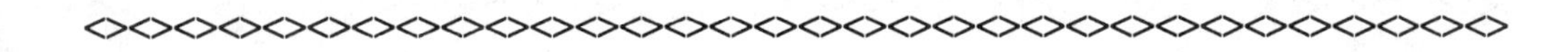

In Section 1-9 we used iteration to solve a Leontief system. We began by setting

$$x^{<0>} := c$$

and setting

$$x^{<n+1>} := D \cdot x^{<n>} + c$$

Substituting in the above we have:

$$x^{<1>} = D \cdot c + c = (D + I) \cdot c$$

$$x^{<2>} = D \cdot (D \cdot c + c) + c = (D^2 + D + I) \cdot c$$

$$x^{<3>} = D \cdot (D^2 \cdot c + D \cdot c + c) + c = (D^3 + D^2 + D + I) \cdot c$$

Write the expression for $\mathbf{x}^{<n>}$.

ANSWER:

Explain why the iteration converges for a matrix **D** satisfying the Leontief Input Constraint.

ANSWER:

<><><><><><><><><><><><><><><><><><><><><><><><><><><><><><><><>

Use iteration to solve the Leontief Models with the industrial demand matrices **A** and **B** and consumer demand vectors **c** and **cc**. (Show your work.)

a) $$A := \begin{pmatrix} .4 & .2 & .1 \\ .2 & .4 & .1 \\ 0 & .2 & .2 \end{pmatrix} \qquad c := \begin{pmatrix} 100 \\ 50 \\ 100 \end{pmatrix}$$

b) $$B := \begin{pmatrix} .2 & .25 & 0 & .25 \\ .3 & .35 & .2 & .5 \\ 0 & 0 & .5 & .2 \\ .4 & .1 & .2 & 0 \end{pmatrix} \qquad cc := \begin{pmatrix} 100 \\ 150 \\ 75 \\ 250 \end{pmatrix}$$

ANSWERS:

<><><><><><><><><><><><><><><><><><><><><><><><><><><><><><><><>

Absorbing Markov Chains

An absorbing state of a Markov chain is, as its name implies, a state that has the property that the probability of leaving the state is zero. For example, in a game that has many states one or more may be designated as winning states and one or more may be designated as losing states. Once these states are entered the game is over. The existence of such states raises questions such as, *what is the average time until absorption?* We review Markov chains, include the concept of an absorbing chain, and discover a way to answer questions such as the one raised above.

Recall that a Markov Chain is a model for the behavior of a system that 'moves' among different 'states.' (See Sections 0-2 and 1-9.) If the system is currently in state j there is a probability $a_{i,j}$ that it will move into state i. (The columns are the **FROM** state and the rows the **TO** state.) Explain why the column sums equal 1. Can you make a similar statement about the row sums?

ANSWER:

The matrix of probabilities, **A**, is called the ***transition matrix*** of the system.

There is a vector **p**, all of whose entries are non-negative and that sum to one, such that $\mathbf{A} \cdot \mathbf{p} = \mathbf{p}$. **p** is called the ***stable probability vector*** or the long run probability vector.

In Section 6-2, Exercise 6 we saw that 1 is an eigenvalue of a transition matrix since $(1,1,...,1)^T$ is an eigenvector of $\mathbf{A}^T$ with eigenvalue 1. Since the eigenvalues (but not the eigenvectors) of **A** are the same as those of $\mathbf{A}^T$ it follows that 1 is an eigenvalue of **A**.

One can show that (a) there can not be an eigenvalue bigger than one, (b) one is a simple eigenvalue (i.e., it is not a multiple root of the characteristic equation) and (c) there exists an eigenvector corresponding to the eigenvalue 1 with all of its entries non-negative. Why does this imply that there is a probability vector that is an eigenvector corresponding to the eigenvalue 1?

ANSWER:

Can there be more than one probability vector with this property? (Hint: What is the dimension of the solution space of $(\mathbf{A} - 1 \cdot \mathbf{I})$?)

ANSWER:

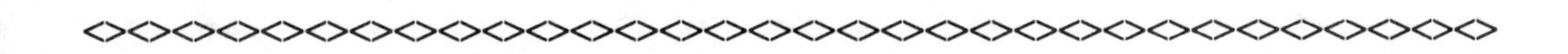

There are various ways of computing stable probability vector **p**.

a) Iterate the system.

b) Solve $\mathbf{A} \cdot \mathbf{p} = \mathbf{p}$ ($(\mathbf{A} - \mathbf{I}) \cdot \mathbf{p} = \mathbf{0}$) with the added condition that the solution be a probability vector; i.e., all the components of p are non-negative and sum to 1.

One must be a little careful here. Suppose we have a two state Markov chain with the property that state one always goes to state two and state two goes to state one. The matrix of this chain is

$$\begin{pmatrix} 0 & 1 \\ 1 & 0 \end{pmatrix}$$

The even powers of the matrix equal the identity and the odd powers equal the original matrix - consequently the iterations of the matrix do not converge. The characteristic equation of the matrix is $(\lambda + 1)(\lambda - 1)$ *and the eigenvectors are* $(.5, .5)^T$ *and* $(-.5, .5)^T$. *The latter is not a probability vector while the former is a stable vector; however, the only way to get to the stable state is to start there. It is an average of the states.*

If some power of the transition matrix has all ***positive*** *entries then this oscillating behavior does not occur. Such Markov chains are called* ***regular****.*

What does it mean for the transition matrix to have the property that for some power all entries are positive? (Hint: Consider moving between states.)

ANSWER:

We have proved that for a transition matrix of a Markov chain, 1 is an eigenvalue. If the Markov chain is regular then the eigenvalue is a simple root of the characteristic equation and there is a related eigenvector all of whose entries are **positive**. The proof of these facts is beyond this course. The statement of these results is called the Frobenius-Perron Theorem.

<><><><><><><><><><><><><><><><><><><><><><><><><><><><><><><>

Example

$$A := \begin{bmatrix} \frac{1}{2} & \frac{1}{4} & 0 & 0 \\ \frac{1}{2} & \frac{1}{2} & \frac{1}{4} & 0 \\ 0 & \frac{1}{4} & \frac{1}{2} & \frac{1}{2} \\ 0 & 0 & \frac{1}{4} & \frac{1}{2} \end{bmatrix} \qquad A^{50} = \begin{bmatrix} 0.16667 & 0.16667 & 0.16667 & 0.16667 \\ 0.33333 & 0.33333 & 0.33333 & 0.33333 \\ 0.33333 & 0.33333 & 0.33333 & 0.33333 \\ 0.16667 & 0.16667 & 0.16667 & 0.16667 \end{bmatrix}$$

<== **Display.**

$$P := \frac{\text{eigenvec}(A, 1)}{\sum \text{eigenvec}(A, 1)} \qquad P = \begin{bmatrix} 0.16667 \\ 0.33333 \\ 0.33333 \\ 0.16667 \end{bmatrix}$$

<== **Display.**

<><><><><><><><><><><><><><><><><><><><><><><><><><><><><><><><>

Exercises

1. Let $B := \begin{bmatrix} 0 & \frac{1}{6} & 1 \\ 0 & \frac{2}{3} & 0 \\ 1 & \frac{1}{6} & 0 \end{bmatrix}$ be the transition matrix of a Markov chain.

Does the process have a stable distribution? Do the iterates converge to the stable distribution? Explain.

ANSWER:

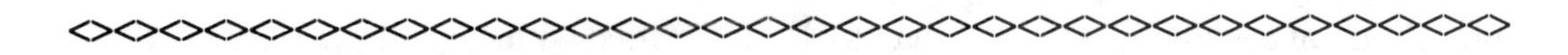

2. For the Markov process with transition matrix **B** ==> compute the stable probability vector by iteration and by finding the stable probability eigenvector.

$$B := \begin{pmatrix} .3 & .2 & .1 & .25 \\ .15 & .35 & .8 & .125 \\ .05 & .4 & 0 & .5 \\ .5 & .05 & .1 & .125 \end{pmatrix}$$

ANSWER:

<><><><><><><><><><><><><><><><><><><><><><><><><><><><><><><><>

Example: Coin Tossing

Assume that a player has k coins, where $1 \le k \le 4$. At each turn in the game the player tosses a coin. If the toss is a head the player wins and receives another coin. If the toss yields a tail the player loses the coin. Assume that the coins are fair; that is, the probability of winning equals the probability of losing = 0.5. The game is over either when the player has either 5 coins on no coins. Represent this game as a Markov chain with 6 states (0, 1, ..., 5) each representing the number of coins that the player has. **A** is the transition matrix for this Markov chain.

$$\mathbf{A} := \begin{bmatrix} 1 & .5 & 0 & 0 & 0 & 0 \\ 0 & 0 & .5 & 0 & 0 & 0 \\ 0 & .5 & 0 & .5 & 0 & 0 \\ 0 & 0 & .5 & 0 & .5 & 0 \\ 0 & 0 & 0 & .5 & 0 & 0 \\ 0 & 0 & 0 & 0 & .5 & 1 \end{bmatrix}$$

Describe in words the states of this model and the probabilities of moving between states.

ANSWER:

$$\mathbf{A}^{100} = \begin{bmatrix} 1 & 0.8 & 0.6 & 0.4 & 0.2 & 0 \\ 0 & 0 & 0 & 0 & 0 & 0 \\ 0 & 0 & 0 & 0 & 0 & 0 \\ 0 & 0 & 0 & 0 & 0 & 0 \\ 0 & 0 & 0 & 0 & 0 & 0 \\ 0 & 0.2 & 0.4 & 0.6 & 0.8 & 1 \end{bmatrix}$$

<== Display.

From the display of $\mathbf{A}^{100}$ above we see that the (6,2)-entry is .2 which implies that the probability of eventually winning (state 6), having started with 1 coin is .2; starting with 2 coins the probability of winning is .4; etc.

The first state (0) and the last state (6) are called absorbing states since a player stays in those states forever. We would like to be able to calculate the probability of winding up in state (0) (resp., state 6) having started in state j, ***directly*** (i.e., not by iteration).

<><><><><><><><><><><><><><><><><><><><><><><><><><><><><><><><><><><><>

Analysis of Markov Chains with absorbing states

The following steps give a procedure for determining the probability of winding up in an absorbing state having started in a non-absorbing state. We begin by arranging the transition matrix so that all the absorbing states precede the non-absorbing states.

$$\mathbf{AA} := \begin{bmatrix} 1 & 0 & .5 & 0 & 0 & 0 \\ 0 & 1 & 0 & 0 & 0 & 0 \\ 0 & 0 & 0 & .5 & 0 & 0 \\ 0 & 0 & .5 & 0 & .5 & 0 \\ 0 & 0 & 0 & .5 & 0 & .5 \\ 0 & 0 & 0 & 0 & .5 & 0 \end{bmatrix}$$

We write this new matrix in block form as follows and name the blocks.

$$\left|\begin{array}{cc|cccc} 1 & 0 & .5 & 0 & 0 & 0 \\ 0 & 1 & 0 & 0 & 0 & 0 \\ \hline 0 & 0 & 0 & .5 & 0 & 0 \\ 0 & 0 & .5 & 0 & .5 & 0 \\ 0 & 0 & 0 & .5 & 0 & .5 \\ 0 & 0 & 0 & 0 & .5 & 0 \end{array}\right| = \begin{pmatrix} I & R \\ 0 & Q \end{pmatrix}$$

Verify for yourself that we may multiply blocks in the 'same way' that we multiply matrices. (Naturally the sizes must be compatible.)
For example,

$$\begin{pmatrix} I & R \\ 0 & Q \end{pmatrix}^2 = \begin{pmatrix} I & R + R\cdot Q \\ 0 & Q^2 \end{pmatrix}$$

Compute the following products by 'hand.' Be careful not to commute matrices.

$$\begin{pmatrix} I & R \\ 0 & Q \end{pmatrix} \cdot \begin{pmatrix} I & R + R\cdot Q \\ 0 & Q^2 \end{pmatrix} = \begin{pmatrix} I & \blacksquare \\ 0 & \blacksquare \end{pmatrix}$$

$$\begin{pmatrix} I & R \\ 0 & Q \end{pmatrix} \cdot \begin{pmatrix} I & R \\ 0 & Q \end{pmatrix} \cdot \begin{pmatrix} I & R + R\cdot Q \\ 0 & Q^2 \end{pmatrix} = \begin{pmatrix} I & \blacksquare \\ 0 & \blacksquare \end{pmatrix}$$

Complete the following;

$$\begin{pmatrix} I & R \\ 0 & Q \end{pmatrix}^k = \begin{pmatrix} I & R + R\cdot Q + \blacksquare \\ 0 & \blacksquare \end{pmatrix}$$

<><><><><><><><><><><><><><><><><><><><><><><><><><><><><><><><><><><><><>

If it is possible to reach an absorbing state from each of the starting states it follows that the sum norm of **Q** is less than one and $\mathbf{Q}^k$ goes to the zero matrix as k gets large.

Likewise, the upper right matrix, $R\cdot(I + Q + Q^2 + \ldots + Q^{k-1}) \rightarrow R\cdot(I - Q)^{-1}$

Set $\quad N = (I - Q)^{-1}$

N is called the ***fundamental matrix*** of the absorbing Markov Chain.

$(\mathbf{R}\cdot\mathbf{N})_{i,j}$ is the probability of winding up in the absorbing state i having started in the non-absorbing state j. (The state numbers are from the rearranged transition matrix.) In the above example we have

$$Q := \begin{pmatrix} 0 & .5 & 0 & 0 \\ .5 & 0 & .5 & 0 \\ 0 & .5 & 0 & .5 \\ 0 & 0 & .5 & 0 \end{pmatrix} \qquad R := \begin{pmatrix} .5 & 0 & 0 & 0 \\ 0 & 0 & 0 & .5 \end{pmatrix}$$

Compute the fundamental matrix **N** of the coin tossing model and **R·N**.

ANSWER:

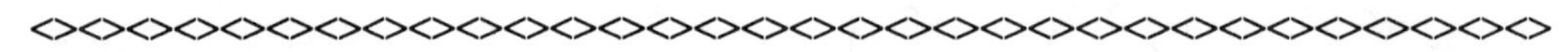

We can modify the above model to consider the effects of a biased coin. Let p be the probability of a win; then 1-p is the probability of a loss.

$$QQ(p) := \begin{pmatrix} 0 & 1-p & 0 & 0 \\ p & 0 & 1-p & 0 \\ 0 & p & 0 & 1-p \\ 0 & 0 & p & 0 \end{pmatrix} \qquad RR(p) := \begin{pmatrix} 1-p & 0 & 0 & 0 \\ 0 & 0 & 0 & p \end{pmatrix}$$

$$p := 0, .05 .. 1$$

f(p) is the probability of winning (state 2 in the rearranged matrix) beginning with 2 coins.

$$f(p) := \left(RR(p)\cdot(\text{identity}(4) - QQ(p))^{-1}\right)_{2,2}$$

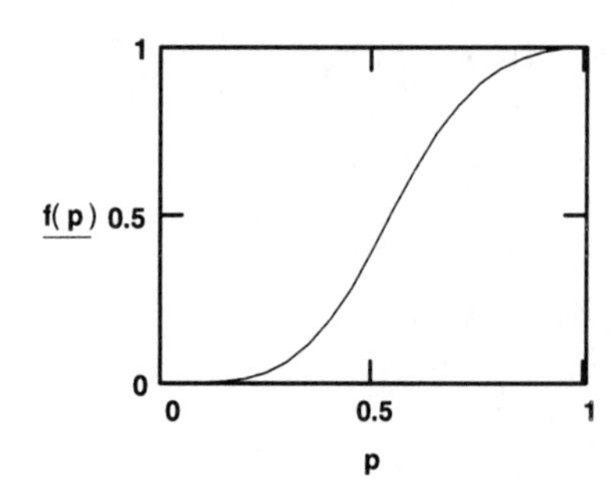

Let $p := .45$

Compute the probabilities of winning and losing starting from each state.
(Call the fundamental matrix **NN** so that it can be used below.)

ANSWER:

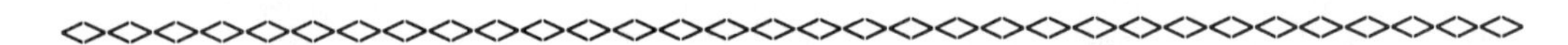

Question 1: What is the average time we visit state i before absorption if we start in state j?

ANSWER:

$$I + \left[Q_{i,j} + (Q^2)_{(i,j)} + (Q^3)_{(i,j)} + (Q^4)_{(i,j)} + \dots \right]$$

This is the (i,j)th element of the fundamental matrix.

Question 2: How many turns does the game last?

The expected time before absorption is then simply the sum of the expected time spent in each non-absorbing state from each starting position. This is

$$((1\ \ 1\ \ 1\ \ 1)\cdot NN)^T =$$

<><><><><><><><><><><><><><><><><><><><><><><><><><><><><>

Another example

Question: Assume that a fair coin is tossed 10 times. What is the probability that there is a string of either 4 heads or 4 tails in this run?

To answer this question set up a Markov chain with the states: absorbing state '4 in a row' (denoted a), h, hh, hhh, t, tt, and ttt. Where hh means two heads in a row and the other symbols are defined similarly.

The transition matrix, **M**, is $(a\ \ h\ \ hh\ \ hhh\ \ t\ \ t\ \ ttt)$

$$M := \begin{bmatrix} 1 & 0 & 0 & .5 & 0 & 0 & .5 \\ 0 & 0 & 0 & 0 & .5 & .5 & .5 \\ 0 & .5 & 0 & 0 & 0 & 0 & 0 \\ 0 & 0 & .5 & 0 & 0 & 0 & 0 \\ 0 & .5 & .5 & .5 & 0 & 0 & 0 \\ 0 & 0 & 0 & 0 & .5 & 0 & 0 \\ 0 & 0 & 0 & 0 & 0 & .5 & 0 \end{bmatrix} \quad \begin{bmatrix} a \\ h \\ hh \\ hhh \\ t \\ tt \\ ttt \end{bmatrix}$$

$$M^9 = \begin{bmatrix} 1 & 0.46484 & 0.55078 & 0.70898 & 0.46484 & 0.55078 & 0.70898 \\ 0 & 0.14453 & 0.12109 & 0.07813 & 0.14648 & 0.12305 & 0.08008 \\ 0 & 0.08008 & 0.06641 & 0.04297 & 0.07813 & 0.06641 & 0.04297 \\ 0 & 0.04297 & 0.03711 & 0.02344 & 0.04297 & 0.03516 & 0.02344 \\ 0 & 0.14648 & 0.12305 & 0.08008 & 0.14453 & 0.12109 & 0.07813 \\ 0 & 0.07813 & 0.06641 & 0.04297 & 0.08008 & 0.06641 & 0.04297 \\ 0 & 0.04297 & 0.03516 & 0.02344 & 0.04297 & 0.03711 & 0.02344 \end{bmatrix}$$ <== **Display.**

Since we can begin either with a head or a tail we see that the probability of absorption is the element in the first row and either second or fifth column; namely, 0.465. Thus the probability of four in a row out of 10 tosses is slightly under a half.

In general, society has a poor understanding of probability. One often hears sports reporters speak of a team as on streak if they win or lose three or four games in a row. As we can see, if we assume that all teams have equal ability, then the probability of a four game 'streak' over a stretch of 10 games is almost one-half.

Try the following experiment. Ask your friends to write a 'random' sequence of ten heads and tails. I guarantee you that very few of your friends will include a string of four in a row of either heads or tails; yet such a string happens 46.5% of the time when the coins are flipped. (Our intuition of what random means is often very poor.)

Assume we continue to toss the coin until we get four in a row. On average, how many tosses will be required?
ANSWER:

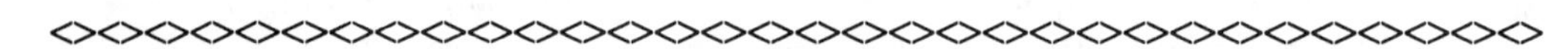

A Random Walk

Consider a Markov chain with n+1 states labeled 1, 2, ..., n + 1.

Transition probabilities are given as follows:

$n := 10$

$T_{2,1} := 1$ $\qquad$ $T_{n,n+1} := 1$

$k := 2..n$

$T_{k-1,k} := 0.5$ $\qquad$ $T_{k+1,k} := 0.5$

Describe this model in words.

ANSWER:

What is the long run probability distribution as a function of n? Experiment with different values of n.

ANSWER:

You are invited to play the following game. You begin in state 1 and the game lasts for n + 1 turns. At the end of that time you are paid k - 1 dollars if your final state is state k. You must pay $4 to play. How large should n be to ensure that your expected winnings are greater than $4? If the probability of being in state k at the end of the game is p_k, then the expected winnings are

$$\sum_{k=1}^{n+1} (k-1) \cdot p_k$$

The payoff vector is given by $i := 1..n+1$ $\quad$ $PAYOFF_i := i - 1$

and the starting vector is given by $START_i := 0$ $\quad$ $START_1 := 1$

ANSWER:

Assume that you begin in state 1, what is the average time until you return to state 1? To answer this question make state 1 an absorbing state, begin in state 2 and compute the average time until absorption. Add 1 to this answer. Examine the cases n = 5, 10, 15, 20, 25.

ANSWER:

Suppose that state 1 is an absorbing state. If you begin in state n+1, what is average time until you are absorbed in state 1?

Examine n = 5, 10, 15, 20, 25.

ANSWER:

<><><><><><><><><><><><><><><><><><><><><><><><><><><><><><>

Exercise: (Individual) Write a summary of the MATHEMATICS that you learned in this section.

CHAPTER 7

LINEAR TRANSFORMATIONS

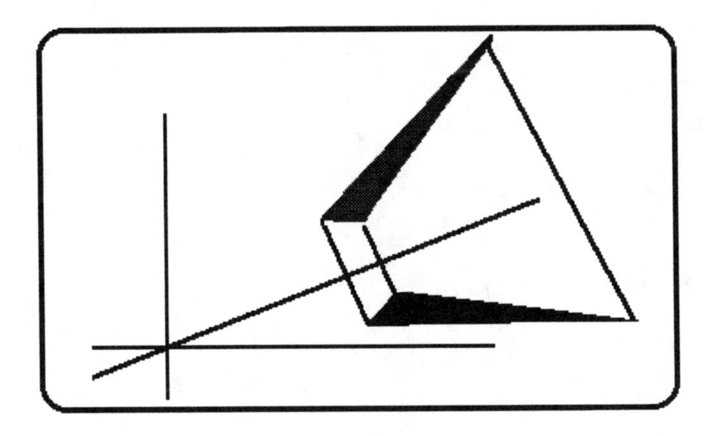

Section 7-1 Linear Transformations

Name:________________

Score:_____________

Overview

In Section 1-7 we discovered the relationship between a matrix and a linear function from $\mathbf{R}^n$ to $\mathbf{R}^m$. We review that relationship in this section and investigate how, given a linear function, we can compute a matrix **A** that corresponds to the function. This matrix can be used to compute values of the linear function when we specify bases in $\mathbf{R}^n$ and $\mathbf{R}^m$. We calculate the matrices corresponding to a rotation and a projection as examples.

<><><><><><><><><><><><><><><><><><><><><><><><><><><><><><><><><><><><>

Let $T:\mathbf{R}^n \longrightarrow \mathbf{R}^m$ be a function.

T is said to be a **linear transformation** if, for **u** and **v** in $\mathbf{R}^n$ and c, a scalar in **R**, it satisfies:

a) $T(\mathbf{u}+\mathbf{v})=T(\mathbf{u})+T(\mathbf{v})$

b) $T(c\cdot\mathbf{u})=c\cdot T(\mathbf{u})$

The vector sum and scalar product on each side of properties a and b above are different. One is performed in $\mathbf{R}^n$ and the other in $\mathbf{R}^m$.

Since T preserves linear combinations, the image of a line is a line; the image of a plane is another plane; and the image of a subspace is a subspace.

<><><><><><><><><><><><><><><><><><><><><><><><><><><><><><><><><><><><>

Example

Let **A** be an m by n matrix. **A** defines a function from $\mathbf{R}^n$ to $\mathbf{R}^m$ given by $\mathbf{A}(\mathbf{v}) = \mathbf{A}\cdot\mathbf{v}$, where **v** is in $\mathbf{R}^n$. The properties of the matrix-vector product ensure that this is a linear transformation. Explain.

ANSWER:

<><><><><><><><><><><><><><><><><><><><><><><><><><><><><><><><><><><><>

If S and T are linear transformations $\mathbf{R}^n \rightarrow \mathbf{R}^m$ then their sum $(S + T):\mathbf{R}^n \rightarrow \mathbf{R}^m$ is defined by $(S + T)(\mathbf{u}) = S(\mathbf{u}) + T(\mathbf{u})$. A direct calculation shows that the sum is also a linear transformation. Likewise if c is a scalar the linear function $c\cdot T:\mathbf{R}^n \rightarrow \mathbf{R}^m$ is defined by $(c\cdot T)(\mathbf{u}) = c\cdot T(\mathbf{u})$.

<><><><><><><><><><><><><><><><><><><><><><><><><><><><><><><><><><><><>

Composition of Linear Transformations

If $T:\mathbf{R}^n \to \mathbf{R}^m$ and $S:\mathbf{R}^m \to \mathbf{R}^p$ the *composite* map $S \circ T:\mathbf{R}^n \to \mathbf{R}^p$ is given by $(S \circ T)(\mathbf{u}) = S(T(\mathbf{u}))$.

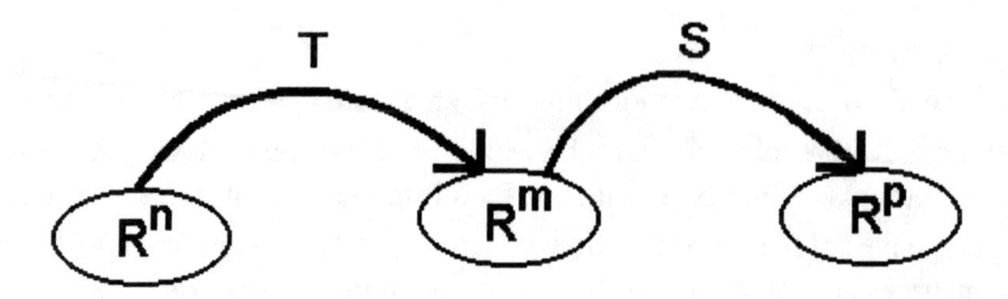

Show that the composition of two linear transformations is a linear transformation. (Hint: Compute $(S \circ T)(\mathbf{u} + \mathbf{v})$ and $(S \circ T)(c \cdot \mathbf{u})$.)

ANSWER:

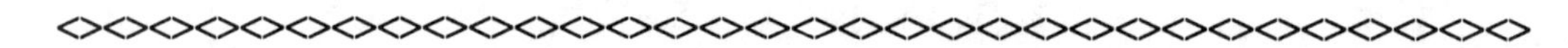

Let T be a linear transformation from $\mathbf{R}^n$ to $\mathbf{R}^m$ and suppose that **ej** is the vector such that $\mathbf{ej}_i = 0$ for $i \neq j$ and $\mathbf{ej}_j = 1$. **e1**, **e2**, ..., **en** is called the standard or natural basis for $\mathbf{R}^n$. Suppose that $T(\mathbf{ej}) = \mathbf{vj}$ for j = 1, ..., n and let $\mathbf{x} = (x1, x2, ..., xn)$. Express the value of $T(\mathbf{x})$ in terms of the scalars, x1, ..., xn and the vectors **v1**, ..., **vn**.

ANSWER:

If you answered the previous question correctly then your answer can be expressed in the language of matrix algebra as follows:

Let **M** be the matrix whose jth column is **vj**. Then $T(\mathbf{x}) = \mathbf{M} \cdot \mathbf{x}$. Explain.

ANSWER:

Summary

Let T be a linear transformation from $\mathbf{R}^n$ to $\mathbf{R}^m$ and let **M** be the m by n matrix whose jth column is $T(\mathbf{ej})$ then for any vector **v** in $\mathbf{R}^n$, $T(\mathbf{v}) = \mathbf{M} \cdot \mathbf{v}$.

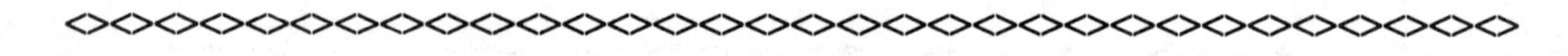

Example: Rotation in the Plane

Let R_θ be rotation through an angle, θ.
Then

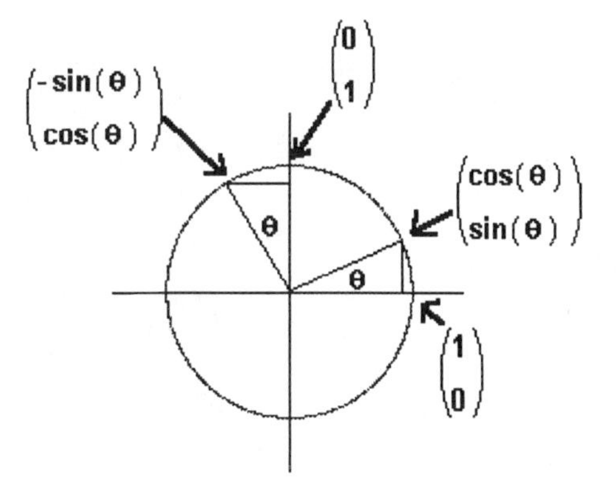

$$R_\theta\cdot\begin{pmatrix}1\\0\end{pmatrix}=\begin{pmatrix}\cos(\theta)\\\sin(\theta)\end{pmatrix}$$

To see this use trigonometry.

$$R_\theta\cdot\begin{pmatrix}0\\1\end{pmatrix}=\begin{pmatrix}-\sin(\theta)\\\cos(\theta)\end{pmatrix}$$

For an arbitrary vector $\begin{pmatrix}x\\y\end{pmatrix}$ in $\mathbf{R}^2$, we have $\begin{pmatrix}x\\y\end{pmatrix}=x\cdot\begin{pmatrix}1\\0\end{pmatrix}+y\cdot\begin{pmatrix}0\\1\end{pmatrix}$

Thus $$R_\theta\cdot\begin{pmatrix}x\\y\end{pmatrix}=x\cdot R_\theta\cdot\begin{pmatrix}1\\0\end{pmatrix}+y\cdot R_\theta\cdot\begin{pmatrix}0\\1\end{pmatrix}=\begin{pmatrix}x\cdot\cos(\theta)-y\cdot\sin(\theta)\\x\cdot\sin(\theta)+y\cdot\cos(\theta)\end{pmatrix}$$

It follows that $$R_\theta\cdot\begin{pmatrix}x\\y\end{pmatrix}=\begin{pmatrix}x\cdot\cos(\theta)-y\cdot\sin(\theta)\\x\cdot\sin(\theta)+y\cdot\cos(\theta)\end{pmatrix}=\begin{pmatrix}\cos(\theta)&-\sin(\theta)\\\sin(\theta)&\cos(\theta)\end{pmatrix}\cdot\begin{pmatrix}x\\y\end{pmatrix}$$

and hence $$R_\theta=\begin{pmatrix}\cos(\theta)&-\sin(\theta)\\\sin(\theta)&\cos(\theta)\end{pmatrix}$$

For ease of reference we change the name of the matrix from R_θ to $R(\theta)$

Let $$R(\theta):=\begin{pmatrix}\cos(\theta)&-\sin(\theta)\\\sin(\theta)&\cos(\theta)\end{pmatrix}$$

Let $w^{<1>}:=\begin{pmatrix}-1\\-1\end{pmatrix}$ $\quad n:=5 \quad j:=1..n+2 \quad w^{<j+1>}:=R\left(\frac{2\cdot\pi}{n}\right)\cdot w^{<j>}$

If we think of $\mathbf{w}^{<j>}$ j = 1, ..., n + 2 as points in the plane, what geometric figure do they determine when connected in sequence? How does this change when we change n?

ANSWER:

Check your answer by viewing the graph below.

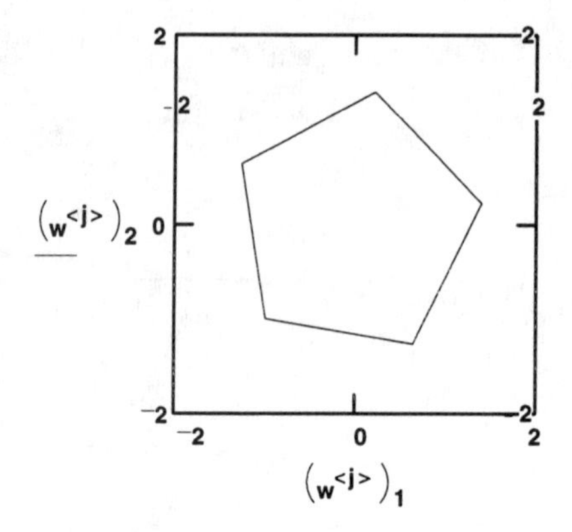

What transformation does the composition R(ϕ)∘R(θ) represent?

$$R(\theta)=\begin{pmatrix}\cos(\theta) & -\sin(\theta)\\ \sin(\theta) & \cos(\theta)\end{pmatrix} \qquad R(\phi)=\begin{pmatrix}\cos(\phi) & -\sin(\phi)\\ \sin(\phi) & \cos(\phi)\end{pmatrix}$$

ANSWER:

Multiply these matrices together to get one representation. State another representation directly from the meaning of R(ϕ)∘R(θ). By equating these two representations you get formulas for what relationships?

ANSWER:

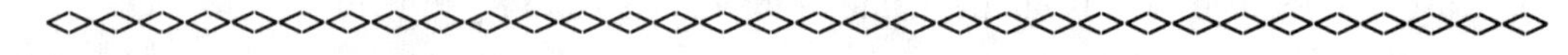

Example: Projection

Let **v** be a vector in $\mathbf{R}^n$. We saw in Chapter 5 that if **x** is any vector in $\mathbf{R}^n$ then the projection, P(**x**), of **x** onto **v** is given by

$$P(x) := \frac{x \cdot v}{v \cdot v} \cdot v$$

Thus to compute the matrix **M** corresponding to the transformation P we need only compute P(**ej**) for each vector in the standard basis.

Let $\mathbf{v} := \begin{pmatrix} 1 \\ -2 \\ 1 \end{pmatrix}$

The first column of **M** equals

$$\frac{\begin{pmatrix} 1 \\ 0 \\ 0 \end{pmatrix} \cdot \begin{pmatrix} 1 \\ -2 \\ 1 \end{pmatrix}}{\begin{pmatrix} 1 \\ -2 \\ 1 \end{pmatrix} \cdot \begin{pmatrix} 1 \\ -2 \\ 1 \end{pmatrix}} \cdot \begin{pmatrix} 1 \\ -2 \\ 1 \end{pmatrix} = \begin{bmatrix} \frac{1}{6} \\ \frac{-1}{3} \\ \frac{1}{6} \end{bmatrix}$$

(This is P(**e1**).)

The second column of **M** equals

$$\frac{\begin{pmatrix} 0 \\ 1 \\ 0 \end{pmatrix} \cdot \begin{pmatrix} 1 \\ -2 \\ 1 \end{pmatrix}}{\begin{pmatrix} 1 \\ -2 \\ 1 \end{pmatrix} \cdot \begin{pmatrix} 1 \\ -2 \\ 1 \end{pmatrix}} \cdot \begin{pmatrix} 1 \\ -2 \\ 1 \end{pmatrix} = \begin{bmatrix} \frac{-1}{3} \\ \frac{2}{3} \\ \frac{-1}{3} \end{bmatrix}$$

(This is P(**e2**).)

The third column of **M** equals

$$\frac{\begin{pmatrix} 0 \\ 0 \\ 1 \end{pmatrix} \cdot \begin{pmatrix} 1 \\ -2 \\ 1 \end{pmatrix}}{\begin{pmatrix} 1 \\ -2 \\ 1 \end{pmatrix} \cdot \begin{pmatrix} 1 \\ -2 \\ 1 \end{pmatrix}} \cdot \begin{pmatrix} 1 \\ -2 \\ 1 \end{pmatrix} = \begin{bmatrix} \frac{1}{6} \\ \frac{-1}{3} \\ \frac{1}{6} \end{bmatrix}$$

(This is P(**e3**).)

To check that **M** is the correct matrix compute **M·x**

$$\begin{bmatrix} \frac{1}{6} & -\frac{1}{3} & \frac{1}{6} \\ -\frac{1}{3} & \frac{2}{3} & -\frac{1}{3} \\ \frac{1}{6} & -\frac{1}{3} & \frac{1}{6} \end{bmatrix} \cdot \begin{pmatrix} x1 \\ x2 \\ x3 \end{pmatrix}$$

<==== **Evaluate symbolically.**

which should be the same as

$$\frac{\begin{pmatrix} 1 \\ -2 \\ 1 \end{pmatrix} \cdot \begin{pmatrix} x1 \\ x2 \\ x3 \end{pmatrix}}{\begin{pmatrix} 1 \\ -2 \\ 1 \end{pmatrix} \cdot \begin{pmatrix} 1 \\ -2 \\ 1 \end{pmatrix}} \cdot \begin{pmatrix} 1 \\ -2 \\ 1 \end{pmatrix}$$

<==== **Evaluate symbolically.**

<><><><><><><><><><><><><><><><><><><><><><><><><><><><><><><><><><><><><><>

Evaluate $\mathbf{M}^2$. Explain why this is equal to $\mathbf{M}$.

ANSWER:

We saw in Chapter 5 that the projection into a subspace can be computed in terms of the pseudo inverse. Use the pseudo inverse to verify that the above matrix calculation is correct.

ANSWER:

If T is a linear transformation, λ is called an ***eigenvalue of T*** if there is a nonzero vector $\mathbf{v}$ such that $T(\mathbf{v}) = \lambda \cdot \mathbf{v}$.

Explain why the eigenvalues of a linear transformation T are the same as the eigenvalues of the matrix, $\mathbf{M}$, corresponding to T.

ANSWER:

Let P be the projection given above, explain why $P^2 = P$. (We use P^2 to stand for $P \circ P$.)

ANSWER:

If T is a linear transformation that satisfies $T^2 = T$, we call T a ***projection***. Let λ be an eigenvalue of a projection and $\mathbf{v}$ a corresponding eigenvector. Then

$$T^2(\mathbf{v}) = T \circ T(\mathbf{v}) = T(\lambda \cdot \mathbf{v}) = \lambda \cdot T(\mathbf{v}) = \lambda^2 \cdot \mathbf{v}$$

Since $T^2 = T$, $T^2(\mathbf{v}) = T(\mathbf{v}) = \lambda \cdot \mathbf{v}$

Therefore the eigenvalues of a projection must satisfy $\lambda^2 = \lambda$. What are the possible eigenvalues of a projection?

ANSWER:

Remember that, for any subspace W of $\mathbf{R}^n$, we can decompose $\mathbf{R}^n$ into the sum of W and its orthogonal complement, $W^\perp$. The **geometric projection, $\mathbf{P_W}$, onto W** is given by $P_W(\mathbf{v})$ = the component of $\mathbf{v}$ that lies in W.

Let T be a projection in the algebraic sense described above (i.e., $T^2 = T$). Explain why this definition of projection is consistent with the definition given in the previous paragraph.

ANSWER:

<><><><><><><><><><><><><><><><><><><><><><><><><><><><>

Exercises

1. Calculate the matrix, **M**, that corresponds to projection in $\mathbf{R}^3$ onto the plane $x - 2y + 4z = 0$. Check your answer by verifying that $\mathbf{M}^2 = \mathbf{M}$ and $\mathbf{M}\cdot(1,-2,4)^T = 0$. (Hint: Use Section 5-1.)

ANSWER:

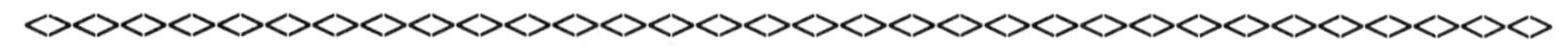

2. Calculate the 4 by 4 matrix that corresponds to projection onto the plane in $\mathbf{R}^4$ spanned by

$$\mathbf{w1} := \begin{pmatrix} 2 \\ 0 \\ 1 \\ -1 \end{pmatrix} \qquad \mathbf{w2} := \begin{pmatrix} 2 \\ 5 \\ -2 \\ 2 \end{pmatrix}$$

ANSWER:

<><><><><><><><><><><><><><><><><><><><><><><><><><><><><><><><><><><><><><><>

3. Let **M** be the matrix given below. Verify that $\mathbf{M}^2 = \mathbf{M}$ and therefore **M** is the matrix of a projection.

$$\begin{bmatrix} \frac{1}{6} & \frac{1}{3} & \frac{-1}{6} \\ \frac{1}{3} & \frac{13}{15} & \frac{1}{15} \\ \frac{-1}{6} & \frac{1}{15} & \frac{29}{30} \end{bmatrix}$$

What is the dimension of the column space of **M**? If it is one, **M** is a projection onto a line; if it is two, **M** is a projection onto a plane. Either find the direction for the line or the equation for the plane (depending upon which is appropriate).

ANSWER:

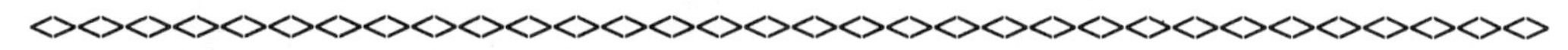

4. (Individual) Write a summary of the MATHEMATICS that you learned in this section.

Section 7-2 Change of Basis

Name:________________

Score:_____________

Overview

In the previous section we computed the matrix of a linear transformation with respect to the standard basis. In this section we extend that calculation to compute the matrix relative to a choice of basis in the domain $\mathbf{R}^n$ and a choice of basis in the range $\mathbf{R}^m$. We examine how matrices of the same transformation relative to different choices of bases are related and exploit this relationship to derive the matrix representation of a projection and a reflection.

<><><><><><><><><><><><><><><><><><><><><><><><><><><><><><><><><><><><><>

Coordinates Relative to an Ordered Basis

> The **coordinates** of a vector **v** relative to an ordered basis **u1, u2, ..., uk** are the coefficients c1, c2, ..., ck such that
>
> $$\mathbf{v} = c1 \cdot \mathbf{u1} + c2 \cdot \mathbf{u2} + \ldots + ck \cdot \mathbf{uk}$$

We write $\mathbf{coord(v)} = \begin{bmatrix} c1 \\ c2 \\ . \\ . \\ ck \end{bmatrix}$

Using the standard basis $\mathbf{e1} = \begin{pmatrix} 1 \\ 0 \end{pmatrix}$ $\mathbf{e2} = \begin{pmatrix} 0 \\ 1 \end{pmatrix}$ in $\mathbf{R}^2$, the coordinates of the vector

$\mathbf{v} = \begin{pmatrix} a \\ b \end{pmatrix}$ are the components of the vector since $\mathbf{v} = a \cdot \mathbf{e1} + b \cdot \mathbf{e2}$.

Geometrically, we start at the origin and step off a units in the **e1** direction followed by b units in the **e2** direction. The values of a and b are the coordinates. (See the diagram below.)

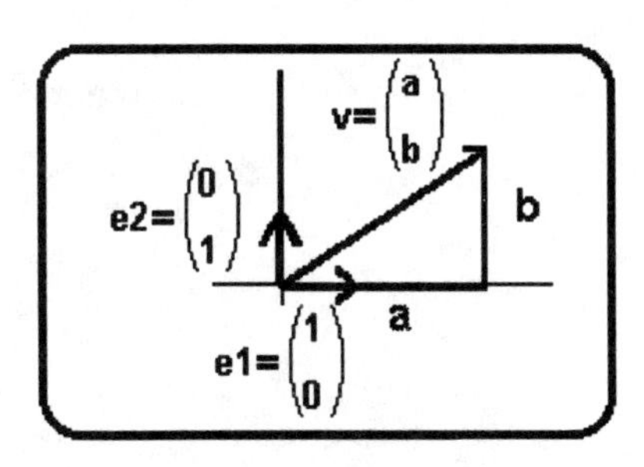

The vectors $\mathbf{u1}=\begin{pmatrix}1\\2\end{pmatrix}$ $\mathbf{u2}=\begin{pmatrix}2\\-3\end{pmatrix}$ are a basis for $\mathbf{R}^2$.

The corresponding geometric picture is

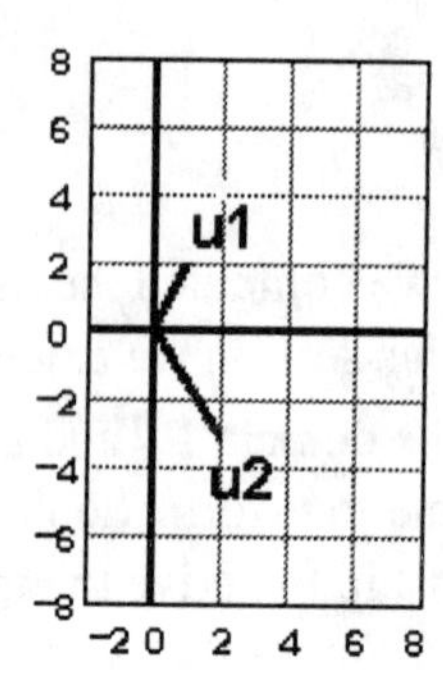

To express a vector $\mathbf{v}=\begin{pmatrix}5\\3\end{pmatrix}$ in terms of a basis consisting of **u1** and **u2** we must find its coordinates with respect to **u1** and **u2**; that is, find coefficients c1 and c2 so that $\mathbf{v} = c1\cdot\mathbf{u1} + c2\cdot\mathbf{u2}$.

Show that the coordinates of **v** relative to the **u1**, **u2** basis are c1 = 3 and c2 = 1.

ANSWER:

The picture on the right shows **v** in terms of the basis vectors, **u1** and **u2**, with the coordinates of **v** used as labels for the vectors in the **u1** and **u2** directions.

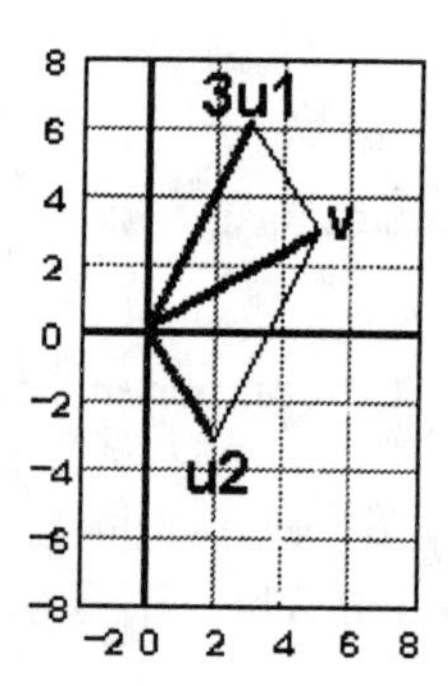

Coordinates of v are $\begin{bmatrix}3\\1\end{bmatrix}$

***Given a basis, the coordinates of a vector relative to that basis are found by solving a linear system whose coefficient matrix has columns corresponding to the basis vectors and whose right side is the vector* v.**

Let **B** be a matrix whose columns form a basis for $\mathbf{R}^n$ and let **bj** be the jth column of the matrix **B**. Let **v** be a vector in $\mathbf{R}^n$. Explain why the coordinates of **v** relative to the **b1, b2, ..., bn** basis are $\mathbf{B}^{-1} \cdot \mathbf{v}$.

ANSWER:

<><><><><><><><><><><><><><><><><><><><><><><><><><><><><><><><><><><><>

Example

Let $\mathbf{B} := \begin{pmatrix} 1 & 0 & 0 & 1 \\ 1 & 0 & 1 & 1 \\ 0 & 1 & 1 & 1 \\ 0 & 1 & 1 & 0 \end{pmatrix}$

Verify that the columns of **B** form a basis for $\mathbf{R}^4$. (We call this the **B**-basis.)

ANSWER:

Let the coordinates relative to the **B**-basis of a vector **v** in $\mathbf{R}^4$ be $\begin{pmatrix} 2 \\ -2 \\ 3 \\ 1 \end{pmatrix}$

Write a sentence that explains how to calculate the coordinates of **v** with respect to the standard basis.

ANSWER:

Do this calculation.

ANSWER:

The correct answer is $\begin{pmatrix} 3 \\ 6 \\ 2 \\ 1 \end{pmatrix}$

Write a sentence that explains how to compute the coordinates of $\begin{pmatrix} 3 \\ 6 \\ 2 \\ 1 \end{pmatrix}$ in terms of the **B**-basis.

ANSWER:

Do the calculation.

ANSWER:

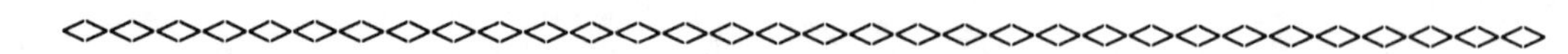

The Matrix of a Linear Transformation Relative to Given Bases

Let $\mathbf{u}_1, \mathbf{u}_2, \ldots, \mathbf{u}_n$ be a basis for $\mathbf{R}^n$. Assume that $T(\mathbf{u}_i) = \mathbf{w}_i$ is known for each i.

(1) Let **u** be any vector in $\mathbf{R}^n$. Explain why T is defined for all vectors **u** in $\mathbf{R}^n$.

ANSWER:

(2) Explain how to compute T(**u**).

ANSWER:

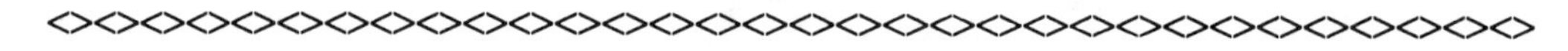

The first step of the procedure to compute T(**u**) must be to find the coordinates of **u** with respect to the given basis. Once we do this we can view the matrix whose jth column is $T(\mathbf{u}_j)$ as a transformation from **u** in $\mathbf{u}_j$-coordinates to T(**u**) in standard coordinates.

Example

Let $\mathbf{u1} := \begin{pmatrix} 1 \\ 1 \end{pmatrix}$ $\quad \mathbf{u2} := \begin{pmatrix} 1 \\ -1 \end{pmatrix}$ be a basis for $\mathbf{R}^2$ and assume that T: $\mathbf{R}^2 \rightarrow \mathbf{R}^3$ is such that

$$T\left(\begin{pmatrix} 1 \\ 1 \end{pmatrix}\right) = \begin{pmatrix} 1 \\ 0 \\ 3 \end{pmatrix} \qquad T\left(\begin{pmatrix} 1 \\ -1 \end{pmatrix}\right) = \begin{pmatrix} 0 \\ 2 \\ -1 \end{pmatrix}$$

We want to compute $T\left(\begin{pmatrix} 5 \\ -2 \end{pmatrix}\right)$

Step 1. What are the coordinates of $\begin{pmatrix} 5 \\ -2 \end{pmatrix}$ in the **u**-basis?

ANSWER:

Step 2. Compute $T\left(\begin{pmatrix} 5 \\ -2 \end{pmatrix}\right)$

ANSWER:

Explain why in general $T\left(\begin{pmatrix} x \\ y \end{pmatrix}\right) = \begin{pmatrix} \frac{1}{2}\cdot x + \frac{1}{2}\cdot y \\ x - y \\ x + 2\cdot y \end{pmatrix}$

ANSWER:

<><><><><><><><><><><><><><><><><><><><><><><><><><><><><><><><><><><>

The matrix $M := \begin{pmatrix} 1 & 0 \\ 0 & 2 \\ 3 & -1 \end{pmatrix}$ represents the transformation relative to the **u**-basis

in the domain and the standard basis in the range. By this we mean that if **w** is the vector of coordinates of a vector **v** in the **u**-basis then **M·w** is the vector of coordinates of T(**v**) in the standard coordinates of the range.

> Corresponding to each linear transformation there is a matrix **A** such that T(**v**) = **A·v**. **A** depends upon the bases used.

<><><><><><><><><><><><><><><><><><><><><><><><><><><><><><><><><><><>

In the preceding investigations we were given a basis in the domain for which we knew the images in the range. We then expressed an arbitrary vector in the domain in terms of its basis vectors and were able to use the matrix algebra to determine a matrix representation for the linear transformation. Implicit in the development was the use of the standard basis in the range. When a basis is specified in the domain and another (non-standard basis) in the range the computations required to determine the matrix representation are a bit more complicated. We develop this procedure below.

<><><><><><><><><><><><><><><><><><><><><><><><><><><><><><><><><><><>

Let $T: \mathbf{R}^n \rightarrow \mathbf{R}^m$ be a linear transformation. Given a basis for $\mathbf{R}^n$, $\mathbf{u}_1, \mathbf{u}_2, \ldots, \mathbf{u}_n$ and a basis for $\mathbf{R}^m$, $\mathbf{v}_1, \mathbf{v}_2, \ldots, \mathbf{v}_m$ there is an m by n matrix, **A**, such that if the coordinates of vector **u** relative to the **u**-basis are $(c_1, \ldots, c_m)$ then the coordinates of T(**u**) relative to the **v**-basis are

$$\mathbf{A} \cdot \begin{bmatrix} c1 \\ c2 \\ . \\ . \\ cm \end{bmatrix}$$

The entries of **A** depend upon the bases - ***choose different bases - get a new A.***

> A matrix representation **A** of a linear transformation gives a function that takes coordinates in the domain to coordinates in the range. The value of this linear transformation is computed by matrix multiplication:
>
> **A** · domain coordinates = range coordinates

We still must determine how to compute the matrix representation **A**.

For the linear transformation $T:\mathbf{R}^n \rightarrow \mathbf{R}^m$ with basis $\mathbf{u}_1, \mathbf{u}_2, \ldots, \mathbf{u}_n$ for $\mathbf{R}^n$ and basis $\mathbf{v}_1, \mathbf{v}_2, \ldots, \mathbf{v}_m$ for $\mathbf{R}^m$, the matrix representation **A** must satisfy

$$\mathbf{A}^{<j>} = \text{coordinates of } T(\mathbf{u}_j)$$

Thus we must compute the image of each of the basis vectors $\mathbf{u}_1, \mathbf{u}_2, \ldots, \mathbf{u}_n$ for the domain $\mathbf{R}^n$ and find its coordinates relative to basis $\mathbf{v}_1, \mathbf{v}_2, \ldots, \mathbf{v}_m$ in the range.

<><><><><><><><><><><><><><><><><><><><><><><><><><><><><><><>

Example: Projections revisited

We determine the matrix representation of the projection onto a line in $\mathbf{R}^2$.

Let L be the line through the origin with direction vector $\begin{pmatrix} u \\ v \end{pmatrix}$

For any vector $\begin{pmatrix} x \\ y \end{pmatrix}$ in $\mathbf{R}^2$, its projection onto L is the vector

$$p=\left[\frac{\begin{pmatrix}u\\v\end{pmatrix}\cdot\begin{pmatrix}x\\y\end{pmatrix}}{\begin{pmatrix}u\\v\end{pmatrix}\cdot\begin{pmatrix}u\\v\end{pmatrix}}\right]\cdot\begin{pmatrix}u\\v\end{pmatrix}$$

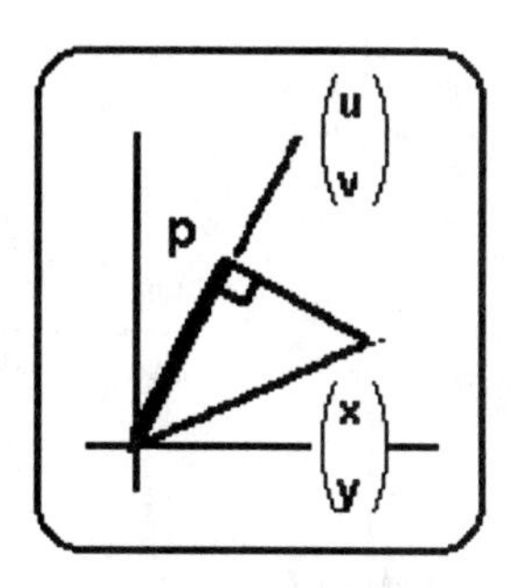

For a fixed line L through the origin, the mapping $T\left(\begin{pmatrix}x\\y\end{pmatrix}\right)=p$ is a linear transformation from $\mathbf{R}^2$ to $\mathbf{R}^2$.

In Section 7-1 we computed the matrix of a projection directly in terms of standard coordinates. In this section we do the same calculation by choosing a convenient basis for the domain and another for the range.

A convenient basis is the direction vector for the line L, $\quad L1=\begin{pmatrix}u\\v\end{pmatrix}$

and a vector orthogonal to L1 $\quad L2=\begin{pmatrix}-v\\u\end{pmatrix}$

Use this basis for both the domain and range. (Call this the **L**-basis, since one vector is in the direction of L and the other is orthogonal to L.)

Explain why the matrix for the projection in terms of the **L**-basis given above is

$$P:=\begin{pmatrix}1&0\\0&0\end{pmatrix}$$

ANSWER:

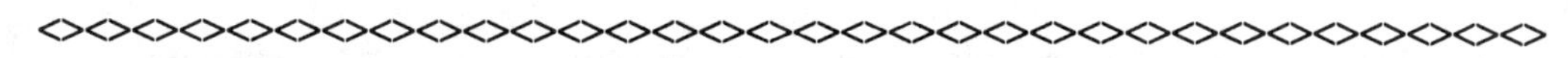

To find the matrix of the projection relative to the natural basis we must first transform coordinates given in terms of the natural basis into coordinates in the **L**-basis. We then transform using the matrix **P** and finally change the **L**-coordinates into standard coordinates.

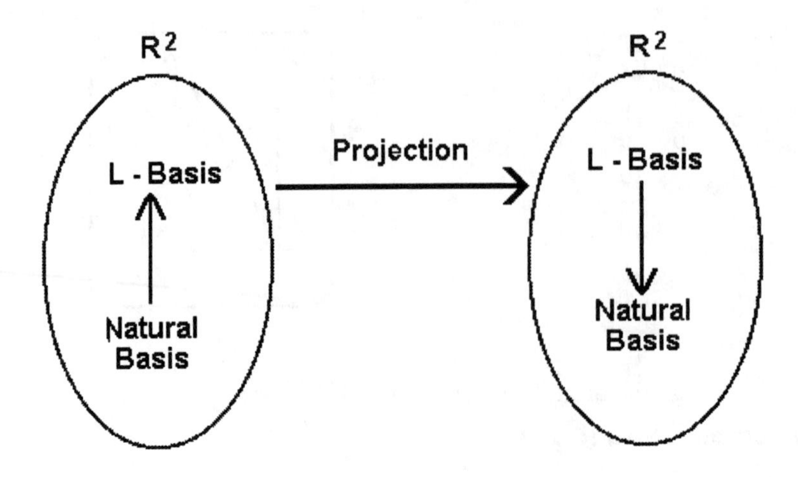

To go from **L**-coordinates to standard coordinates is the easier task. For if the coordinates relative to the **L**-basis are (x,y) then the coordinates relative to natural basis are simply $x \cdot \mathbf{L1} + y \cdot \mathbf{L2}$. Let the **L** be the matrix whose columns are the **L**-basis vectors.

If (x,y) are the L-coordinates then $\mathbf{L} \cdot \begin{pmatrix} x \\ y \end{pmatrix}$ is the vector of the coordinates relative to the natural basis. If the matrix **L** takes **L**-coordinates to natural coordinates, what matrix takes natural coordinates to **L**-coordinates?

ANSWER:

The matrix of the projection relative to standard coordinates is then $\mathbf{L} \cdot \mathbf{P} \cdot \mathbf{L}^{-1}$. In terms of the above example we then have:

$$\mathbf{PS} := \begin{pmatrix} u & -v \\ v & u \end{pmatrix} \cdot \begin{pmatrix} 1 & 0 \\ 0 & 0 \end{pmatrix} \cdot \begin{pmatrix} u & -v \\ v & u \end{pmatrix}^{-1}$$

which is

$$\mathbf{PS} := \begin{bmatrix} \dfrac{u^2}{(u^2 + v^2)} & \dfrac{u \cdot v}{u^2 + v^2} \\ \dfrac{u \cdot v}{u^2 + v^2} & \dfrac{v^2}{(u^2 + v^2)} \end{bmatrix}$$

Check that this is correct by calculating the matrix directly (i.e., the first column is the projection of (1,0) and the second is the projection of (0,1)).

ANSWER:

<><><><><><><><><><><><><><><><><><><><><><><><><><><><><><><><><><><><><><>

Let $S:\mathbf{R}^n \to \mathbf{R}^m$ and $T:\mathbf{R}^m \to \mathbf{R}^p$ be linear transformations. Let $\mathbf{u}_1, \mathbf{u}_2, ..., \mathbf{u}_n$ be a basis for $\mathbf{R}^n$, $\mathbf{v}_1, \mathbf{v}_2, ..., \mathbf{v}_m$ a basis for $\mathbf{R}^m$, and $\mathbf{w}_1, \mathbf{w}_2, ..., \mathbf{w}_p$ a basis for $\mathbf{R}^p$ and let $\mathbf{M}_S$ and $\mathbf{M}_T$ be the matrices for S and T corresponding to these bases. What is the matrix of $(T \circ S)$ relative to the **u** and **w** bases?

ANSWER: $\mathbf{M}_T \cdot \mathbf{M}_S$

It is enough to show that this is correct for each basis vector, $\mathbf{u}_j$.

The coordinates of $S(\mathbf{u}_j)$ with respect to the $\mathbf{v}_j$'s is the jth column of $\mathbf{M}_S$. But then the coordinates of $T(S(\mathbf{u}_j))$ is $\mathbf{M}_T$ times the jth column of $\mathbf{M}_S$, this is the jth column of the product $\mathbf{M}_T \cdot \mathbf{M}_S$.

<> <><><><><><><><><><><><><><><><><><><><><><><><><><><><><><><><><><><>

Let $T:\mathbf{R}^n \to \mathbf{R}^m$ be a linear transformation. Let $\mathbf{u}_1, \mathbf{u}_2, ..., \mathbf{u}_n$ be a basis for $\mathbf{R}^n$ and let $\mathbf{v}_1, \mathbf{v}_2, ..., \mathbf{v}_m$ be a basis for $\mathbf{R}^m$.

Let **M** be the matrix of T with respect to these two bases. We denote this as follows:

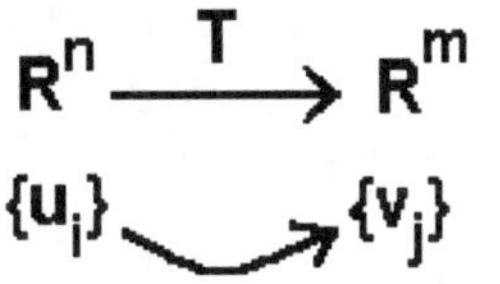

$$\mathbf{R}^n \xrightarrow{T} \mathbf{R}^m \qquad \{u_i\} \to \{v_j\}$$

(This means **M** takes u-coordinates to v-coordinates.)

Now suppose that $\mathbf{w}_1, \mathbf{w}_2, ..., \mathbf{w}_n$ is another basis for $\mathbf{R}^n$. What is the matrix of T with respect to the **w**- and **v**-bases?

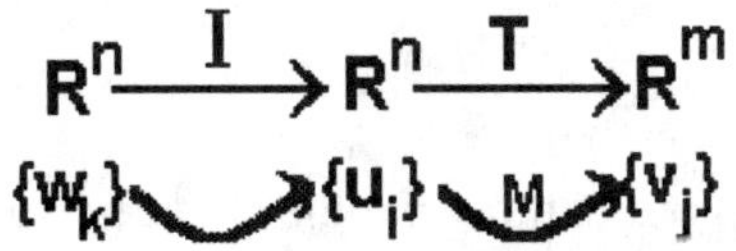

$$\mathbf{R}^n \xrightarrow{I} \mathbf{R}^n \xrightarrow{T} \mathbf{R}^m \qquad \{w_k\} \to \{u_i\} \xrightarrow{M} \{v_j\}$$

What is the matrix of the Identity Mapping with respect to the **u**'s and **w**'s? Call this matrix **C**. If $a_1, a_2, ..., a_n$ are the coordinates of **x** with respect to the **w**-basis, then $\mathbf{C} \cdot \mathbf{a}$ is the vector of coordinates with respect to the **w**-basis. In particular, the jth column of **C** is the vector of coordinates of $\mathbf{w}_j$ with respect to the **u**'s. **C** is invertible and its inverse changes basis from the **u**-basis to the **w**-basis.

In the same way that we can change basis in $\mathbf{R}^n$ by premultiplying by a 'change-of-basis' matrix, we can change basis in $\mathbf{R}^m$ by postmultiplying by a 'change-of-basis' matrix.

Let $\mathbf{s1} := \begin{pmatrix} 1 \\ 1 \\ 0 \end{pmatrix}$ $\mathbf{s2} := \begin{pmatrix} 1 \\ 0 \\ 1 \end{pmatrix}$ $\mathbf{s3} := \begin{pmatrix} 0 \\ 1 \\ 1 \end{pmatrix}$

be a basis for $\mathbf{R}^3$. What are the coordinates of $\mathbf{u} := \begin{pmatrix} 1 \\ -1 \\ 2 \end{pmatrix}$ with respect to the **s**-basis?

ANSWER:

What is the matrix **C** that represents the change of basis from the **s**-basis to the standard basis?

ANSWER:

Experiment: Reflection about a Line in R^2

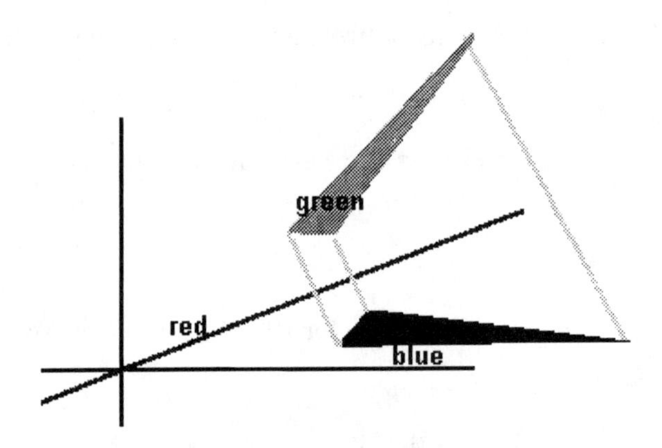

The green triangle is a reflection of the blue triangle about the red line.

Explain why a reflection about a line through the origin is a linear transformation.

ANSWER:

Consider reflecting a vector $\mathbf{v} = \begin{pmatrix} x \\ y \end{pmatrix}$ about the line whose direction vector is $\begin{pmatrix} a \\ b \end{pmatrix}$

Let $\mathbf{u1} := \begin{pmatrix} a \\ b \end{pmatrix}$ and $\mathbf{u2} := \begin{pmatrix} -b \\ a \end{pmatrix}$ **u1**, **u2** is a basis for $\mathbf{R}^2$.

Call the reflection S(**v**). Pictorially we have:

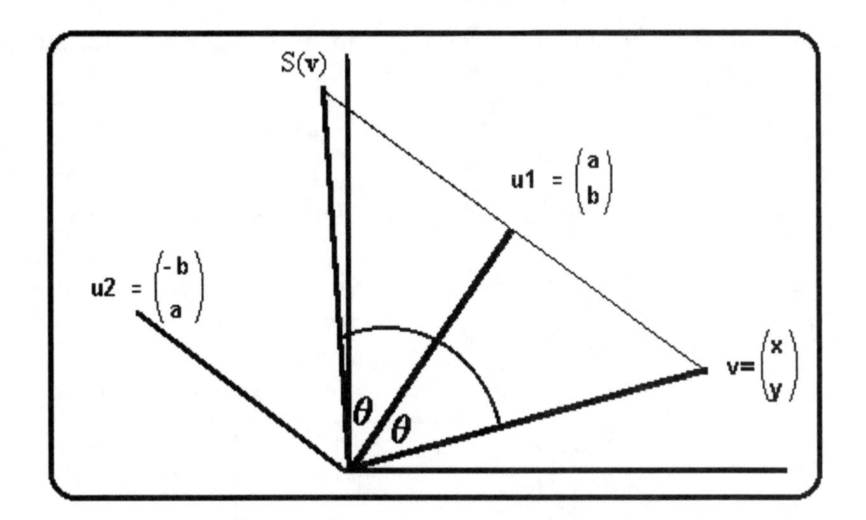

If the coordinates of **v** are (p,q) relative to the **u1**, **u2** basis, what are the coordinates of S(**v**) relative to this basis. (Hint: **v** and S(**v**) are two sides of an isosceles triangle in which **u1** is the perpendicular bisector of the base.)

ANSWER:

What is the matrix of the reflection relative to the **u**-basis?

ANSWER:

To get the matrix for the reflection about the line whose direction vector is **u1** in terms of the natural basis we must first change basis from the natural basis to the **u**-basis, do the reflection, then change back.

The matrix $\mathbf{U} := \begin{pmatrix} a & -b \\ b & a \end{pmatrix}$ is the matrix from the **u**-basis to the natural basis.

What is $\mathbf{U}^{-1}$?

ANSWER:

<><><><><><><><><><><><><><><><><><><><><><><><><><><><><><><><><><><><><><><><>

Let $\mathbf{R(a,b)}$ denote the matrix that reflects a vector about the line with direction vector **u1** given above.

$\mathbf{R(a,b)} :=$ ▪ **<====== Complete this expression**

You must compute the matrix **R(a, b)** to finish the lesson.

Let $\mathbf{v} := \begin{pmatrix} 0 & 5 \\ 0 & 2 \end{pmatrix}$ $\quad \mathbf{j} := 1..2 \quad \mathbf{k} := 1..4$

Vector v ==>

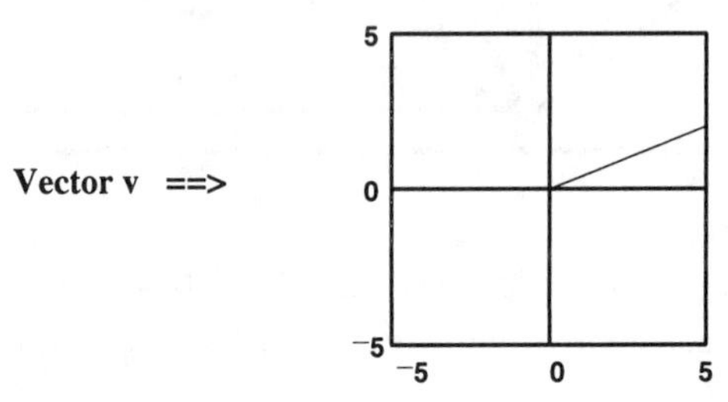

$\mathbf{a} := 1$
$\mathbf{b} := 3$ **<== Direction vector of line L**

$\mathbf{L} := \begin{pmatrix} -5\cdot a & 0 & 5\cdot a & 0 \\ -5\cdot b & 0 & 5\cdot b & 0 \end{pmatrix}$ **<== Points on the line.**

$\mathbf{w} := \mathbf{R(a,b)\cdot v}$ **<== Compute the reflection.**

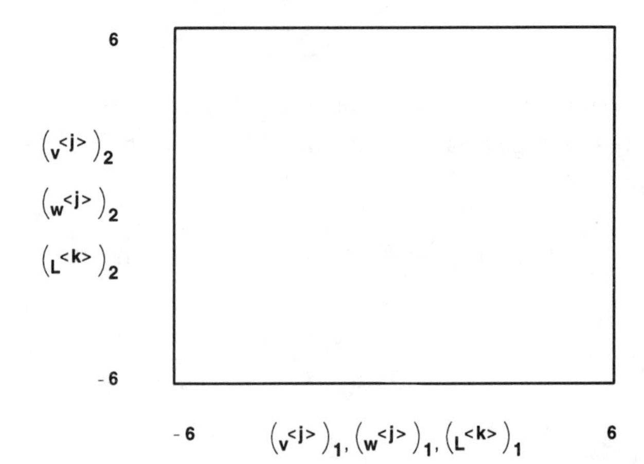

<><><><><><><><><><><><><><><><><><><><><><><><><><><><><><><><><><><><>

Exercise: (Individual) Write a summary of the MATHEMATICS that you learned in this sectio

Section 7-3
The Kernel & Image of a Linear Transformation

Name:________________

Score:______________

Overview

For an m by n matrix **A**, the rank of **A** is the number of linearly independent rows and the dimension of the solution space of the related homogeneous system of linear equations is n - rank(**A**). Our goal in this section is to prove that the dimension of the column space of the matrix equals the dimension of the row space. This may appear obvious from the reduced row echelon form of the matrix; however, we must be careful since row operations change the column space even though they preserve the row space.

Example

Consider $\mathbf{A} := \begin{pmatrix} 1 & 1 & 0 \\ 0 & 0 & 0 \\ 1 & 0 & 1 \end{pmatrix}$

Explain why the second and third columns span the column space of **A**.

ANSWER:

Find the rref of **A**. Show that you get the following matrix. $\begin{pmatrix} 1 & 0 & 1 \\ 0 & 1 & -1 \\ 0 & 0 & 0 \end{pmatrix}$

ANSWER:

Now the first and second columns span the column space and their span is different from the span of the original columns of **A**; i.e., the column space has been changed by row reduction.

To prove that the dimension of the column space of the matrix equals the dimension of the row space, we consider the more general situation of linear transformations and study two related subspaces, the kernel of the transformation and the image of the transformation. The dimensions of these subspaces will be related to the dimensions of the row space and column space of a matrix.

In the following discussion, assume that U = $\mathbf{R}^m$ and V = $\mathbf{R}^n$ for some m and n.

Kernel of T

Let T:U → V be a linear transformation. The **kernel of T**, denoted Ker(T), is the set of all vectors **u** in U such that T(**u**) = **0**, the zero vector. (The kernel of T is also called the **null space of T**.)

Show that Ker(T) is a subspace of U.

ANSWER:

The dimension of Ker(T) is called the **nullity of T**.

If T is a matrix mapping, the kernel is simply the solution space of the associated homogeneous system of linear equations and the nullity is the dimension of that solution space.

Image of T

Let T:U → V be a linear transformation. The **image space** of T, denoted by Image(T), is the set of all vectors **v** in V such that **v** = T(**u**) for some vector **u** in U. (The **image space** of T is also called the **range** of T, denoted Range(T).)

Show that Image(T) is a subspace of V. (Hint: You must show that if **v1** and **v2** are vectors in Image(T), then **v1** + **v2** and c·**v1** are also vectors in Image(T).)

ANSWER:

The dimension of Image(T) is called the **rank of T**.

Explain why, in the case that T is a matrix mapping the rank of T is the dimension of the column space.

ANSWER:

We say that T is **one-one** (pronounced 'one to one'), if T(**x**) = T(**y**) implies that **x** = **y**.

Explain why it is true that the following three statements are equivalent. (Equivalent means that whenever one is true the others are also true.)

a) T is one-one.

b) Ker(T) = **0**, the zero vector.

c) nullity(T) = 0.

ANSWER:

We say that T is **onto**, if whenever **v** is an element of V there is **u** in U such that T(**u**) = **v**.

Explain why the following three statements are equivalent.

a) T is onto.
b) Image(T) = V.
c) Rank(T) = dim(V).

ANSWER:

If T is 1-1 and onto we say that T is an **isomorphism**.

If **A** is a matrix and T(**u**) = **A**·**u**, what does it mean (in terms of **A**) to say that

a) T is one-one.
b) T is onto.
c) T is an isomorphism.

ANSWER:

<><><><><><><><><><><><><><><><><><><><><><><><><><><><><><><><><><><>

Let T:U → V be a linear transformation. Then T is an isomorphism if and only if there is a linear transformation S:V → U such that S∘T = I_U and T∘S = I_V. (I_U is the identity map on space U and I_V the identity map on V.)

Assume that S exists. We show that T is one-one and onto.

If T(**u**) = **0** then **u** = S∘T(**u**) = S(**0**) = **0**. This says Ker(T) contains only the zero vector, thus by the previously established equivalences we have that T is one-one.

If **v** in V, then T(S(**v**)) = (T∘S)(**v**) = **v**. This says that for every **v** in V there is a vector **u** in U such that T(**u**) = **v**. Therefore T is onto.

Assume that T is an isomorphism. We show that there exists a linear transformation S:V → U such that S∘T = I_U and T∘S = I_V.

Define S(**v**) by noting that since T is onto there is a vector **u** such that T(**u**) = **v** and moreover that **u** is unique since T is one-one. Set S(**v**) = **u**. To see that S is a linear transformation, note that T(S(**u** + **w**) - S(**u**) - S(**w**)) = **0** and T is one-one.

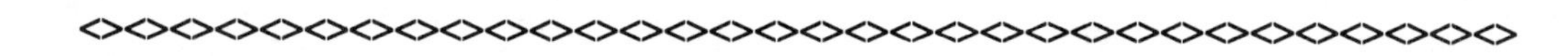

Let T:U $\rightarrow$ V and assume that dim(U) = n, then rank(T) is less than or equal to n.

Assume that dim(Image(T)) > n. That means that there is a basis for Image(T) that has more that n elements. Denote this basis as $\mathbf{v}_1, \mathbf{v}_2, \ldots, \mathbf{v}_m$ where $m > n$.

Since T is onto the image, there are vectors $\mathbf{u}_1, \mathbf{u}_2, \ldots, \mathbf{u}_m$ such that $T(\mathbf{u}_j) = \mathbf{v}_j$ for j = 1, ..., m. But the **u**'s must be linearly dependent since dim(U) = n. Thus there are constants $a_1, \ldots, a_m$ not all zero such that

$$a_1 \cdot u_1 + a_2 \cdot u_2 + \ldots + a_m \cdot u_m = 0$$

Thus $$T(a_1 \cdot u_1 + a_2 \cdot u_2 + \ldots + a_m \cdot u_m) = 0$$

or $$a_1 \cdot v_1 + a_2 \cdot v_2 + \ldots + a_m \cdot v_m = 0$$ with not all the a's = 0.

But this is a contradiction of the assumption that the v's were a basis.

<><><><><><><><><><><><><><><><><><><><><><><><><><><><><><><>

The main result of this section is:

> **Let T:U $\rightarrow$ V be a linear transformation and assume that dim(U) = n.**
>
> **Then rank(T) + nullity(T) = n.**

Before proving this, we shall use it to conclude that if **A** is a matrix, the dimension of the column space equals the dimension of the row space.

<><><><><><><><><><><><><><><><><><><><><><><><><><><><><><><>

Let **A** be an m by n matrix and $\mathbf{A} \cdot \mathbf{x} = \mathbf{0}$ the corresponding homogeneous system of linear equations.

The **rank of A** is the dimension of its image. The image space is spanned by the columns of **A**, that is the column space of **A**.

The row space of **A** is the space spanned by the rows. The dimension of the row space is the number of non-zero rows in the reduced row echelon form. This is true since the elementary operations don't change the span of the rows and the resulting rows are linearly independent.

If T:U $\rightarrow$ V is defined by $T(\mathbf{x}) = \mathbf{A} \cdot \mathbf{x}$, then the row space is a subspace of __________ and the column space is a subspace of _____________. (Fill in the blanks.)

By our row reduction algorithm we know that the solution space has a basis vector for each column that is not a pivot column after row reduction. Thus

$$\textbf{nullity(T) = dim(solution space A) = n - dim(row space)}$$

The columns of $\mathbf{A}$ span the image of $\mathbf{A}$. Thus

$$\textbf{dim(column space) = rank(A) = rank(T)}$$

Putting these together we have:

$$n = \text{rank}(T) + \text{nullity}(T) = \text{dim(column space)} + n - \text{dim(row space)}$$

which implies that $0 = \text{dim(column space)} - \text{dim(row space)}$ and hence

$$\text{dim(column space)} = \text{dim(row space)}.$$

We return to the proof that rank(T) + nullity(T) = n. To show that this is true,

1. Choose a basis $\mathbf{u}_1, \mathbf{u}_2, \ldots, \mathbf{u}_k$ for the Kernel of T.
2. Choose a basis $\mathbf{v}_1, \mathbf{v}_2, \ldots, \mathbf{v}_m$ for the Image of T.
3. Choose $\mathbf{w}_1, \mathbf{w}_2, \ldots, \mathbf{w}_m$ such that $T(\mathbf{w}_j) = \mathbf{v}_j$.
4. It then suffices to show that $\mathbf{u}_1, \mathbf{u}_2, \ldots, \mathbf{u}_k, \mathbf{w}_1, \mathbf{w}_2, \ldots, \mathbf{w}_m$ is a basis for U.

Explain why we can find the **w**'s in step 3 and explain why step 4 suffices.

ANSWER:

To show that $\mathbf{u}_1, \mathbf{u}_2, \ldots, \mathbf{u}_k, \mathbf{w}_1, \mathbf{w}_2, \ldots, \mathbf{w}_m$ form a basis for U we must show that (a) they are linearly independent and (b) they span U.

(a) **Showing linear independence:**

Assume: $\mathbf{a_1 \cdot u_1 + \ldots + a_k \cdot u_k + b_1 \cdot w_1 + \ldots + b_m \cdot w_m = 0}$

then $\mathbf{T(a_1 \cdot u_1 + \ldots\ldots + a_k \cdot u_k + b_1 \cdot w_1 + \ldots + b_m \cdot w_m) = 0}$

this implies that $\mathbf{b_1 \cdot v_1 + \ldots + b_m \cdot v_m = 0}$ (1)

which implies all the b's are zero since the v's are linearly independent.

Explain why (1) is true.

ANSWER:

Therefore $\mathbf{a_1 \cdot u_1 + \ldots + a_k \cdot u_k = 0}$

Why does this imply that the a's are zero?

ANSWER:

(b) **Showing spanning:**

Let **u** be in U. Then $\mathbf{T(u)=c_1 \cdot v_1 + \ldots + c_m \cdot v_m}$

Explain why.

ANSWER:

and hence $\mathbf{u - (c_1 \cdot w_1 + \ldots + c_m \cdot w_m)}$ is in Ker(T),

Explain why.

ANSWER:

Since $\mathbf{u}_1, \ldots, \mathbf{u}_k$ span the kernel,

$$\mathbf{u - (c_1 \cdot w_1 + \ldots + c_m \cdot w_m) = b_1 \cdot u_1 + \ldots + b_k \cdot u_k}$$

so $\mathbf{u = c_1 \cdot w_1 + \ldots + c_m \cdot w_m + b_1 \cdot u_1 + \ldots + b_k \cdot u_k}$

and it follows that the u's and w's span.

<><><><><><><><><><><><><><><><><><><><><><><><><><><><><><><>

Exercises

1. Let $\mathbf{u} := \begin{bmatrix} 1 \\ 0 \\ -2 \\ 1 \\ 0 \end{bmatrix}$ and $\mathbf{v} := \begin{bmatrix} 0 \\ 3 \\ 1 \\ -1 \\ 0 \end{bmatrix}$

Define $T:\mathbf{R}^5 \to \mathbf{R}^5$ by $T(\mathbf{x}) = \mathbf{x} - (\mathbf{x} \cdot \mathbf{u}) \cdot \mathbf{u} - (\mathbf{x} \cdot \mathbf{v}) \cdot \mathbf{v}$.

What is

(a) Rank (T)?

(b) Nullity (T)?

ANSWERS:

<><><><><><><><><><><><><><><><><><><><><><><><><><><><><><>

2. Give an example of a linear transformation $T:\mathbf{R}^6 \to \mathbf{R}^3$ such that Rank (T) = 2.
What is Nullity(T)?

ANSWER:

<><><><><><><><><><><><><><><><><><><><><><><><><><><><><><>

3. (Individual) Write a summary of the MATHEMATICS that you learned in this section.

CHAPTER 8

VECTOR SPACES

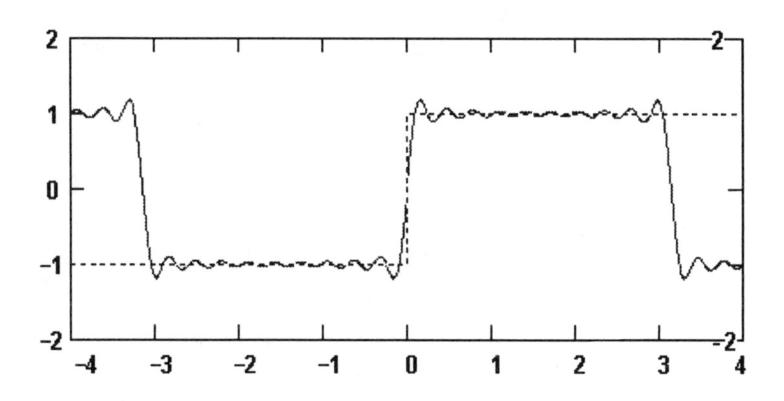

Section 8-1 Vector Spaces

Name:________________

Score:_____________

Overview

Throughout this text we have focused our discussion on vectors in $\mathbf{R}^n$, maps between $\mathbf{R}^n$ and $\mathbf{R}^m$, and dot products of these vectors, although in Section 6-3 we did expand this study to include complex vector spaces. It is our goal in this chapter to extend this discussion to a more general concept of **vector space** that includes infinite dimensional vector spaces and vector spaces of functions.

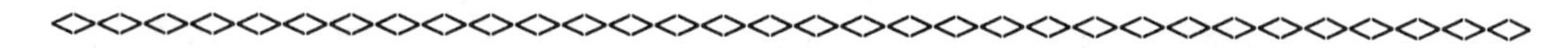

Mathematics is in many ways the study of structures. The structure that we have focused on is the solution space of a system of linear equations. Let us pause and look back at our work and ask the general question:

What are the properties of $\mathbf{R}^n$ that make things 'work'?

This is an imprecise question. But that shouldn't keep us from making a short list of properties.

1. We can add vectors.
2. We can multiply a vector by a scalar.
3. Vector addition and scalar multiplication behave nicely (i.e., they satisfy properties such as commutativity and associativity, are distributive, etc.).
4. The dot product enables us to define length, angles, and distance.

We use properties 1-3 to define what we mean by a **vector space**. Once we do that the concepts of subspace, basis, dimension, linear transformation, and eigenvalue/eigenvector can be generalized from our earlier work.

If in addition there is a operation on pairs of vectors that obeys the properties of the dot product, we will be able to define an **inner product** on the vector space and extend the definitions of length, angles, orthogonality, projection, and distance.

In Section 8-2 we discuss the vector space of infinite sequences and in Section 8-3 we discuss vector spaces of functions.

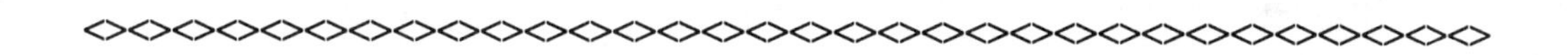

Definition (Vector Space)

A **vector space**, V, is a set of elements (called vectors) and two operations, vector addition (denoted by +) and scalar multiplication that satisfy the following conditions. (We will take the scalars to be the real numbers.)

For $\mathbf{u}$, $\mathbf{v}$, and $\mathbf{w}$ any vectors and any scalars a and b, $\mathbf{u} + \mathbf{v}$ is in V (closure of +), $a \cdot \mathbf{v}$ is in V (closure of scalar multiplication) and

1. $\mathbf{v} + \mathbf{w} = \mathbf{w} + \mathbf{v}$ (Commutativity of +.)
2. $\mathbf{u} + (\mathbf{v} + \mathbf{w}) = (\mathbf{u} + \mathbf{v}) + \mathbf{w}$ (Associativity of +.)
3. There exists a vector $\mathbf{0}$ such that $\mathbf{0} + \mathbf{v} = \mathbf{v}$, for any vector $\mathbf{v}$. (Existence of additive identity.)
4. For each vector $\mathbf{v}$ there is another vector, denoted $-\mathbf{v}$, such that $\mathbf{v} + (-\mathbf{v}) = \mathbf{0}$. (Existence of additive inverse.)
5. $a \cdot (b \cdot \mathbf{v}) = (a \cdot b) \cdot \mathbf{v}$
6. $a \cdot (\mathbf{u} + \mathbf{v}) = a \cdot \mathbf{u} + a \cdot \mathbf{v}$
7. $(a + b) \cdot \mathbf{v} = a \cdot \mathbf{v} + b \cdot \mathbf{v}$
8. $1 \cdot \mathbf{v} = \mathbf{v}$
9. $0 \cdot \mathbf{v} = \mathbf{0}$ (The 0 on the left is the scalar zero, while the $\mathbf{0}$ on the right is the additive identity.)

More precisely, we have defined a **REAL** vector space. We could have defined a vector space with scalars in the complex numbers or more generally in an object that mathematicians call a field.

We now list definitions and properties of vector spaces that are immediate extensions of our study of solution spaces.

- A **subspace** W of a vector space V is a non-empty subset that is closed under addition and scalar multiplication.

- A set of vectors, $\mathbf{v}_1, \mathbf{v}_2, \ldots$ in a vector space V is called **linearly independent** if

  $a_1 \cdot \mathbf{v}_1 + a_2 \cdot \mathbf{v}_2 + \ldots + a_k \cdot \mathbf{v}_k = \mathbf{0}$ implies that $a_1 = a_2 = \ldots = a_k = 0$.

  (If the set of vectors is infinite, then we require that the property given above holds for all finite subsets.)

- A set of vectors, $\mathbf{v}_1, \mathbf{v}_2, \ldots$ is said to **span** V if every vector in V can be written as a linear combination of a finite number of the $\mathbf{v}$'s.

- A set of vectors, $\mathbf{v}_1$, $\mathbf{v}_2$, ... is said to be a **basis** for V if it spans V and is linearly independent. (Note that, in general, the number of elements in a basis may be infinite.)

- The **dimension** of a subspace is the number of elements in a basis for the subspace.

- A vector space is said to be **finite dimensional** if its dimension is finite.

  If we consider the vector space that consists of ***all polynomials***, then we see that there is a basis consisting of the polynomials 1, x, x^2, x^3, ... , x^n, ... This vector space is infinite dimensional and the basis has an infinite number of elements. We note that the infinite series $1 + x + x^2 + x^3 + \ldots + x^n + \ldots$ is **NOT** a polynomial.

- A mapping T:U → V (U and V are vector spaces) is called a **linear transformation** if $T(a \cdot \mathbf{x} + \mathbf{y}) = a \cdot T(\mathbf{x}) + T(\mathbf{y})$ for all **x** and **y** that are elements of U and all scalars a.

- The **kernel** of a linear transformation, T:U → V is the set of all vectors **u** in U such that $T(\mathbf{u}) = \mathbf{0}$.

- The **image** of a linear transformation, T:U → V is the set of all vectors **v** in V such that $\mathbf{v} = T(\mathbf{u})$ for some **u** in U.

- We say that λ is an **eigenvalue** for a linear transformation T:U → U if there is a non-zero vector **u** (called an associated **eigenvector**) such that $T(\mathbf{u}) = \lambda \cdot \mathbf{u}$.

- If T:U → V is a linear transformation of **finite** dimensional vector spaces and $\mathbf{u}_1, \mathbf{u}_2, \ldots, \mathbf{u}_k$ is a basis of U and $\mathbf{v}_1, \mathbf{v}_2, \ldots, \mathbf{v}_m$ is a basis of V then the **matrix of T relative to** the **u**'s and **v**'s, **M**, maps the coordinates of **u** relative to the **u**-basis to the coordinates of T(**u**) relative to the **v**-basis. The jth column of **M** is the vector of coordinates of $T(\mathbf{u}_j)$ in the **v**-basis.

Explain how to define the **rank** of T and the **nullity** of T.

ANSWER:

We have indicated how definitions generalize in a natural way. The same is true for the theorems that we have stated and/or proved. For example:

- ***If T: U → V is a linear transformation of finite dimensional vector spaces, then rank(T) + nullity(T) = dim(U).***

We give examples to illustrate these concepts in the next two sections.

<><><><><><><><><><><><><><><><><><><><><><><><><><><><><><><>

Definition (Inner Product)

Let V be a real vector space (i.e., the scalars are real numbers). An operation on a pair of vector **u** and **v** from V, denoted **<u,v>**, that produces a real scalar is called an **inner product** if it satisfies the following properties:

1. **<u,v>** = **<v,u>**.
2. <r·**u**,**v**> = r·**<u,v>**, for any real scalar r.
3. **<u+v,w>** = **<u,w>** + **<v,w>**.
4. **<u,u>** = 0, if and only if **u** = **0**.

When there is no confusion we will use **u**·**v** to mean the same as **<u,v>**.

In an inner product space:

- The **length** of the vector, **v** is defined to be sqrt(**<v,v>**), this is denoted by |**v**|.
- The **angle**, θ, between two vectors, **u** and **v**, is defined by cos(θ) = **<u,v>/(|u|·|v|)**.
- Two vectors are said to be orthogonal or perpendicular if the angle between them is π/2.
- **Theorem**: **u** is orthogonal to **v** if and only if **<u,v>** = 0.
- A set of vectors **v1**, **v2**, ... is said to be an **orthogonal** set if **<vi,vj>** = 0 for i ≠ j.
- Given a set of vectors, $\mathbf{v}_1$, $\mathbf{v}_2$, ..., $\mathbf{v}_n$, the **Gram-Schmidt** algorithm will yield a set of vectors $\mathbf{w}_1$, $\mathbf{w}_2$, ..., $\mathbf{w}_n$ such that the **w**'s are an orthogonal set and Span($\mathbf{v}_1$, ..., $\mathbf{v}_k$) = Span($\mathbf{w}_1$, .., $\mathbf{w}_k$).

These concepts are illustrated in the next two sections.

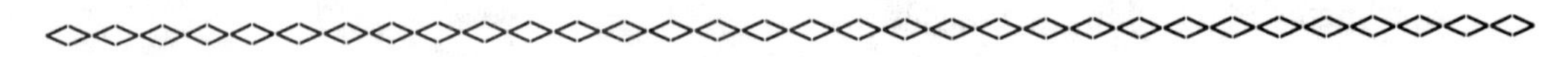

Exercise: (Individual) Write a summary of the MATHEMATICS in this section.

Section 8-2 The Vector Space of Infinite Sequences

Name:________________

Score:______________

Overview

Let $\mathscr{Inf}$ be the set of all infinite sequences. We show that $\mathscr{Inf}$ is a vector space, investigate some subspaces of $\mathscr{Inf}$, and examine certain linear transformations on $\mathscr{Inf}$.

<><><><><><><><><><><><><><><><><><><><><><><><><><><><><><><><>

Let $\mathbf{a} = (a_1, a_2, a_3, \ldots)$ and $\mathbf{b} = (b_1, b_2, b_3, \ldots)$ be two members of $\mathscr{Inf}$. We define the sum of **a** and **b** by constructing the infinite sequence whose entries are the sum of the corresponding entries of **a** and **b**:

$$\mathbf{a} + \mathbf{b} = (a_1 + b_1, a_2 + b_2, a_3 + b_3, \ldots)$$

Scalar multiplication is given by

$$k \cdot \mathbf{a} = (k \cdot a_1, k \cdot a_2, k \cdot a_3, \ldots)$$

<><><><><><><><><><><><><><><><><><><><><><><><><><><><><><><><>

Verify that addition in $\mathscr{Inf}$ is commutative.

ANSWER:

Verify that addition in $\mathscr{Inf}$ is associative.

ANSWER:

Determine the additive identity in $\mathscr{Inf}$.

ANSWER:

Describe the additive inverse of $\mathbf{b} = (b_1, b_2, b_3, \ldots)$.

ANSWER:

Let r and s be real scalars below.

Verify that $r \cdot (s \cdot \mathbf{b}) = (r \cdot s) \cdot \mathbf{b}$.

ANSWER:

Verify that $r\cdot(\mathbf{a} + \mathbf{b}) = r\cdot\mathbf{a} + r\cdot\mathbf{b}$.

ANSWER:

Verify that $(r + s)\cdot\mathbf{b} = r\cdot\mathbf{b} + s\cdot\mathbf{b}$.

ANSWER:

Verify that $1\cdot\mathbf{b} = \mathbf{b}$.

ANSWER:

Verify that $0\cdot\mathbf{b}$ = the additive identity.

ANSWER:

It follows that $\mathscr{Inf}$ with the operations defined above is a real vector space. The vectors in $\mathscr{Inf}$ can be thought of as columns (or rows) with infinitely many entries.

Let **ej** be the infinite sequence whose jth term is 1 and whose ith term is 0 for $j \neq i$.

Show that **e1**, **e2**, **e3**, ... , **en**, ... are a linearly independent set of vectors.

ANSWER:

Explain why **e1**, **e2**, **e3**, ... , **en**, ... are **NOT** a basis for the vector space $\mathscr{Inf}$?

ANSWER:

It is true that there is a basis for $\mathscr{Inf}$ that contains **e1**, **e2**, **e3**, ... , **en**, ... as a subset. However the proof that there is such a basis, known as a Kümmer basis, is not constructive. This means that although we know there is such a basis we can not explicitly describe it.

What must the dimension of $\mathscr{Inf}$ be?

ANSWER:

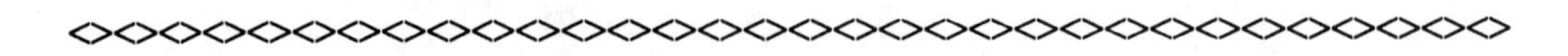

Some subspaces of $\mathscr{I}nf$.

(a) Verify that the set C of all convergent sequences is a subspace of $\mathscr{I}nf$.

ANSWER:

(b) Let F be the set of sequences in which all but a finite number of terms are 0.

List five vectors that belong to F.

ANSWER:

Show that F is a subspace of $\mathscr{I}nf$.

ANSWER:

Find a basis for F.

ANSWER:

Discuss why the usual inner product in $\mathbf{R}^n$ generalizes naturally to the subspace F.

ANSWER:

(c) Let K be the set of sequence in which the kth term is zero. Show that K is a subspace of $\mathscr{I}nf$.

ANSWER:

Since infinite sums may diverge there is no obvious way to define an inner product on the vector space $\mathscr{I}nf$ or on some its subspaces.

<><><><><><><><><><><><><><><><><><><><><><><><><><><><><><><><><><><><>

Some linear transformations $\mathscr{Inf} \to \mathscr{Inf}$.

(a) Left shift, $LS(x_1, x_2, x_3, \ldots.) = (x_2, x_3, \ldots)$

Verify that LS is a linear transformation.
ANSWER:

What is the kernel of LS?
ANSWER:

What is the image of LS?
ANSWER:

<><><><><><><><><><><><><><><><><><><><><><><><><><><><><><><><><><><><><><>

(b) Right shift, $RS(x_1, x_2, \ldots.) = (0, x_1, x_2, \ldots)$

Verify that RS is a linear transformation.
ANSWER:

What is the kernel of RS?
ANSWER:

What is the image of RS?
ANSWER:

<><><><><><><><><><><><><><><><><><><><><><><><><><><><><><><><><><><><><><>

A linear transformations from $\mathcal{Inf} \to \mathbf{R}^n$.

$P_n(x_1, x_2, x_3, ...) = (x_1, ..., x_n)$

For n = 5, $P_5(x_1, x_2, x_3, ...)$ is the first five members of sequence **x**.

Verify that P_5 is a linear transformation.

ANSWER:

What is the kernel of P_5?

ANSWER:

What is the image of P_5?

ANSWER:

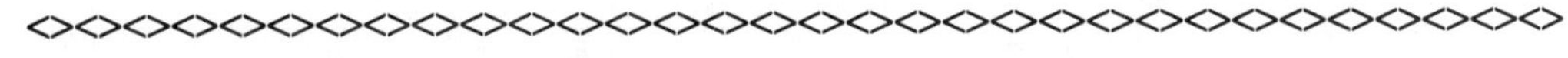

A linear transformations from $\mathbf{R}^n \to \mathcal{Inf}$.

$J_n(x_1, ..., x_n) = (x_1, ..., x_n, 0, 0, ...)$

Verify that J_n is a linear transformation.

ANSWER:

What is the kernel of J_n?

ANSWER:

What is the image of J_n?

ANSWER:

<><><><><><><><><><><><><><><><><><><><><><><><><><><><><><><>

Which of the maps given above are one-one and which are onto?

ANSWER:

What are the eigenvalues and corresponding eigenvectors of LS and RS?

ANSWER:

<><><><><><><><><><><><><><><><><><><><><><><><><><><><><><><>

Exercise: (Individual) Write a summary of the MATHEMATICS that you learned in this section.

Section 8-3 Vector Spaces of Functions

Name:__________________

Score:______________

Overview

In the same way that we used linear algebra to study the sets of solutions to a system of linear equations we can use linear algebra to study the set of solutions to a linear differential equation. The conclusions are almost immediate once we notice that the set of all continuous and infinitely differentiable functions from the reals to the reals form a vector space and that the derivative is a linear transformation on this space. We discuss these concepts in this section.

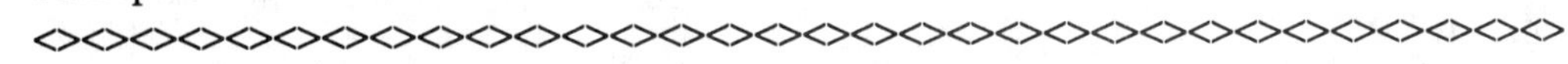

The vector space of polynomials

$\mathcal{P}_n$ is the set of all polynomials of degree less than or equal to n. We add two polynomials p(x) and q(x) by setting

$$(p+q)(x) = p(x) + q(x)$$

and define scalar multiplication by

$$(c \cdot p)(x) = c \cdot p(x).$$

With these two operations it is easy to prove that $\mathcal{P}_n$ is a vector space.

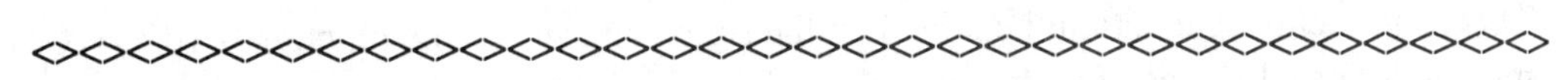

Carefully describe how the addition of two polynomials in $\mathcal{P}_n$ is performed.

ANSWER:

In $\mathcal{P}_3$, $2 \cdot x^2 + 3 \cdot x + 5$ is a vector. What is the zero vector?

ANSWER:

In $\mathcal{P}_4$, what is the additive inverse of $-5 \cdot x^4 + 3 \cdot x^2 - 12$?

ANSWER:

Describe the vector space $\mathcal{P}_0$.

ANSWER:

Describe the vector space $\mathcal{P}_1$.

ANSWER:

Discuss why $\mathcal{P}_1$ is a subspace of $\mathcal{P}_2$.

ANSWER:

How are the vector spaces $\mathcal{P}_0$, $\mathcal{P}_1$, $\mathcal{P}_2$, and $\mathcal{P}_3$ related?

ANSWER:

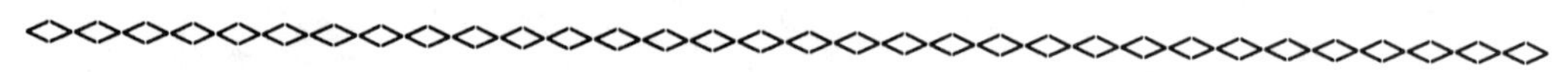

In $\mathcal{P}_n$ are the vectors $1, x, x^2, \ldots, x^n$ linearly independent?

We investigate this question as follows.

Suppose $a_0 + a_1 \cdot x + a_2 \cdot x^2 + \ldots + a_n \cdot x^n = 0$

This is a functional equation; i.e., the 0 on the right hand side is the function 0 $(0(x) = 0)$ so it must hold for all values of x. In particular, when $x = 0$, it implies that $a_0 = 0$.

Since it is a functional equation we can take the derivative of each side and maintain equality.

$$a_1 + 2 \cdot a_2 \cdot x + \ldots + n \cdot a_n \cdot x^{n-1} = 0$$

Letting $x = 0$ we see that $a_1 = 0$.

Continuing in this way we can conclude that **all** the coefficients = 0 and that $1, x, x^2, \ldots, x^n$ are linearly independent.

Let $p_1 = 2 \cdot x^2 + x - 3$

$p_2 = x + 2$

$p_3 = x^2 + 1$

Are $\mathbf{p}_1, \mathbf{p}_2, \mathbf{p}_3$ linearly independent? (Hint: Use the same procedure as above, but instead of the simple set of equations $a_k = 0$, we get a homogeneous system that must have only the trivial solution if the vectors are to be linearly independent.)

ANSWER: (Show your work.)

You should get a system equivalent to

$$\begin{aligned} -3 \cdot a_1 + 2 \cdot a_2 + a \cdot_3 &= 0 \\ a_1 + a_2 &= 0 \\ 4 \cdot a_1 + 2 \cdot a_3 &= 0 \end{aligned}$$

Which of the following are subspaces of $\mathcal{P}_3$? For those that are not subspaces, explain why they are not; for those that are, find a basis and compute the dimension.

a) polynomials of degree 2 or less

b) polynomials with constant term = 0

c) polynomials such that the coefficient of x is positive

d) polynomials of degree exactly two

e) polynomials p such that p(2) = 0

f) polynomials with constant term 1

ANSWERS:

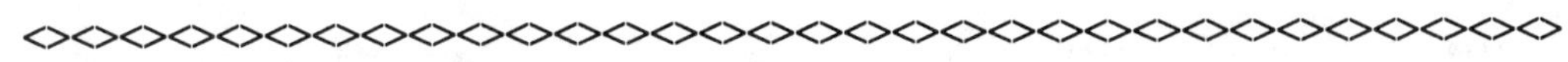

Which of the following are linear transformations?

$T(a + b \cdot x + c \cdot x^2 + d \cdot x^3) =$

(i) $b + d \cdot x$

(ii) $a \cdot x + b \cdot x^2 + c \cdot x^3$

(iii) $ab + c \cdot d \cdot x$

(iv) $a + b + c + d$

(v) $b + 2 \cdot c \cdot x + 3 \cdot d \cdot x^2$

ANSWERS:

<><><><><><><><><><><><><><><><><><><><><><><><><><><><><><><>

The vector space of real valued functions

Let **F** be the set of **all** functions from the reals to the reals. This is a very large set. Define addition of two functions by $(f + g)(x) = f(x) + g(x)$ and scalar multiplication by $(c \cdot f)(x) = c \cdot (f(x))$.

F is a vector space under these two operations.

What is the zero vector?

ANSWER:

Which of the following subsets of **F** are subspaces?

a) continuous functions

b) differentiable functions

c) all f such that $f(2) = 0$

d) all f such that $f(2) = 0$ and $f(5) = 0$

e) all f such that $f(2) = 0$ or $f(5) = 0$

f) all f such that $f(2) = f(5)$

g) all f such that $f(2) = 1$

h) The set of functions which are periodic of period 1.

i) all f such that $f(x) > 0$ for all x

j) all f such that $f'' - 5 \cdot f' + 6 \cdot f = 0$ (f' is the derivative of f; f'', the second derivative)

ANSWERS:

There is no natural basis for **F** and most of the sets given above that are subspaces do not have a finite basis. There is one exception however. Determine which subspace has a finite basis and find a basis.

ANSWER:

<><><><><><><><><><><><><><><><><><><><><><><><><><><><>

near transformations on function spaces

Differentiation

Let **DIFF** be the vector space of differentiable functions. Then

$$D: \mathbf{DIFF} \to \mathbf{DIFF}$$

defined by D(f) = f ' is a linear transformation since the derivative of the sum is the sum of the derivatives and the derivative of a constant times a function is the constant times the derivative.

Integration

Let $\boldsymbol{C}$ be the vector space of continuous functions. Then

$$\mathrm{Int}(a): \boldsymbol{C} \to \boldsymbol{C}$$

defined by

$$\mathrm{Int}(a) = \int_a^x f(t)\,dt$$

is a linear transformation by the properties of integration.

What is D∘Int(a)? $$\frac{d}{dx}\int_a^x f(t)\,dt = f(x)$$

This is the **FUNDAMENTAL THEOREM OF CALCULUS.**

Since D∘Int(a) = I, D is a left inverse to Int(a) (and Int(a) is a right inverse to D).

What is Int(a)∘D? $$\int_a^x \frac{d}{dt} f(t)\,dt = f(x) - f(a)$$

In general, Int(a)∘D is not the identity.

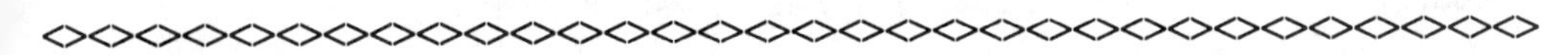

We know that the composition of linear transformations is a transformation, so D^k, the kth derivative function, is a linear transformation. (We should be a little careful here since the derivative of a differentiable function need not be differentiable. To correct this we should consider D as a linear transformation from k-times differentiable functions to (k-1)-times differentiable functions. Since we know that we can do that, we will be sloppy and ignore that problem.)

What is the kernel of D? What is its dimension and what is a basis?

ANSWER:

What is the kernel of D^k? What is its dimension and what is a basis?

ANSWER:

What is the kernel of Int(a)? What is its dimension and what is a basis?

ANSWER:

What is the kernel of $D^2 - 5 \cdot D + 6 \cdot I$? (I is the identity.) What is its dimension and what is a basis?

ANSWER:

What are the eigenvalues and eigenvectors of D?

ANSWER:

What are the eigenvalues and eigenvectors of D^2?

ANSWER:

Consider the subspace that is spanned by the functions in the first column below

$\exp(x)\cdot\sin(x)$	$\exp(x)\cdot\sin(x) + \exp(x)\cdot\cos(x)$
$\exp(x)\cdot\cos(x)$	$\exp(x)\cdot\cos(x) - \exp(x)\cdot\sin(x)$
$\exp(x)\cdot\sin(x)^2$	$\exp(x)\cdot\sin(x)^2 + 2\cdot\exp(x)\cdot\sin(x)\cdot\cos(x)$
$\exp(x)\cdot\cos(x)^2$	$\exp(x)\cdot\cos(x)^2 - 2\cdot\exp(x)\cdot\sin(x)\cdot\cos(x)$
$\exp(x)\cdot\sin(x)\cdot\cos(x)$	$\exp(x)\cdot\sin(x)\cdot\cos(x) + \exp(x)\cdot\cos(x)^2 - \exp(x)\cdot\sin(x)^2$

The second column contains the derivatives of the corresponding functions in the first column. As you can see, this subspace is closed under differentiation. The functions in the first column are linearly independent and hence are a basis.

What is the matrix for D relative to this basis?

It is

$$D := \begin{bmatrix} 1 & -1 & 0 & 0 & 0 \\ 1 & 1 & 0 & 0 & 0 \\ 0 & 0 & 1 & 0 & -1 \\ 0 & 0 & 0 & 1 & 1 \\ 0 & 0 & 2 & -2 & 1 \end{bmatrix}$$

Carefully explain how this matrix represents D relative to the given basis.

ANSWER:

Show that the matrix D is nonsingular and hence invertible.

ANSWER:

To compute

$$\int_0^x \exp(t)\cdot\sin(t)^2\,dt$$

all we need to do is use the inverse matrix and compute

$$D^{-1}\cdot\begin{bmatrix} 0 \\ 0 \\ 1 \\ 0 \\ 0 \end{bmatrix} = \begin{bmatrix} 0 \\ 0 \\ 0.6 \\ 0.4 \\ -0.4 \end{bmatrix}$$

Explain the preceding expression.

ANSWER:

The answer to $\int_0^x \exp(t)\cdot\sin(t)^2\,dt$ is

$$\frac{3\cdot(\exp(x)\cdot\sin(x)^2)}{5} + \frac{2\cdot(\exp(x)\cdot\cos(x)^2)}{5} - \frac{2\cdot(\exp(x)\cdot\sin(x)\cdot\cos(x))}{5}$$

Use symbolic differentiation to check the answer. (Box an 'x' in the expression, then choose Differentiate on Variable.)

Use similar reasoning to integrate $x^7 \exp(x)$. (Hint: A basis is $x^k \exp(x)$ for $k = 0, \ldots, 7$.)

ANSWER:

<><><><><><><><><><><><><><><><><><><><><><><><><><><><><><><><><><><>

Exercise: (Individual) Write a summary of the MATHEMATICS that you learned in this section

Section 8-4 Inner Products in Function Spaces

Name:________________

Score:_____________

Overview

The statement that ***the polynomial x is orthogonal to x^2 on the interval*** **[-1,1]** has no intuitive meaning. Yet the idea of orthogonality in function spaces is useful because (1) we know that if a subspace has an orthogonal basis, a vector's coordinates with respect to that basis can be easily calculated using the inner product and (2) least square approximation can be computed using projections. We use these ideas in this section to study Fourier Series and Legendre Polynomials which are both based on the idea of orthogonal sets of functions.

<><><><><><><><><><><><><><><><><><><><><><><><><><><><><><><><><><><><><><>

Let $\boldsymbol{C}$ be the space of continuous functions on the interval [a,b]. Define an inner product as follows:

$$<f,g> = \int_a^b f(x)\cdot g(x)\, dx$$

It is easy to show that <f,g> satisfies the properties required of an inner product. The only verification that requires some work is the fact that <f,f> = 0 implies that f = 0. This follows from the definition of a continuous function.

<><><><><><><><><><><><><><><><><><><><><><><><><><><><><><><><><><><><><>

This is a natural generalization of the inner product in $\mathbf{R}^n$ which is defined by

$$u\cdot v = \sum_{j=1}^{n} u_j \cdot v_j$$

<><><><><><><><><><><><><><><><><><><><><><><><><><><><><><><><><><><><>

Consider the subspace of polynomials, $\mathcal{P}$, within $\boldsymbol{C}$ and choose a = -1 and b = 1.

$<x, x> = \int_{-1}^{1} x^2\, dx = \frac{2}{3}$ so $|x| = \sqrt{\frac{2}{3}}$ is the **norm** or **length** of x in this inner product

$<x, x^2> = \int_{-1}^{1} x^3\, dx = 0$ so, x is orthogonal to x^2 over [-1, 1]

A function, f, is called **even** if f(-x) = f(x) and **odd** if f(-x) = - f(x). Show that if f is even and g is odd then f and g are orthogonal over the interval [-a,a] for any value of a > 0.

ANSWER:

If the inner product were taken over [0,1] then

$$\langle x, x^2 \rangle = \int_0^1 x^3\,dx = \frac{1}{4}$$

So orthogonality depends upon which inner product one chooses.

What is the norm of x over [0,1]?

ANSWER:

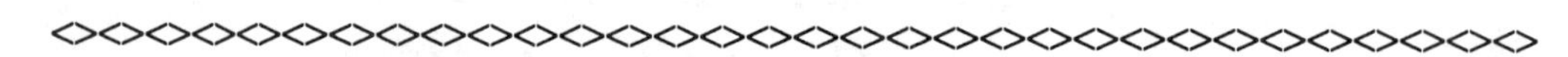

Legendre Polynomials

Let f be a function defined on [a,b] (for example) and let $\mathcal{P}_n$ be the subspace of polynomials of degree less than or equal to n. Our goal is to find the nth degree polynomial that best approximates f. Once again we are faced with the problem of defining what we mean by best approximates. This is the same situation that we discussed when we computed the least square approximations in Chapter 5. Here the answer will again be that polynomial p that minimizes $|f - p|$. The major difference is that the inner product is different. Otherwise the solution is the same and the answer is that p is the projection of f into the subspace $\mathcal{P}_n$.

As we have seen earlier, to compute the projection it is useful to have an orthogonal (or orthonormal) basis for the subspace into which we are projecting. We begin with the basis 1, x, x^2, ..., x^n and use the Gram-Schmidt process to find an orthogonal basis. We do this below for the case n = 3 over the interval [-1,1].

<><><><><><><><><><><><><><><><><><><><><><><><><><><><><><><><><><>

Find an orthogonal basis for the space $\mathcal{P}_3$, i.e. replace 1, x, x^2, x^3 with an orthogonal basis under the inner product.

$$\int_{-1}^{1} f(x) \cdot g(x)\,dx$$

<><><><><><><><><><><><><><><><><><><><><><><><><><><><><><><><><><>

The Gram-Schmidt method begins with vectors **v1, v2, ..., vn** and replaces them one at a time to get an orthogonal basis **w1, w2, ..., wn** with the property that Span(**v1, v2, ..., vj**) = Span(**w1, w2, ..., wj**) for each j. If we have already computed **w1, ..., w(k-1)** then **wk** equals

$$wk := vk - \sum_{j=1}^{k-1} \frac{vk \cdot wj}{wj \cdot wj} \cdot wj$$

<><><><><><><><><><><><><><><><><><><><><><><><><><><><><><>

We will call the resulting orthogonal basis **g1**, **g2**, **g3**, and **g4** and L_i will be **<gi,gi>**.

STEP 1:

$$g1(x) := 1 \qquad L_1 := \int_{-1}^{1} 1\,dx \qquad L_1 = 2$$

<><><><><><><><><><><><><><><><><><><><><><><><><><><><><><>

STEP 2:

$$g2(x) = x - \frac{\langle g1, x \rangle}{L_1} \cdot g1$$

Use Maple to compute

$$x - \left(\frac{\int_{-1}^{1} x\,dx}{2} \right) \cdot 1$$

Set your answer equal to **g2(x)**

$g2(x) := \blacksquare$

Use Maple to compute

$$L_2 := \int_{-1}^{1} g2(x)^2\,dx$$

(You must substitute for **g2(x)** to use Maple.) Set your answer equal to L_2.

If your work is correct you should have $L_2 := \frac{2}{3}$

<><><><><><><><><><><><><><><><><><><><><><><><><><><><><><>

STEP 3:

$$g3 = x^2 - \frac{\langle g1, x^2 \rangle}{L_1} \cdot g1 - \frac{\langle g2, x^2 \rangle}{L_2} \cdot g2$$

Substitute the values for **g1**, **g2**, L_1 and L_2 in the previous expression and use Maple to evaluate it. Call the result **g3**.

g3 := ▮

Substitute for **g3(x)** and use Maple to compute

$$L_3 := \int_{-1}^{1} (g3(x))\cdot(g3(x))\,dx$$

L_3 := ▮

If your work is correct you should have $L_3 := \frac{8}{45}$

<><><><><><><><><><><><><><><><><><><><><><><><><><><><><><><>

STEP 4:

$$g4 = x^3 - \frac{\langle g1, x^3 \rangle}{L_1}\cdot g1 - \frac{\langle g2, x^3 \rangle}{L_2}\cdot g2 - \frac{\langle g3, x^3 \rangle}{L_3}\cdot g3$$

Substitute the values for **g1**, **g2**, **g3**, L_1, L_2 and L_3 in the previous expression and use Maple to evaluate it. Call the result **g4**.

g4 := ▮

Substitute for **g4(x)** and use Maple to compute

$$L_4 := \int_{-1}^{1} (g4(x))\cdot(g4(x))\,dx$$

L_4 := ▮

If your work is correct you should have $L_4 := \frac{8}{175}$

<><><><><><><><><><><><><><><><><><><><><><><><><><><><><><><>

If your work above is correct you should have that an orthogonal basis for $\mathcal{P}_3$ is

$g1(x) := 1 \qquad g2(x) := x \qquad g3(x) := \left(x^2 - \frac{1}{3}\right) \qquad g4(x) := \left(x^3 - \frac{3}{5} \cdot x\right)$

$L_1 := 2 \qquad L_2 := \frac{2}{3} \qquad L_3 := \frac{8}{45} \qquad L_4 := \frac{8}{175}$

(We use an orthogonal basis instead of an orthonormal basis to avoid the square roots of the lengths.)

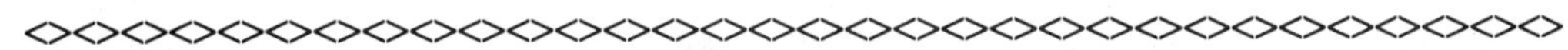

Find the polynomial **f(x)** of degree 3 that best approximates **sin(π·x)** using this orthogonal basis. From our work with projections we know that

$$f(x) = \mathrm{proj}_{\mathcal{P}_3}(\sin(\pi \cdot x))$$

That is, **f(x)** is projection onto subspace $\mathcal{P}_3$. This projection is given by

$$\sum_{j=1}^{4} \frac{\langle \sin(\pi \cdot x) \cdot gj(x) \rangle}{L_j} \cdot gj(x)$$

<><><><><><><><><><><><><><><><><><><><><><><><><><><><><><>

We denote the coefficients of **gj(x)** in this expression by f1, f2, f3, and f4, respectively.

$f1(x) := \int_{-1}^{1} \sin(\pi \cdot x) \cdot \frac{1}{2} \, dx$ **Evaluate symbolically to obtain**

$f1(x) := \blacksquare$

<><><><><><><><><><><><><><><><><><><><><><><><><><><><><><>

$f2(x) := \int_{-1}^{1} \sin(\pi \cdot x) \cdot \left[\left(\frac{3}{2}\right) \cdot x\right] dx$ **Evaluate symbolically to obtain**

$f2(x) := \blacksquare$

<><><><><><><><><><><><><><><><><><><><><><><><><><><><><><>

$$f3(x) := \int_{-1}^{1} \sin(\pi \cdot x) \cdot \left[\left(\frac{45}{8}\right) \cdot \left(x^2 - \frac{1}{3}\right) \right] dx$$

Evaluate symbolically to obtain

$$f3(x) := \blacksquare$$

<><><><><><><><><><><><><><><><><><><><><><><><><><><><><><>

$$f4(x) := \left[\int_{-1}^{1} \sin(\pi \cdot x) \cdot \left[\left(\frac{175}{8}\right) \cdot \left(x^3 - \frac{3}{5} \cdot x\right) \right] dx \right]$$

Evaluate symbolically to obtain

$$f4(x) := \blacksquare$$

If you completed the above computations correctly you should have that

$$f(x) := \frac{3}{\pi} \cdot x + \frac{35}{\left(2 \cdot \pi^3\right)} \cdot \left(-15 + \pi^2\right) \cdot \left(x^3 - \frac{3}{5} \cdot x\right)$$

Why do all the even power terms have coefficient zero?

ANSWER:

The polynomial f(x) is called the **Legendre polynomial approximation** to **sin($\pi \cdot$x)**.

The above answer is very different from the Taylor Series Approximation. We plot the curve, the Legendre approximation and the Taylor Series Approximation.

$t := -1, -.95 .. 1$

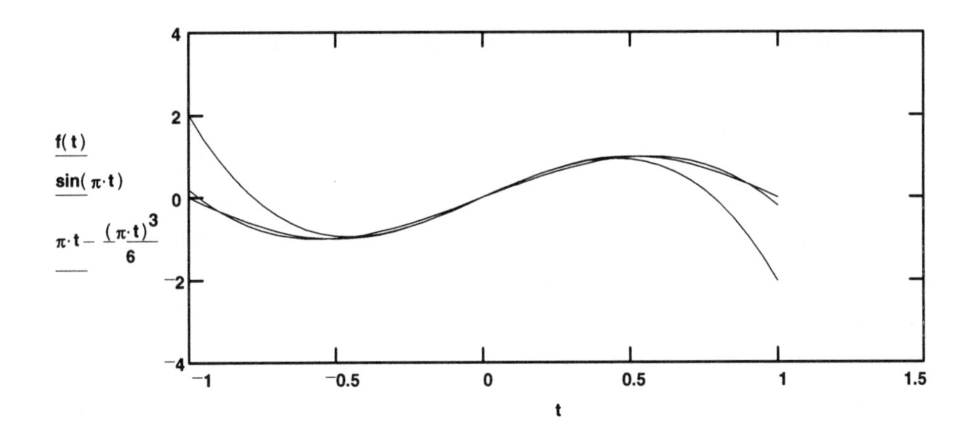

Explain in what sense the Legendre polynomial is a best cubic approximation and in what sense the Taylor Series is a best cubic approximation.
ANSWER:

Compute the best 3rd order (cubic) Legendre approximation to x^4 over [-1,1].

ANSWER:

What would be the best 4th order Legendre approximation to x^4 over [-1,1]. (Hint: No computations are required.)

ANSWER:

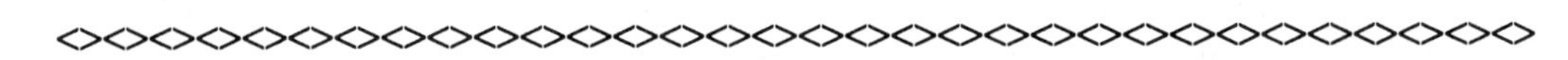

Fourier Series

Claim: The functions $\frac{1}{\sqrt{\pi}} \cdot \sin(j \cdot x)$ $\qquad \frac{1}{\sqrt{\pi}} \cdot \cos(j \cdot x)$ $\qquad$ and $\frac{1}{\sqrt{2 \cdot \pi}}$

form an orthonormal set where j = 1, 2, ..., n, ...

Verification

$$\int_{-\pi}^{\pi} \sin(j\cdot x)\cdot\cos(k\cdot x)\,dx$$ **Evaluate symbolically to obtain**

$$0$$

$$Is(j) := \int_{-\pi}^{\pi} \sin(j\cdot x)\cdot\sin(j\cdot x)\,dx$$ **Evaluate symbolically to obtain**

$$Is(j) := \frac{(-\cos(j\cdot\pi)\cdot\sin(j\cdot\pi) + j\cdot\pi)}{j}$$

$$Ic(j) := \int_{-\pi}^{\pi} \cos(j\cdot x)\cdot\cos(j\cdot x)\,dx$$ **Evaluate symbolically to obtain**

$$Ic(j) := \frac{(\cos(j\cdot\pi)\cdot\sin(j\cdot\pi) + j\cdot\pi)}{j}$$

Simplify the above expression for Is(1)? Ic(1)? Is(2)? Ic(2)?
ANSWER:

Let $\mathcal{J}$ be the subspace spanned by the family of functions given above. For any function **f(x)** the projection into $\mathcal{J}$ is called the **Fourier series of f(x)**. We note that in this case, unlike the previous case, the dimension of the subspace is infinite. This explains why we may get an infinite series of functions rather than a finite sum.

Expand the square wave function $F(x) := \text{if}(x>0, 1, -1)$
in a Fourier series. That is, for a particular value of j, obtain the projection of **F(x)** onto the subspace with orthonormal basis

$$\frac{1}{\sqrt{\pi}}\cdot\sin(j\cdot x) \qquad \frac{1}{\sqrt{\pi}}\cdot\cos(j\cdot x) \qquad \text{and} \qquad \frac{1}{\sqrt{2\cdot\pi}}$$

$t := -\pi, -\pi + .01 .. \pi$

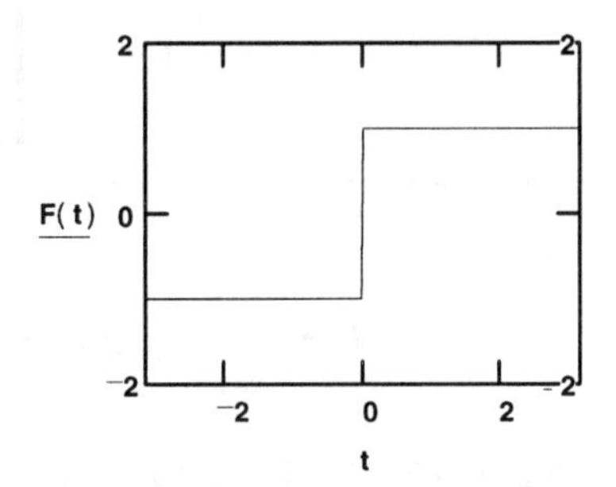

We calculate the coefficients in terms of the inner products.

$$c = \left\langle F(x), \frac{1}{\sqrt{2\cdot\pi}} \right\rangle$$

$$c := \frac{1}{\sqrt{2\cdot\pi}} \cdot \left(\int_{-\pi}^{0} -1 \, dx + \int_{0}^{\pi} 1 \, dx \right)$$

$$a_i = \left\langle F(x), \frac{1}{\sqrt{\pi}} \cdot \sin(i\cdot x) \right\rangle$$

$i := 1 .. 20$

$$a_i := \frac{1}{\sqrt{\pi}} \cdot \left(\int_{-\pi}^{0} -\sin(i\cdot x) \, dx + \int_{0}^{\pi} \sin(i\cdot x) \, dx \right)$$

Using Maple, we compute the above expression to get

$$a_i := \frac{1}{\sqrt{\pi}} \cdot \left(\frac{2}{i} - 2 \cdot \frac{\cos(i\cdot\pi)}{i} \right)$$

$$a_i := -2 \cdot \frac{(-1 + \cos(i\cdot\pi))}{\left(i\cdot\sqrt{\pi}\right)}$$

$a^T =$

	1	2	3	4	5	6	7	8	9
1	2.257	0	0.752	0	0.451	0	0.322	0	0.251

$$b_i = \left\langle F(x), \frac{1}{\sqrt{\pi}} \cdot \cos(i \cdot x) \right\rangle$$

$$b_i := \frac{1}{\sqrt{\pi}} \cdot \left(\int_{-\pi}^{0} -\cos(i \cdot x)\, dx + \int_{0}^{\pi} \cos(i \cdot x)\, dx \right)$$

$b^T =$

	1	2	3	4	5	6	7	8	9	10	11	12
1	0	0	0	0	0	0	0	0	0	0	0	0

$i := 1 .. 20$

$$G(x) := \frac{c}{\sqrt{2 \cdot \pi}} + \sum_i \left(a_i \cdot \frac{\sin(i \cdot x)}{\sqrt{\pi}} + b_i \cdot \frac{\cos(i \cdot x)}{\sqrt{\pi}} \right)$$

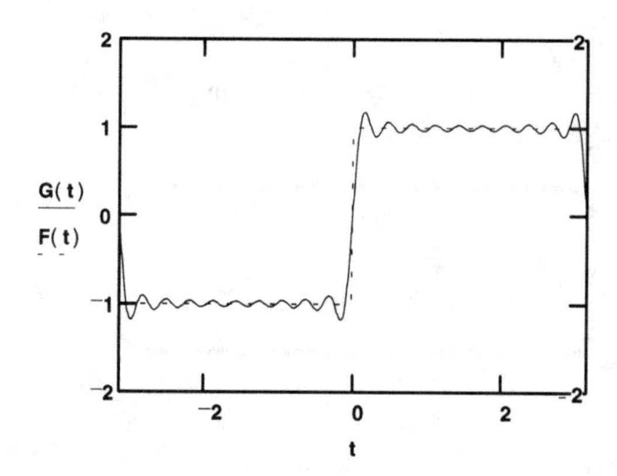

Vary the number of terms, i, and report on the results. In particular note what happens at the point of discontinuity.

ANSWER:

Compute the Fourier series approximation of f(x) = x with i = 5.

ANSWER:

◇◇

Exercise: (Individual) Write a summary of the MATHEMATICS that you learned in this section

INDEX

A

B

C

D

E

F

G

N

O

P

Q

R

S

stable state .... Section 1-6 page 9